Bruder
User's Guide to Plastic

Bonus Download Material:

On the website **www.brucon.se** (under "Our Books") you can download functional versions of the Excel worksheets described in the book.
This material can also be downloaded from
http://files.hanser.de/Files/Article/ARTK_GCL_9781569907344_0001.zip

Ulf Bruder

User's Guide to Plastic

2nd Edition

Hanser Publishers, Munich

HANSER

Hanser Publications, Cincinnati

The Author:

Ulf Bruder, Bruder Consulting AB, Barkassgatan 9, SE-371 32 Karlskrona, Sweden

Distributed in the Americas by:
Hanser Publications
414 Walnut Street, Cincinnati, OH 45202 USA
Phone: (800) 950-8977
www.hanserpublications.com

Distributed in all other countries by:
Carl Hanser Verlag
Postfach 86 04 20, 81631 Munich, Germany
Fax: +49 (89) 98 48 09
www.hanser-fachbuch.de

The use of general descriptive names, trademarks, etc., in this publication, even if the former are not especially identified, is not to be taken as a sign that such names, as understood by the Trade Marks and Merchandise Marks Act, may accordingly be used freely by anyone. While the advice and information in this book are believed to be true and accurate at the date of going to press, neither the authors nor the editors nor the publisher can accept any legal responsibility for any errors or omissions that may be made. The publisher makes no warranty, express or implied, with respect to the material contained herein.

Disclaimer:
We have done our best to ensure that the information found in this publication is both useful and accurate. However, please be aware that errors may exist in this publication, and that neither the author nor Bruder Consulting AB guarantee the accuracy of the information. All the tradenames and registered trademarks occurring in this book are for identification purposes only and are the property of their respective owners.

The final determination of the suitability of any information for the use contemplated for a given application remains the sole responsibility of the user.

Cataloging-in-Publication Data is on file with the Library of Congress

All rights reserved. No part of this book may be reproduced or transmitted in any form or by any means, electronic or mechanical, including photocopying or by any information storage and retrieval system, without permission in writing from the publisher.

© Carl Hanser Verlag, Munich 2019
Editor: Dr. Mark Smith
Production Management: Jörg Strohbach
Coverdesign: Max Kostopoulos
Typesetting: Kösel Media, Krugzell
Printed and bound by Appl, Wemding
Printed in Germany

ISBN 978-1-56990-734-4
E-Book ISBN 978-1-56990-735-1

Contents

Foreword XVII

CHAPTER 1
Polymers and Plastics 1
1.1   Thermosets 3
1.2   Thermoplastics 4
1.3   Amorphous and Semi-crystalline Plastics 5

CHAPTER 2
Commodities 7
2.1   Polyethylene (PE) 7
    2.1.1   Classification 8
    2.1.2   Properties of Polyethylene 8
    2.1.3   Recycling 9
    2.1.4   Application Areas 9
2.2   Polypropylene (PP) 11
    2.2.1   Properties of Polypropylene 12
    2.2.2   Recycling 13
2.3   Polyvinylchloride (PVC) 13
    2.3.1   Properties of PVC 14
    2.3.2   Recycling 14
2.4   Polystyrene (PS) 15
    2.4.1   Classification 16
    2.4.2   Properties of Polystyrene 16
    2.4.3   Recycling 17
    2.4.4   Application Areas 17
2.5   Styrene-Acrylonitrile (SAN) 17
2.6   Acrylonitrile-Butadiene-Styrene (ABS) 18
    2.6.1   ABS Blends 19
    2.6.2   Properties of ABS 19
    2.6.3   Recycling 19
    2.6.4   Application Areas 20
2.7   Polymethyl Methacrylate (PMMA) 21
    2.7.1   Properties of PMMA 22
    2.7.2   Recycling 22
    2.7.3   Application Areas 22

CHAPTER 3
Engineering Polymers 23
3.1   Polyamide or Nylon 23
    3.1.1   Classification 23
    3.1.2   Properties of Polyamide 24
    3.1.3   Recycling 25
    3.1.4   Application Areas 25

V

Contents

	3.2	Acetal		26
		3.2.1	Properties of Acetal	27
		3.2.2	Recycling	28
		3.2.3	Application Areas	28
	3.3	Polyester		29
		3.3.1	Properties of Polyester PBT and PET	31
		3.3.2	Recycling	31
		3.3.3	Application Areas	31
	3.4	Polycarbonate		33
		3.4.1	Properties of Polycarbonate	34
		3.4.2	Recycling	34
		3.4.3	Application Areas	34
CHAPTER 4	**Thermoplastic Elastomers**			**36**
	4.1	TPE-O		36
		4.1.1	Properties of TPE-O	36
		4.1.2	Application Areas	37
	4.2	TPE-S		38
		4.2.1	Properties of TPE-S	38
		4.2.2	Application Areas	39
	4.3	TPE-V		39
		4.3.1	Properties of TPE-V	40
		4.3.2	Application Areas	40
	4.4	TPE-U		41
		4.4.1	Properties of TPE-U	41
		4.4.2	Application Areas	42
	4.5	TPE-E		42
		4.5.1	Properties of TPE-E	42
		4.5.2	Application Areas	43
	4.6	TPE-A		44
		4.6.1	Properties of TPE-A	44
		4.6.2	Application Areas	45
CHAPTER 5	**High-Performance Polymers**			**46**
	5.1	Advanced Thermoplastics		46
		5.1.1	Recycling	47
	5.2	Fluoropolymers		47
		5.2.1	Properties of PTFE	48
		5.2.2	Application Areas	48
	5.3	"High-Performance" Nylon – PPA		48
		5.3.1	Properties of PPA	49
		5.3.2	Application Areas	50
	5.4	"Liquid Crystal Polymer" – LCP		50
		5.4.1	Properties of LCP	50
		5.4.2	Application Areas	51

5.5	Polyphenylene Sulfide – PPS		52
	5.5.1 Properties of PPS		52
	5.5.2 Application Areas		52
5.6	Polyether Ether Ketone – PEEK		53
	5.6.1 Properties of PEEK		53
	5.6.2 Application Areas		54
5.7	Polyetherimide – PEI		54
	5.7.1 Properties of PEI		55
	5.7.2 Application Areas		55
5.8	Polysulfone – PSU		56
	5.8.1 Properties of PSU		56
	5.8.2 Application Areas		57
5.9	Polyphenylsulfone – PPSU		57
	5.9.1 Properties of PPSU		57
	5.9.2 Application Areas		58

Bioplastics and Biocomposites — 59 — CHAPTER 6

6.1	Definition	59
	6.1.1 What Do We Mean by Bioplastic?	60
6.2	The Market	60
6.3	Bioplastics	62
6.4	Biopolymers	62
6.5	Biobased Polymers: Biopolyester	63
6.6	Biobased Polymers: Biopolyamides	65
6.7	Biobased Polymers from Microorganisms	66
6.8	Bioethanol or Biomethanol	67
6.9	Biocomposites	68
6.10	More Information about Bioplastics	69

Plastic and the Environment — 70 — CHAPTER 7

7.1	Plastic is Climate-Friendly and Saves Energy	70
7.2	Environmental Effects on Plastic	72
7.3	Recycling Plastic	73
	7.3.1 Plastic Recycling in the EU	74

Modification of Polymers — 76 — CHAPTER 8

8.1	Polymerization	76
8.2	Additives	78
	8.2.1 Stiffness and Tensile Strength	79
	8.2.2 Surface Hardness	79
	8.2.3 Wear Resistance	79
	8.2.4 Toughness	80

	8.3	Physical Properties	80
		8.3.1 Appearance	80
		8.3.2 Crystallinity	81
		8.3.3 Weather Resistance	81
		8.3.4 Friction	82
		8.3.5 Density	82
	8.4	Chemical Properties	83
		8.4.1 Permeability	83
		8.4.2 Oxidation Resistance	83
		8.4.3 Hydrolysis Resistance	84
	8.5	Electrical Properties	84
	8.6	Thermal Properties	85
		8.6.1 Heat Stabilization	85
		8.6.2 Heat Deflection Temperature	86
		8.6.3 Flame Retardant Classification	86
	8.7	Material Price	87

CHAPTER 9 Material Data and Measurements — 88

9.1	Tensile Strength and Stiffness	89
9.2	Impact Strength	92
9.3	Maximum Service Temperature	93
	9.3.1 UL Service Temperature	93
	9.3.2 Heat Deflection Temperature	93
9.4	Flammability Tests	95
	9.4.1 HB Rating	95
	9.4.2 V Rating	95
9.5	Electrical Properties	96
9.6	Flow Properties: Melt Index	97
9.7	Shrinkage	97

CHAPTER 10 Material Databases on the Internet — 98

10.1	CAMPUS	98
	10.1.1 Properties of CAMPUS 5.2	99
10.2	Material Data Center	99
	10.2.1 Properties of Material Data Center	100
10.3	Prospector Plastics Database	100

CHAPTER 11 Test Methods for Plastic Raw Materials and Moldings — 102

11.1	Quality Control during Raw Material Production	102
11.2	Visual Quality Control of Plastic Granules	103
11.3	Visual Inspection of Plastic Parts	104
11.4	Tests That Can Be Performed by the Molder	105
11.5	Advanced Testing Methods	107

Injection-Molding Methods 110   CHAPTER 12

12.1	History	110
12.2	Properties	111
	12.2.1 Limitations	111
12.3	The Injection-Molding Machine	112
	12.3.1 The Injection Unit	112
	12.3.2 Locking Unit	113
	12.3.3 Injection-Molding Cycle	114
12.4	Alternative Injection-Molding Methods	115
	12.4.1 Multi-component Injection Molding	115
	12.4.2 Gas or Water Injection	116

Post-molding Operations 117   CHAPTER 13

13.1	Surface Treatment of Moldings	117
	13.1.1 Printing	117
	13.1.2 "Hot Stamp" Printing	118
	13.1.3 Tampon Printing	119
	13.1.4 Screen Printing	119
	13.1.5 IMD: In-Mold Decoration	120
	13.1.6 Laser Marking	121
	13.1.7 Painting	121
	13.1.8 Metalizing/Chroming	122

Different Types of Molds 123   CHAPTER 14

14.1	Two-Plate Molds	123
14.2	Three-Plate Molds	124
14.3	Molds with Slides	124
14.4	Molds with Rotating Cores	125
14.5	Stack Molds	125
14.6	Molds with Ejection from the Fixed Half	126
14.7	Family Molds	126
14.8	Multi-component Molds	127
14.9	Molds with Melt Cores	128

Structure of Molds 129   CHAPTER 15

15.1	The Function of the Mold	130
15.2	Runner Systems – Cold Runners	130
15.3	Runner Systems – Hot Runners	132
15.4	Cold Slug Pockets/Pullers	133
15.5	Tempering or Cooling Systems	134
15.6	Venting Systems	136
15.7	Ejector Systems	137
15.8	Draft Angles	138

Contents

CHAPTER 16   **Mold Design and Product Quality** **139**

     16.1   Mold-Related Problems 139
           16.1.1   Too-Weak Mold Plates 139
           16.1.2   Incorrect Sprue and Nozzle Design 140
           16.1.3   Incorrect Runner Design 141
           16.1.4   Incorrectly Designed, Located, or Missing Cold Slug Pocket 141
           16.1.5   Incorrect Gate Design 142
           16.1.6   Incorrect Venting 143
           16.1.7   Incorrect Mold Temperature Management 144

CHAPTER 17   **Prototype Molds and Mold Filling Analysis** **145**

     17.1   Prototype Molds 145
     17.2   Mold Filling Analysis 146
     17.3   Workflow 147
           17.3.1   Mesh Model 147
           17.3.2   Material Selection 148
           17.3.3   Process Parameters 148
           17.3.4   Selection of Gate Location 148
           17.3.5   Simulations 149
           17.3.6   Results Generated by Simulations 149
           17.3.7   Filling Sequence 150
           17.3.8   Pressure Distribution 150
           17.3.9   Clamping Force 150
           17.3.10   Cooling Time 151
           17.3.11   Temperature Control 151
           17.3.12   Shrinkage and Warpage 151
           17.3.13   Glass Fiber Orientation 152
           17.3.14   Warpage Analysis 152
           17.3.15   Gate Location 152
           17.3.16   Material Replacement 153
           17.3.17   Simulation Software 153

CHAPTER 18   **Rapid Prototyping and Additive Manufacturing** **154**

     18.1   Prototypes 154
     18.2   Rapid Prototyping (RP) 155
           18.2.1   SLA – Stereolithography 156
           18.2.2   SLS – Selective Laser Sintering 159
           18.2.3   The FDM Method 161
           18.2.4   3DP Printing 162
           18.2.5   3D Printing 163
           18.2.6   PolyJet 164
     18.3   Additive Manufacturing 166

CHAPTER 19

Cost Calculations for Moldings 168
- 19.1 Part Cost Calculator 169
- 19.2 Part Cost Scenarios 173
- 19.2 Replacement Cost 174

CHAPTER 20

Extrusion 177
- 20.1 The Extrusion Process 177
  - 20.1.1 Advantages (+) and Limitations (−) 177
- 20.2 Materials for Extrusion 179
- 20.3 The Extruder Design 180
  - 20.3.1 The Cylinder 180
  - 20.3.2 Single-Screws 181
  - 20.3.3 Barrier Screws 181
  - 20.3.4 Straight Twin-Screws 182
  - 20.3.5 Conical Twin-Screws 182
  - 20.3.6 Rotational Direction 183
  - 20.3.7 Comparison of Single-Screws and Twin-Screws 183
  - 20.3.8 Tool/Die 184
  - 20.3.9 Calibration 184
  - 20.3.10 Corrugation 185
  - 20.3.11 Cooling 185
  - 20.3.12 Feeding 186
  - 20.3.13 Marking 186
  - 20.3.14 Further Processing 186
  - 20.3.15 Cutting 187
  - 20.3.16 Winding 188
- 20.4 Extrusion Processes 188
  - 20.4.1 Straight Extrusion 189
  - 20.4.2 Extrusion with Angle Tool/Coating 189
  - 20.4.3 Extrusion of Plates and Sheets 190
  - 20.4.4 Co-extrusion 191
  - 20.4.5 Film Blowing 191
    - 20.4.5.1 Advantages (+) and Limitations (−) 192
  - 20.4.6 Cable Production 193
  - 20.4.7 Monofilament 194
  - 20.4.8 Compounding 195
- 20.5 Design for Extrusion 196
  - 20.5.1 Ribbing – Stiffening 197
  - 20.5.2 Cavity 197
  - 20.5.3 Sealing Lip 197
  - 20.5.4 Hinge 198
  - 20.5.5 Guide 198
  - 20.5.6 Sliding Joint 198
  - 20.5.7 Snap-Fit Joint 199
  - 20.5.8 Bellow 199
  - 20.5.9 Insert/Reinforcement 199
  - 20.5.10 Friction Surface 200

Contents

	20.5.11	Printing/Stamping	200
	20.5.12	Decoration Surface	200
	20.5.13	Drilled Side Holes	201
	20.5.14	Irregular Holes	201
	20.5.15	Corrugation	201
	20.5.16	Spiral Forming	202
	20.5.17	Foaming	202
	20.5.18	Extruded Screw Holes	202
	20.5.19	Muffing and Hot Plate Welding	203

CHAPTER 21 — Alternative Processing Methods for Thermoplastics 204

21.1	Blow Molding	204
21.2	Rotational Molding	206
21.3	Vacuum Forming	207

CHAPTER 22 — Material Selection Process 209

22.1	How Do You Select the Right Material in Your Development Project?	209
22.2	Development Cooperation	210
22.3	Establishing the Requirement Specifications	210
22.4	MUST Requirements	211
22.5	WANT Requirements	212
22.6	Specify and Sort the Material Candidates	213
22.7	Make a Detailed Cost Analysis	214
22.8	Establish a Meaningful Test Program	215

CHAPTER 23 — Requirements and Specification for Plastic Products 216

23.1	Background Information	216
23.2	Batch Size	217
23.3	Part Size	218
23.4	Tolerance Requirements	218
23.5	Part Design	220
23.6	Assembly Requirements	223
23.7	Mechanical Load	223
23.8	Chemical Resistance	224
23.9	Electrical Properties	225
23.10	Environmental Impact	226
23.11	Color	227
23.12	Surface Properties	228
23.13	Other Properties	230
23.14	Regulatory Requirements	231
23.15	Recycling Requirements	232
23.16	Cost Requirements	233

23.17	Requirement Specification – Checklist	234	
	23.17.1 Background information	234	
	23.17.2 Batch Size	234	
	23.17.3 Part Size	235	
	23.17.4 Tolerance Requirements	235	
	23.17.5 Part Design	235	
	23.17.6 Assembly Requirements	235	
	23.17.7 Mechanical Load	235	
	23.17.8 Chemical Resistance	235	
	23.17.9 Electrical Properties	235	
	23.17.10 Environmental Impact	236	
	23.17.11 Color	236	
	23.17.12 Surface Properties	236	
	23.17.13 Other Properties	236	
	23.17.14 Regulatory Requirements	237	
	23.17.15 Recycling	237	
	23.17.16 Costs	237	

Design Rules for Thermoplastic Moldings — 238 **CHAPTER 24**

24.1	Rule 1 – Remember That Plastics Are Not Metals	239	
24.2	Rule 2 – Consider the Specific Characteristics of Plastics	240	
	24.2.1 Anisotropic Behavior	241	
	24.2.2 Temperature-Dependent Behavior	241	
	24.2.3 Time-Dependent Stress-Strain Curve	242	
	24.2.3.1 Creep	242	
	24.2.3.2 Relaxation	242	
	24.2.4 Speed-Dependent Characteristics	243	
	24.2.5 Environmentally Dependent Characteristics	244	
	24.2.6 Easy to Design	244	
	24.2.7 Easy to Color	244	
	24.2.8 Easy to Assemble	245	
	24.2.9 Recycling	245	
24.3	Rule 3 – Design with Regard to Future Recycling	246	
	24.3.1 Dismantling	246	
	24.3.2 Reused Materials	248	
	24.3.3 Coding	248	
	24.3.4 Cleaning	249	
24.4	Rule 4 – Integrate Several Functions into One Component	249	
24.5	Rule 5 – Maintain an Even Wall Thickness	251	
24.6	Rule 6 – Avoid Sharp Corners	252	
24.7	Rule 7 – Use Ribs to Increase Stiffness	254	
	24.7.1 Limitations when Designing Ribs	254	
	24.7.2 Material-Saving Design	255	
	24.7.3 Avoid Sink Marks at Rib Joints	255	

	24.8	Rule 8 – Be Careful with Gate Location and Dimensions	256
		24.8.1 Weld Lines	257
	24.9	Rule 9 – Avoid Tight Tolerances	258
	24.10	Rule 10 – Choose an Appropriate Assembly Method	259

CHAPTER 25 Assembly Methods for Thermoplastics — 260

25.1	Assembly Methods That Facilitate Disassembly		260
25.2	Integrated Snap-Fits		261
25.3	Permanent Assembly Methods		262
	25.3.1	Ultrasonic Welding	262
	25.3.2	Vibration Welding	263
	25.3.3	Rotational Welding	264
	25.3.4	Hot Plate Welding	265
	25.3.5	Infrared Welding	266
	25.3.6	Laser Welding	266
	25.3.7	Riveting	268
	25.3.8	Gluing	269

CHAPTER 26 The Injection-Molding Process — 270

26.1	Molding Processing Analysis	270
26.2	Contact Information	272
26.3	Information Pane	272
26.4	Material Information	273
26.5	Information about the Machine	274
26.6	Information about the Mold	276
26.7	Drying	278
26.8	Processing Information	280
26.9	Temperatures	281
26.10	Pressure, Injection Speed, and Screw Rotation Speed	286
26.11	Hold Pressure	287
26.12	Injection	289
26.13	Screw Rotation Speed	290
26.14	Time and Length Settings	292

CHAPTER 27 Injection Molding Process Parameters — 297

CHAPTER 28 Problem Solving and Quality Management — 301

28.1	Increased Quality Demands		301
28.2	Analytical Troubleshooting – ATS		301
	28.2.1	Definition of the Problem	302
	28.2.2	Deviation Definition	302
28.3	Defining a Problem		303
	28.3.1	Classification of Problems	304
	28.3.2	Problem Analysis	306

		28.3.3 Brainstorming	308
		28.3.4 Verification of Causes	308
		28.3.5 Planning of Actions to Take	309
28.4	Statistical Design of Experiments – DOE		309
		28.4.1 Factorial Experiments	310
28.5	Failure Mode Effect Analysis – FMEA		313
		28.5.1 General Concepts of FMEA	314

Troubleshooting – Causes and Effects 316   CHAPTER 29

29.1	Molding Problems	316
29.2	Fill Ratio	318
	29.2.1 Short Shots – The Part Is Not Completely Filled	318
	29.2.2 Flashes	319
	29.2.3 Sink Marks	319
	29.2.4 Voids or Pores	320
29.3	Surface Defects	321
	29.3.1 Burn Marks	321
	29.3.1.1 Discoloration, Dark Streaks, or Degradation ...	321
	29.3.1.2 Black Specks	321
	29.3.1.3 Splays or Silver Streaks (Partly over the Surface)	322
	29.3.1.4 Diesel Effect – Entrapped Air	323
	29.3.2 Splays or Silver Streaks (All over the Surface)	324
	29.3.3 Color Streaks – Bad Color Dispersion	324
	29.3.4 Color Streaks – Unfavorable Pigment Orientation	325
	29.3.5 Surface Gloss – Matte/Shiny Surface Variations	325
	29.3.6 Surface Gloss – Corona Effect	326
	29.3.7 Splays, Stripes, and Blisters	326
	29.3.8 Glass Fiber Streaks	327
	29.3.9 Weld-Lines (Knit-Lines)	327
	29.3.10 Jetting	328
	29.3.11 Delamination	329
	29.3.12 Record Grooves (Orange Peel)	329
	29.3.13 Cold Slug	330
	29.3.14 Ejector Pin Marks	330
	29.3.15 Oil Stain – Brown or Black Specks	331
	29.3.16 Water Stain	331
29.4	Poor Mechanical Strength	332
	29.4.1 Bubbles or Voids inside the Part	332
	29.4.2 Cracks	332
	29.4.3 Unmelts (Also Called Pitting)	333
	29.4.4 Brittleness	334
	29.4.5 Crazing	334
	29.4.6 Problems with Regrind	335

		29.5	Dimensional Problems	335
			29.5.1 Incorrect Shrinkage	335
			29.5.2 Unrealistic Tolerances	336
			29.5.3 Warpage	337
		29.6	Production Problems	338
			29.6.1 Part Sticks in the Cavity	338
			29.6.2 Part Sticks on the Core	338
			29.6.3 Part Sticks on the Ejector Pins	339
			29.6.4 Sprue Sticks in the Mold	340
			29.6.5 Stringing	341
CHAPTER 30		**Statistical Process Control (SPC)**		**342**
	30.1	Why SPC?		342
	30.2	Definitions in SPC		343
		30.2.1 Normal Distribution (Gaussian Dispersion)		343
	30.3	Standard Deviations		343
		30.3.1 One Standard Deviation		343
		30.3.2 Six Standard Deviations (Six Sigma)		344
		30.3.3 Control Limits		344
		30.3.4 Target Value		346
		30.3.5 Target Value Centering (TC)		347
		30.3.6 Capability Machine (Cm)		347
		30.3.7 Capability Machine Index (Cmk)		348
		30.3.8 Capability Process (Cp)		348
		30.3.9 Capability Process Index (Cpk)		349
		30.3.10 Six Important Factors		349
		30.3.11 Machine Capability		350
		30.3.12 Process Capability		350
	30.4	How SPC Works in Practice		351
		30.4.1 Software		351
		30.4.2 Process Data Monitoring		352
CHAPTER 31		**Internet Links**		**354**
		Index		**355**

Foreword

For many years, I had the idea of writing a book about injection molding, as I have spent over 45 years of my working life on this subject.

When I retired in 2009 I was given great support by my friends Katarina Elner-Haglund and Peter Schulz of the Swedish plastics magazine Plastforum, who asked me to write a series of articles about thermoplastics and their processing for the magazine.

I was also hired at this time to work with educational programs at the Lund University of Technology, the Royal University of Technology in Stockholm, and a number of industrial companies in Sweden, as a result of which this book was developed.

My aim has been to write in such a way that this book can be understood by everyone, regardless of prior knowledge about plastics. The book has a practical approach with lots of pictures and is intended to be used in secondary schools, universities, industrial training, and self-study. In some of the chapters there are references to worksheets in Excel that can be downloaded free from my website: www.brucon.se.

In addition to the above-mentioned persons, I would like to extend a warm thanks to my wife Ingelöv, who has been very patient when I've been totally absent in the "wonderful world of plastics" and then proofread the book; my brother Hans-Peter, who has spent countless hours on adjustments of all the images etc; and my son-in-law Stefan Bruder, who has checked the contents of the book and contributed with many valuable comments.

I would also like to thank my previous employer, DuPont Performance Polymers and especially my friends and former managers Björn Hedlund and Stewart Daykin, who encouraged the development of my career as a trainer until I reached my ultimate goal and dream job of "global technical training manager". They have also contributed with a lot of information and many valuable images in this book.

I also want said a big thank you to my friends and business partners in all educational programs in recent years, who have supported me and contributed with many valuable comments, information and images for this book, and a special thanks to those who have made this printing possible thanks to the ads in the beginning. The whole list would be very long but you will find some of them in the list of internet links in Chapter 31.

In this Second Edition:

This edition contains greatly expanded coverage of extrusion, collected into a new Chapter 20. There are also a number of new and updated figures, with numerous small improvements and corrections throughout the text. These are complemented by an all-new professional layout and structure, which I hope will help readers to navigate the book comfortably.

I would like to thank Mark Smith at Carl Hanser Verlag for all the support that I received during the last years with my book in various languages.

Ulf Bruder

Karlskrona, Sweden

CHAPTER 1
Polymers and Plastics

Sometimes you get the question: What is the difference between polymer and plastic? The answer is simple: there is no difference, it's the same thing. The word "polymer" comes from the Greek "poly", which means many, and "more" or "meros", which means unity.

The online encyclopedia Wikipedia (www.wikipedia.org) states the following: "Polymers are chemical compounds that consist of very long chains composed of small repeating units, monomers. Polymer chains are different from other chain molecules in organic chemistry because they are much longer than, for example, chains of alcohols or organic acids. The reaction that occurs when the monomers become a polymer is called polymerization. Polymers in the form of engineering materials are known in daily speech as plastics.

By plastic, we mean that the engineering material is based on polymers, generally with various additives to give the material the desired properties, such as colors or softeners. Polymeric materials are usually divided into rubber materials (elastomers), thermosets and thermoplastics."

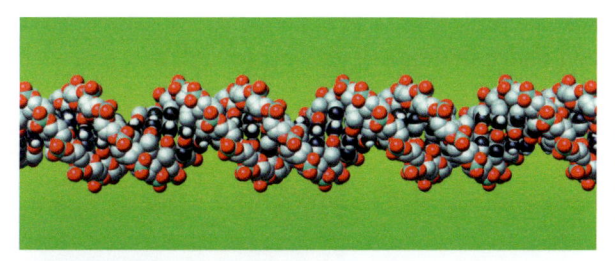

Figure 1.1 Polymers are large macromolecules where monomer molecules bind to each other in long chains. There may be several thousand monomer molecules in a single polymer chain.

Figure 1.2 Amber is a natural polymer. The mosquito in this stone got stuck in the resin of a conifer more than 50 million years ago—something to think about when considering the decomposition of certain polymers in nature.

Chapter 1 — Polymers and Plastics

Most polymers are synthetically produced, but there are also natural polymers such as natural rubber and amber that have been used by mankind for thousands of years.

Other natural polymers include proteins, nucleic acids, and DNA. Cellulose, which is the major component in wood and paper, is also a natural polymer.

In other words, plastic is a synthetically manufactured material composed of monomer molecules that bind to each other in long chains.

If the polymer chain is made up solely of one monomer it is called polymer homopolymer.

If there are several kinds of monomers in the chain, the polymer is called copolymer.

An example of a plastic that can occur both as homopolymer and copolymer is acetal.

Acetal is labeled POM (polyoxymethylene) and is mostly up-built of a monomer known as formaldehyde. The building blocks (atoms) in formaldehyde are composed of carbon, hydrogen, and oxygen.

Most plastic materials are composed of organic monomers but may in some cases also be composed of inorganic acids. One example of an inorganic polymer is a silicone resin consisting of polysiloxanes, where the chain is built up of silicon and oxygen atoms.

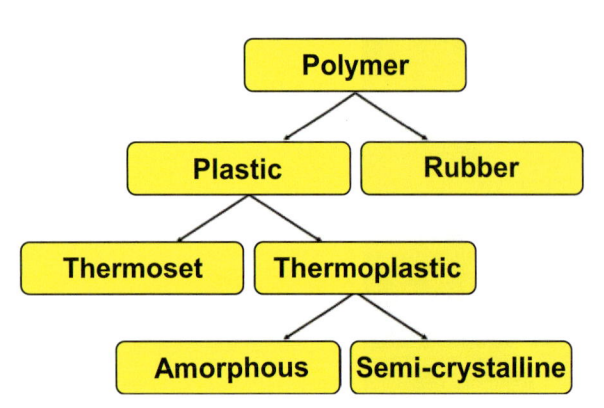

Figure 1.3 Here you can see how you usually divide the synthetic polymers into rubber and plastic, with subgroups thermoset and thermoplastic. Thermoplastic is in turn divided into amorphous and semi-crystalline plastics.

Figure 1.4 Kautschuk or natural rubber is a natural polymer used by man for thousands of years.
In 1839 the American Charles Goodyear invented vulcanization, a cross-linking process in which natural rubber is mixed with sulfur wherein molecular chains are cross-linked under heat and pressure. This process refined rubber's properties significantly.

Carbon and hydrogen are the other dominant elements in plastics. In addition to the aforementioned elements carbon (C), hydrogen (H), oxygen (O), and silicon (Si), plastics typically consist of another five elements: nitrogen (N), fluorine (F), phosphorus (P), sulfur (S), and chlorine (Cl).

It is extremely rare to work with a pure polymer. As a rule, different additives (modifiers) are used to affect a material's properties. Common additives include:

- Surface lubricants (facilitate ejection)
- Heat stabilizers (improve the process window)
- Color pigments
- Reinforcement additives such as glass or carbon fiber (increase stiffness and strength)
- Impact or toughness modifiers
- UV modifiers (e.g. to protect against UV light)
- Fire retardants
- Antistatic agents
- Foaming agents (e.g. EPS, expanded polystyrene)

1.1 Thermosets

In thermosets as well as in rubber, binding can occur between the molecular chains, which is described "cross-linking." These cross-links are so strong that they do not break when heated–thus the material cannot be melted.

Figure 1.5 Plastic padding, or so-called two-component adhesive, is present in many homes. Here two components are mixed with each other, starting a chemical cross-linking reaction causing the material to harden. One of the components is therefore called "hardener." In this case, the reaction occurs at atmospheric pressure and is referred to as a low-pressure reaction.

Figure 1.6 Polyurethane can exist as both a thermoset and a thermoplastic. It can also be rigid or soft as seen in the foam blocks shown here.

Thermosets occur in both liquid and solid form, and in some cases can be processed with high-pressure methods. Some common thermosets include:

- Phenolic plastic (used in saucepan handles)
- Melamine (used in plastic laminates)
- Epoxy (used in two-component adhesives)
- Unsaturated polyester (used in boat hulls)
- Vinyl ester (used in automobile bodywork)
- Polyurethane (used in shoe soles and foam)

Many thermosets have excellent electrical properties and can withstand high operating temperatures. They can be made extremely stiff and strong with glass, carbon, or Kevlar fibers. The main disadvantages are a slower machining process and difficulties of material or energy recycling.

1.2 Thermoplastics

Thermoplastics have the advantage that they melt when heated. They are easy to process with a variety of methods, such as:

- Injection molding (the most common process method for thermoplastics)
- Blow molding (for making bottles and hollow products)
- Extrusion (for pipes, tubes, profiles, and cables)
- Film blowing (e.g. for plastic bags)
- Rotational molding (for large hollow products such as containers, buoys and traffic cones)
- Vacuum forming (for packaging, panels, and roof boxes)

Figure 1.7 Many households now sort their garbage so that the plastic bottles, bags, film, and other plastic products can be recycled.

Figure 1.8 Discarded thermoplastic products can be recycled. These acoustic screens from Polyplank AB are an excellent example.
[Photo: Polyplank AB]

Thermoplastic can be re-melted several times. It is therefore important to recycle plastic products after use. Commodities can usually be recycled up to seven times before the properties become too poor. In the case of engineering and advanced plastics, a maximum of 30% regrind is usually recommended so that the mechanical properties of the new material are not significantly affected. If you cannot use recycled plastics in new products, energy recycling through incineration is often a suitable choice. There is however another option called chemical recycling, although this process has not yet become popular due to the high costs involved versus virgin manufactured material.

1.3 Amorphous and Semi-crystalline Plastics

As shown in Figure 1.3 plastics can be divided into two main groups depending on the plastic structure, i.e. amorphous or semi-crystalline. Glass is another common amorphous material in our environment, and metals have a crystalline structure. An amorphous plastic softens as glass does if you raise the temperature and can therefore be thermoformed.

Amorphous materials have no specific melting point—instead we use the so-called glass transition temperature (T_g), when the molecular chains begin to move. Semi-crystalline plastics do not soften in the same way—instead they change from solid to liquid at the melting point (T_m).

Figure 1.9 Thermoplastic polyester (PET) is a plastic that can be either amorphous, as in soft drink bottles, or semi-crystalline, as in the iron.

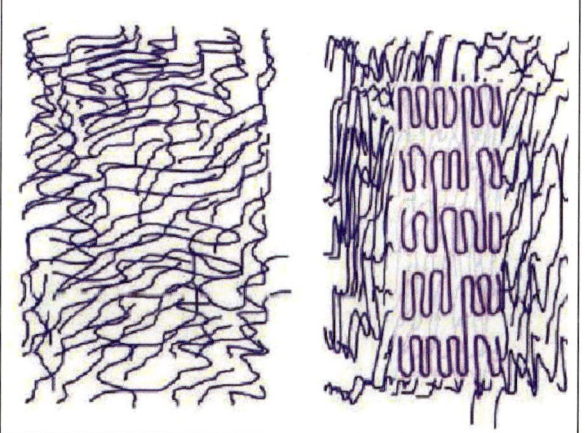

Figure 1.10 The amorphous structure is completely disordered, whereas in the semi-crystalline plastic the molecular chains align themselves in orderly layers (lamellae).

As a rule, semi-crystalline plastics cope better with elevated temperatures than amorphous plastics and have better fatigue resistance and chemical resistance. They are also not sensitive to stress-cracking. Semi-crystalline plastics are more like metal and have better spring properties than amorphous resins. Amorphous plastics can be completely transparent and can be thermoformed. They generally have less mold- and post-shrinkage and have less warpage than semi-crystalline plastics.

It is important that designers and processors of plastic products are aware of the type of material being used since amorphous and semi-crystalline materials behave differently when heated and require different process parameters.

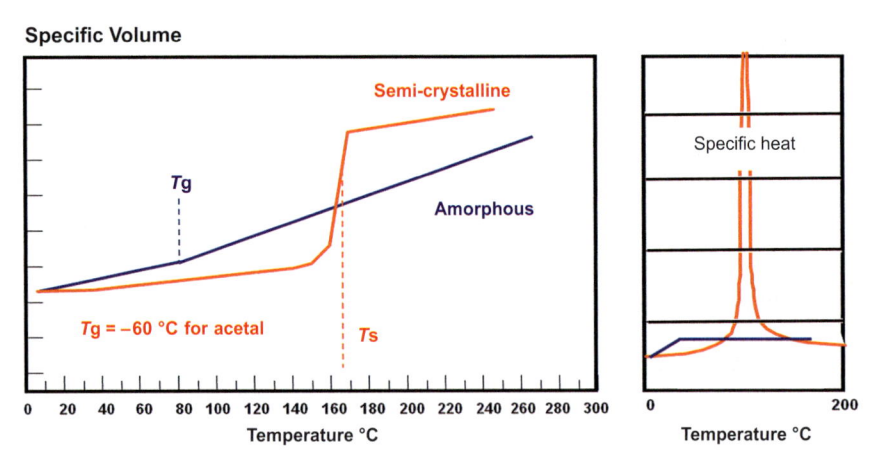

Figure 1.11 Heating increases the specific volume linearly above and below the glass transition temperature (T_g) of the amorphous material. The semi-crystalline material also has a glass transition temperature as there are no plastics with 100% crystallinity. Around the melting point (T_m) the specific volume increases significantly. For acetal this is about 20%, which explains the high shrinkage with injection molding. Amorphous materials have no melting point and significantly less shrinkage. The energy required to raise the temperature one degree remains constant above the T_g of the amorphous material, as shown in the right-hand figure. The semi-crystalline material requires a significant increase in energy to achieve the melting point, the so-called specific heat, to convert the material from a solid to a liquid state. This causes problems for the injection molding processor, as it requires a large energy input when semi-crystalline plastics freeze in the nozzle or hot runners in the mold. Sometimes you have to take a blowtorch to melt the frozen slugs in the cylinder nozzle.

CHAPTER 2
Commodities

2.1 Polyethylene (PE)

Polyethylene or polyethene is a semi-crystalline commodity, denoted as PE. It is the most common plastic, and more than 60 million tons are manufactured each year worldwide. "Low-density" polyethylene (LDPE) was launched on the market by the British chemicals group ICI in 1939.

> **Chemical facts:**
>
> Polyethylene has a very simple structure and consists only of carbon and hydrogen. It belongs to a class of plastics called olefins. These are characterized by their monomers having a double bond, and they are very reactive. The chemical symbol for ethylene, the monomer in PE, is C_2H_4 or $CH_2 = CH_2$, where the "=" sign symbolizes the double bond. Polyethylene can be graphically described as:
>
> $$-\overset{\overset{\displaystyle H}{|}}{\underset{\underset{\displaystyle H}{|}}{C}}-\overset{\overset{\displaystyle H}{|}}{\underset{\underset{\displaystyle H}{|}}{C}}-\overset{\overset{\displaystyle H}{|}}{\underset{\underset{\displaystyle H}{|}}{C}}-\overset{\overset{\displaystyle H}{|}}{\underset{\underset{\displaystyle H}{|}}{C}}-\overset{\overset{\displaystyle H}{|}}{\underset{\underset{\displaystyle H}{|}}{C}}-\overset{\overset{\displaystyle H}{|}}{\underset{\underset{\displaystyle H}{|}}{C}}-$$

Figure 2.1 One reason that PE has become the main commodity is its extensive usage as a packaging material. Plastic bags are made of LDPE.

2.1.1 Classification

Polyethylene can be classified into different groups depending on its density and the lateral branches on the polymer chains:

- UHMWPE Ultrahigh molecular weight
- HDPE   High density
- MDPE   Medium density
- LLDPE   Linear low density
- LDPE   Low density
- PEX   Cross-linked

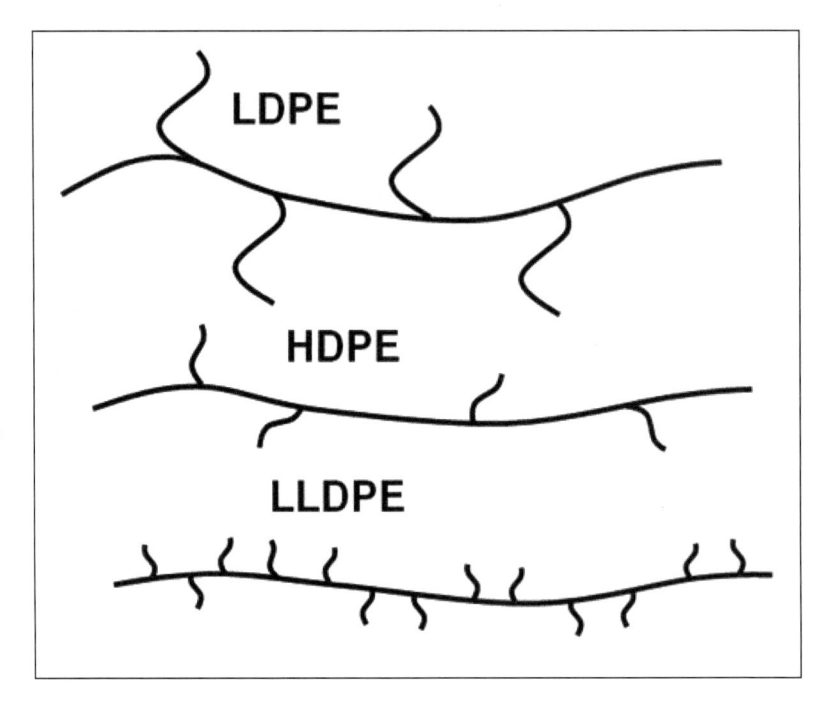

Figure 2.2 When polymerizing ethylene to polyethylene, there are various processes resulting in more or less lateral branches on the molecular chains. A smaller number of lateral branches give a higher crystallinity, molecular weight, and density, since the chains can thus be packed more densely.
HDPE has few or no lateral branches and is also called linear polyethylene.

2.1.2 Properties of Polyethylene

+ Low material price and density
+ Excellent chemical resistance
+ Negligible moisture absorption
+ Food-approved grades are available
+ High elasticity down to < −50 °C

+ Excellent wear resistance (UHMWPE)
+ Easy to color
− Stiffness and tensile strength
− Cannot handle temperatures above 80 °C
− Difficult to paint

The mechanical properties depend largely on the presence of lateral branches, crystallinity, and density, i.e. the type of polyethylene.

2.1 Polyethylene (PE)

2.1.3 Recycling

Polyethylene is one of the most recycled plastic materials. Many of the bags, garbage bags, and dog bags that we use are made from recycled polyethylene. If you use the recycled materials in energy production, the energy content is on par with oil.

When it comes to recycling, the following coding is used:

2.1.4 Application Areas

1) UHMWPE is processed mainly by extrusion into pipes, film, or sheets.

Figure 2.3 Slide rail.
UHMWPE has excellent friction and wear properties and is used in demanding industrial applications, such as this white slide rail for a conveyor belt in gray acetal.

Figure 2.4 Rubbish bins.
HDPE is low cost to produce and easy to mold, even in great detail.

2) HDPE is used for injection molding, blow molding, extrusion, film blowing, and rotational molding.

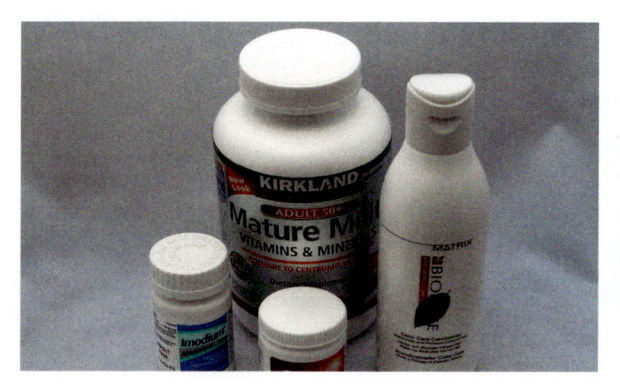

Figure 2.5 Tubs and bottles.
HDPE is appropriate for blow molding and meets food industry standards.

Figure 2.6 Hosepipes.
HDPE is suitable for extrusion. A water hose is tough and strong, approved for drinking water, and can handle the pressure of the mains water supply for the foreseeable future.

3) LDPE is used for film blowing and extrusion.

A large part of all the polyethylene produced is used for film blowing. If the film is soft and flexible, it is either made of LDPE or LLDPE. If it has the rustle of the free bags at the grocery store, it is probably made of HDPE. LLDPE is also used to improve the strength of LDPE film.

Figure 2.7 Garbage bags.
LDPE is excellent for film blowing and is the most common material used in bags, plastic sacks, and construction film.

Figure 2.8 Cable jacketing.
LDPE is used in the extrusion of jacketing for high voltage cables

4) PEX

Cross-linked polyethylene is mainly used in the extrusion of tubes. The cross-linking provides improved creep resistance and better high-temperature properties.

Figure 2.9 Tubes in PEX resist both high temperatures (120°C) and pressure and are used for the hot water supply of cleaning or washing machines.

You can even copolymerize ethylene with polar monomers and get everything from viscous products (e.g. melting glue) to tough films and impact-resistant hard shells such as golf balls.

A common copolymer is EVA (ethylene-vinyl-acetate). By varying the concentration of vinyl acetate (VA) from 2.5 to 95%, you can control the properties and produce a range of different types of material. Increased VA content leads to higher transparency and toughness.

Adhesives, carpet underlay, cable insulation, carriers of color masterbatches, stretch film, and coating film for cardboard and paper are typical uses of EVA.

2.2 Polypropylene (PP)

> **Chemical facts:**
>
> PP has a simple structure and is made up, like PE, only of carbon and hydrogen. It also belongs to the category of plastics called olefins.
>
> Polypropylene is made up of a chain of carbon atoms, where every other carbon atom is bonded to two hydrogen atoms and every other to a hydrogen atom and a methyl group. The monomer formula is:
>
> $$H_2C = CH - CH_3$$
>
> Graphically you describe polypropylene:
>
> $$\begin{array}{c} HHHHHH \\ |||||| \\ -C-C-C-C-C-C- \\ |||||| \\ CH_3HCH_3HCH_3H \end{array}$$

Polypropylene is a semi-crystalline commodity, denoted by–and commonly referred to as–PP. It is also known as "polypropylene." It is the second-largest plastic on the market, after LDPE.

Polypropylene was discovered in 1954, almost simultaneously by two independent researchers Ziegler and Natta, who went on to share the Nobel Prize in 1963.

The Italian chemical company Montecatini launched the material on the market in 1957.

The polymerization of polypropylene can control both crystallinity and molecule size. One can also copolymerize polypropylene with other monomers (e.g. ethylene).

Polypropylene can occur as a homopolymer, random or block copolymer depending on the polymerization method. Polypropylene can also be mixed with elastomers (e.g. EPDM), filled with talc (chalk), or reinforced with glass fiber. In this way it is possible to obtain more grades with widely differing characteristics than can be achieved for any other plastic. Certain grades of polypropylene can handle a continuous temperature of 100 °C plus peaks of up to 140 °C and can therefore be classified as engineering plastics.

2.2.1 Properties of Polypropylene

+ Low material cost and density
+ Excellent chemical resistance
+ Does not absorb moisture
+ Food-approved grades are available
+ Fatigue resistance
− Poor UV resistance (unmodified)
− Brittle at low temperatures (unmodified)
− Poor scratch resistance

Figure 2.10 Buckets and plastic jars.
Household products such as bowls, jars, plastic jars, and buckets are manufactured advantageously in different colors.

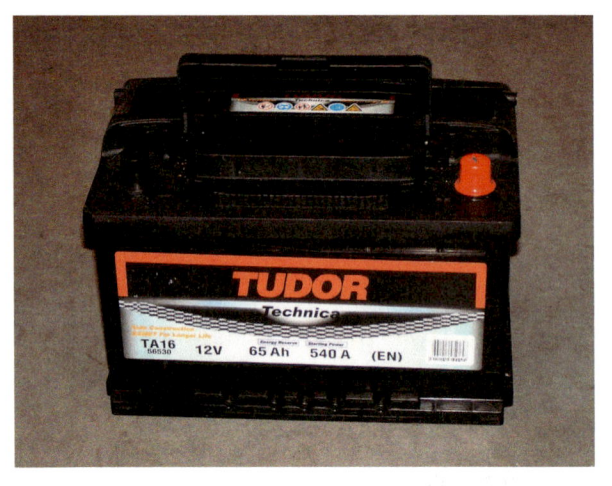

Figure 2.11 Car batteries.
Polypropylene has excellent chemical resistance and can withstand strong acids, making it an excellent material for use in the shell casing for car batteries.

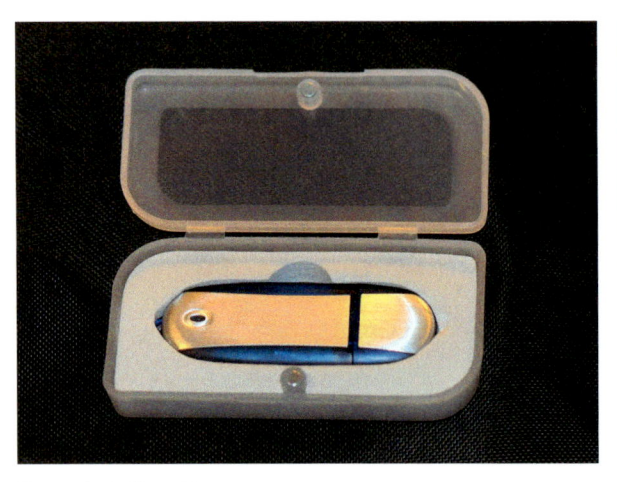

Figure 2.12 Hinged boxes.
PP is widely used in boxes, tubs, and plastic crates. PP hinges are virtually indestructible.

Figure 2.13 Stock and butt for hunting rifles.
Glass fiber reinforced PP has the rigidity and impact strength of polyamide, but lacks the temperature resistance of polyamide.
[Source: Plastinject AB]

2.2.2 Recycling

Recycling of polypropylene will preferably be made by material recycling and secondly by incineration for energy extraction. The recycling code for PP is the triangular recycling symbol with the number 5 inside or > PP < in terms of technical moldings:

2.3 Polyvinylchloride (PVC)

> **Chemical facts:**
>
> PVC has a simple structure but differs from the other basic plastics in that, in addition to carbon and hydrogen, it also has chlorine in the chain. PVC is made up of a chain of carbon atoms bonded alternately—one to two hydrogen atoms, the next to a hydrogen atom and a chlorine atom, and so on, so that the monomer has the formula:
>
> $$H_2C = CH-Cl$$
>
> Graphical view of PVC:
>
> $$-\overset{H}{\underset{H}{C}}-\overset{H}{\underset{Cl}{C}}-\overset{H}{\underset{H}{C}}-\overset{H}{\underset{Cl}{C}}-\overset{H}{\underset{H}{C}}-\overset{H}{\underset{Cl}{C}}-$$

Polyvinylchloride is an amorphous commodity, denoted—and commonly referred to—as PVC. It is the third largest type of plastic, with more than 20 million tonnes produced each year.

PVC was discovered in the 1800s, but did not come into commercial production until 1936 when Union Carbide in the United States launched the material as a substitute for rubber in cable manufacturing.

In the production of PVC, you can use different polymerization methods, and at the compounding stage you can influence the properties more than with any other plastic, ranging from very soft (e.g. garden hoses) to rigid and tough (e.g. waste pipes).

PVC is commonly distinguished as one of three different types: rigid, plasticized, and latex.

2.3.1 Properties of PVC

+ Low material cost and density
+ Excellent chemical resistance
+ Does not absorb moisture
+ Resistant to microorganisms
+ Good long-term strength

+ Food-approved grades are available
+ Self-extinguishing (when not plasticized)
+ Good UV resistance
− Hydrochloric acid is formed during thermal decomposition (fire/burning)

Figure 2.14 Waste pipes.
PVC has excellent chemical resistance and long term durability. Approximately 80% of all the PVC manufactured is used in construction.

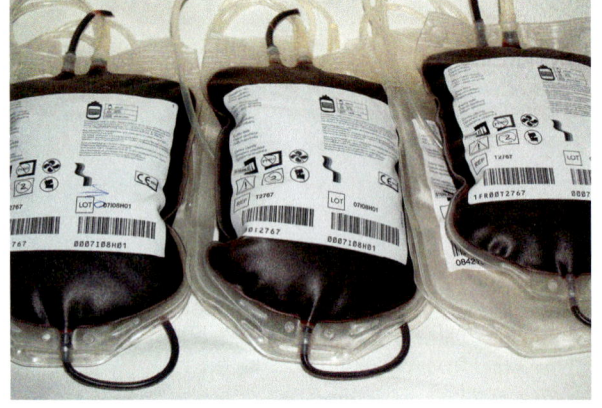

Figure 2.15 Bags for blood transfusion.
Many disposable products in the health service are manufactured from flexible PVC.

2.3.2 Recycling

Recycling of PVC will preferably be made by material recycling and secondly by incineration for energy extraction, and the plastics industry is investing heavily to increase the recycled volumes.

The recycling code for PVC is the triangular recycling symbol with the number 3 inside:

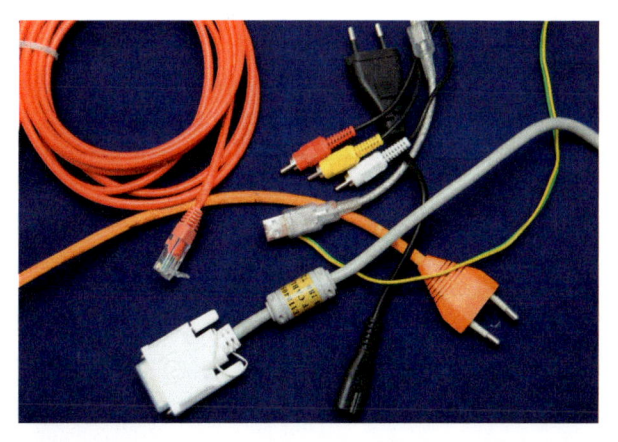

Figure 2.16 Cables.
Plasticized PVC is the predominant material used in the sheathing for cables.

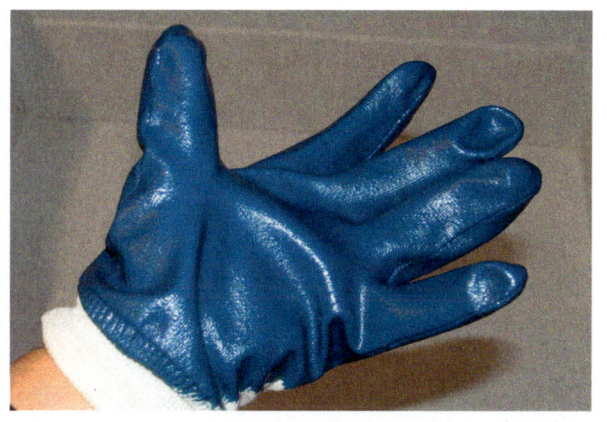

Figure 2.17 Rubber gloves.
Protective gloves and rainproof clothing are often made of latex PVC.

2.4 Polystyrene (PS)

> **Chemical facts:**
>
> PS is manufactured from the monomer styrene, a liquid hydrocarbon produced from oil. The chemical designation of the styrene monomer in PS is:
>
> $H_2C = CH-C_6H_5$
>
> where "=" is a double bond and the hexagon is a so-called benzene ring consisting of six carbon atoms. Each of the ring's carbon atoms is also bound to a hydrogen atom. Polystyrene has an irregular structure in which the polymer chain formula is:
>
> $-CH_2-CH(C_6H_5)-CH_2-CH(C_6H_5)-$

Polystyrene is a crystal-clear amorphous commodity denoted by PS. Polystyrene has traditionally been the cheapest plastic to produce and is widely used in disposable products.

Polystyrene was discovered in 1839 but was not produced on a commercial scale until 1931 when it was launched by IG Farben in Germany.

In 1959 expanded styrene was developed, denoted as EPS. Styrofoam by Dow is the best-known brand of expanded styrene.

2.4.1 Classification

The polymerization of styrene results in a transparent, rigid, and hard plastic with a high gloss finish. Unfortunately, it is very brittle.

Sacrificing the transparency and stiffness, it can be mixed with 5–10% butadiene rubber (BR) to get what is called high-impact polystyrene (HIPS), with an impact strength of up to five times higher than standard polystyrene.

Figure 2.18 Disposable cups.
Many disposable articles are made of polystyrene.

Figure 2.19 CD cases.
CD cases are a typical product made from polystyrene.

Besides mixing polystyrene with other polymers, styrene can also be copolymerized with other monomers to improve properties such as heat resistance, impact strength, stiffness, processability, and chemical resistance. Some common styrene plastics are:

- Styrene-butadiene plastic (SB)
- Acrylonitrile-styrene-acrylate (ASA)
- Styrene-acrylonitrile (SAN)
- Acrylonitrile-butadiene-styrene (ABS)

2.4.2 Properties of Polystyrene

+ Low material cost
+ High transparency (88%)
+ Negligible moisture absorption
+ Food-approved grades are available
+ High hardness and surface gloss

− Brittle
− Poor chemical resistance
− Low softening temperature
− Turns yellow if left outdoors

2.4.3 Recycling

Polystyrene is an easy material to recycle, and is coded:

2.4.4 Application Areas

Polystyrene can be injection molded and extruded. The extruded sheets can be vacuum formed.

Figure 2.20 Parts of polystyrene foam (EPS).
This material is about 80 times higher in volume than conventional polystyrene and is often used as an insulation material in the construction industry, or for disposable cups, shock absorbing packaging for electronic products, etc., as well as floats.
EPS can also be extruded in films. Thicker films can be hot stamped and used for egg cartons, meat trays, and other food packaging.

2.5 Styrene-Acrylonitrile (SAN)

Chemical facts:

SAN is a copolymer of two monomers and typically contains 24% acrylonitrile (the group on the right).

SAN is an amorphous plastic belonging to the styrene family. The material has higher strength, significantly improved chemical resistance, e.g. to fats and oils, and is less sensitive to stress-cracking than polystyrene. SAN has a slightly higher operating temperature compared to PS and also has a better outdoor durability.

It is used to replace glass in some products and is often used in packaging for the cosmetics industry. Other application areas include household products, toothbrush handles, refrigerator interiors, and disposable medical products. The correct recovery code is > SAN <.

Figure 2.21 SAN's good chemical resistance makes it an excellent substitute for glass and is widely used in transparent cosmetic jars.

Figure 2.22 Drawers, shelves, and other transparent refrigerator interior parts are manufactured from SAN. Its original color is slightly yellowish but this is fixed by adding a blue pigment.

2.6 Acrylonitrile-Butadiene-Styrene (ABS)

Acrylonitrile-butadiene-styrene is an amorphous copolymer, abbreviated to and commonly referred to as ABS. ABS was introduced onto the market in 1948.

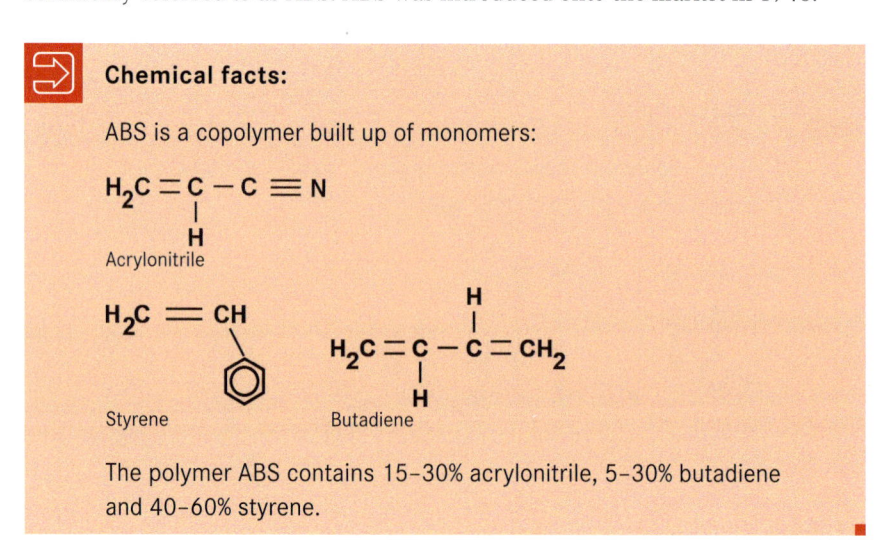

> **Chemical facts:**
>
> ABS is a copolymer built up of monomers:
>
> $$H_2C = C - C \equiv N$$
> $$\quad\quad\quad |$$
> $$\quad\quad\quad H$$
> Acrylonitrile
>
> $$H_2C = CH - C_6H_5$$
> Styrene
>
> $$H_2C = C - C = CH_2$$
> $$\quad\quad | \quad\quad |$$
> $$\quad\quad H \quad\quad H$$
> Butadiene
>
> The polymer ABS contains 15–30% acrylonitrile, 5–30% butadiene and 40–60% styrene.

ABS is manufactured by copolymerizing acrylonitrile and styrene in the presence of polybutadiene (latex rubber). Higher levels of acrylonitrile result in higher strength and better chemical resistance, but mean less butadiene particles resulting in lower impact strength.

The styrene contributes with a high surface gloss and good processing properties, and results in ABS fetching an attractive price.

2.6.1 ABS Blends

In addition to being able to control the properties of ABS by varying the concentration of the monomers, its properties can be further improved by blending with certain engineering plastics. Polycarbonate + ABS (PC/ABS) or polyester + ABS (PBT/ABS) are standard mixtures, known as "plastic alloys." These mixtures fetch a lower price compared to pure polycarbonate or polyester PBT and can even be made flame-resistant.

PC/ABS blends combine the advantages of both plastics and result in a material with better flow properties and better temperature and UV resistance than that of pure ABS. Furthermore, PBT/ABS blends provide better resistance to chemicals (including gasoline) and dimensional stability at elevated temperatures than pure ABS. In the automotive industry blends of PBT/ABS are replacing ABS, PP, and PC/ABS, due to the matte surface providing a better replication of textile surfaces than other plastics, which is much appreciated in interior paneling, etc.

2.6.2 Properties of ABS

+ Combines stiffness, strength, and toughness
+ Dimensionally stable under stress
+ Does not absorb moisture
+ Good surface gloss
+ Easy to color and to paint
+ Excellent for chrome plating

+ Good electrical insulation
+ Can be made transparent
− Heat resistance
− Sensitive to stress-cracking
− Poor UV resistance
− Solvent resistance

2.6.3 Recycling

ABS is an ideal material for recycling. The correct recycling code is > **ABS** <. Sometimes the packaging code is shown as ♻ but this does not indicate the type of polymer used.

2.6.4 Application Areas

ABS can be injection molded and extruded. The extruded sheets can be vacuum formed.

ABS is the most suitable plastic to chrome. During the chrome plating process, etching is used to remove small nitrile particles on the surface, resulting in the formation of small craters. The chromium then penetrates down into the craters, and you get an excellent adhesion between the metal surface and the ABS surface. Besides the aesthetic value, the scratch resistance is much improved. Even PC/ABS blends can be chrome plated. Car door handles, among other things, are made of chrome-plated PC/ABS blends.

Figure 2.23 Chrome-plated showerhead.
Many products made of chrome-plated ABS look as if they are made of metal.

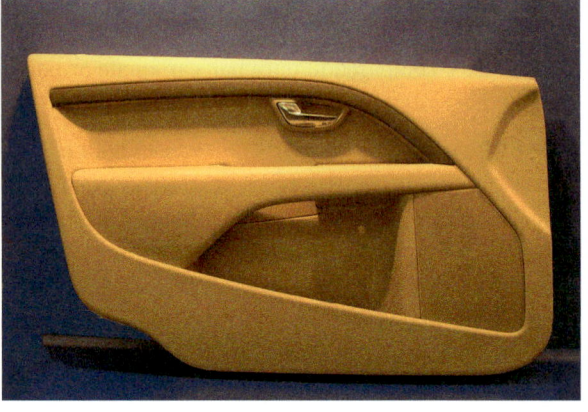

Figure 2.24 Door panel.
Large parts such as door panels and instrument panels for cars are manufactured from ABS.

Figure 2.25 Office machinery.
ABS is a common material for the housings on office equipment, computers, and TVs.

Figure 2.26 Lego bricks.
ABS is used in Lego and other brightly colored plastic toys.

2.7 Polymethyl Methacrylate (PMMA)

Most people don't know the term "PMMA", but if you say "Plexiglas", which is the most famous brand name, everyone will know what you mean.

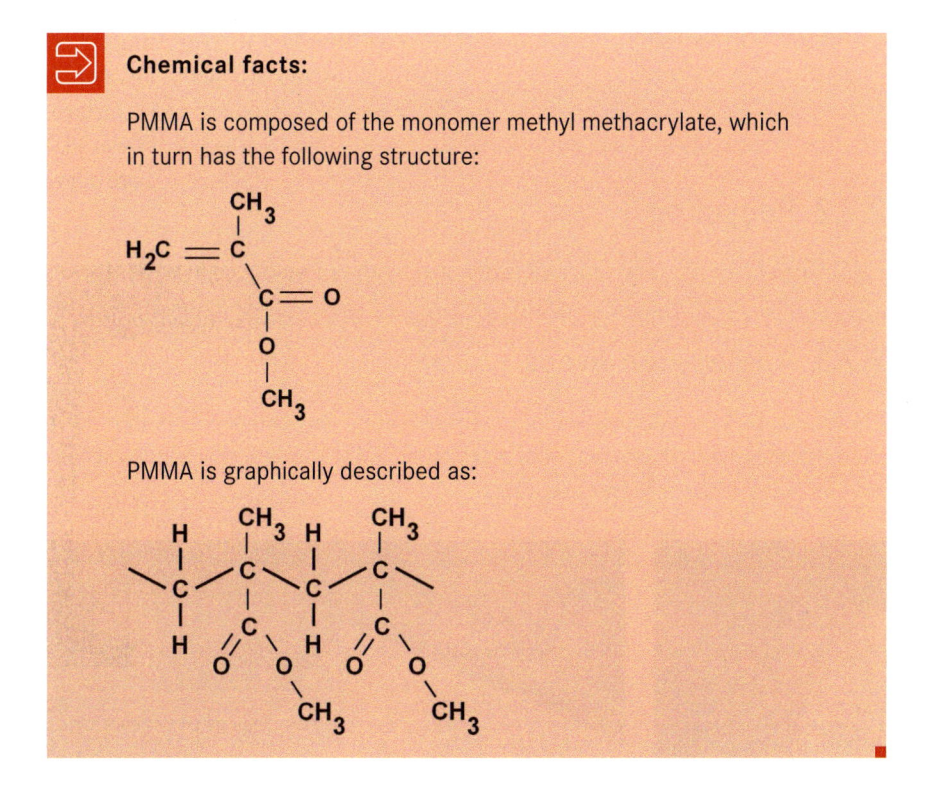

> **Chemical facts:**
>
> PMMA is composed of the monomer methyl methacrylate, which in turn has the following structure:
>
> [structure of methyl methacrylate monomer]
>
> PMMA is graphically described as:
>
> [structure of PMMA polymer]

PMMA is an amorphous crystal-clear acrylic plastic. It was launched as a replacement for glass in 1933 by Rohm & Haas in Germany under the name "Plexiglas".

PMMA has a density of 1.15 to 1.19 g/cm^3, which is less than half the density of glass. The material made a breakthrough during World War II in aircraft cockpit canopies.

Usually PMMA is not used in its pure form, but rather with various additives to improve such properties as:

- Heat stability and processability
- Toughness
- Higher operating temperature
- UV stability

Compared with polystyrene, PMMA has better impact resistance and UV resistance. Compared with polycarbonate, PMMA has lower impact strength but offers greater value for the money. Compared with glass, PMMA is as high in transparency, lighter in weight, and has better impact resistance but lower scratch resistance.

PMMA is supplied as granules for injection molding and extrusion, or in semi-manufactured forms, i.e. sheets, rods, or tubes.

2.7.1 Properties of PMMA

+ Very high transparency (98%)
+ High rigidity and surface hardness
+ Very good UV resistance
+ Good optical properties
+ Can be used in implants

− High thermal expansion coefficient
− Scratch resistance
− Low resistance to stress-cracking
− Solvent resistance
− High melt viscosity (difficult to fill thin walls)

2.7.2 Recycling

PMMA can be easily recycled, and is denoted by the recycling code > PMMA <.

2.7.3 Application Areas

PMMA can be injection molded and extruded. Semi-finished products in PMMA can be processed with conventional machining. PMMA is superior to polycarbonate and polystyrene for laser marking.

Figure 2.27 PMMA works really well in reflective items.

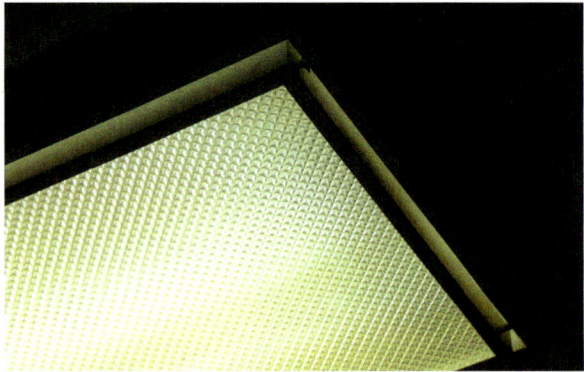

Figure 2.28 PMMA is much used by the lighting industry, e.g. as a screen for fluorescent tubes.

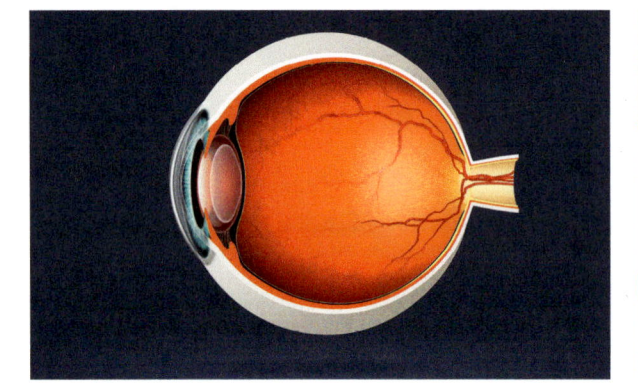

Figure 2.29 Ophthalmic lenses.
PMMA is highly compatible with the human body and is therefore used in implants. Due to its extremely good optical properties, PMMA is used in artificial lenses that are surgically inserted into the eye.

Figure 2.30 Safety glass at sports arenas.
The protective glass shields around hockey rinks are usually made of PMMA as the material has high transparency and sufficient toughness.

CHAPTER 3
Engineering Polymers

3.1 Polyamide or Nylon

Polyamide is a semi-crystalline engineering plastic, denoted by PA. There are several different types of polyamide, of which PA6 and PA66 are the most common. Polyamide was the first engineering polymer launched on the market. It is also the largest in volume since it is widely used in the automotive industry.

Polyamide was invented by DuPont in the United States in 1934 and was first launched as a fiber in parachutes and women's stockings under the trade name Nylon.

A few years later, the injection-molding grades were launched. Nylon became a general term; DuPont lost the trademark and currently markets its polyamides under the trade name Zytel. Ultramid from BASF, Durethan from Lanxess, and Akulon from DSM are some of the other famous trade names on the market.

3.1.1 Classification

The development of polyamide has focused on improving the high-temperature properties and reducing water absorption. This has led to a number of variants where in addition to PA6 and PA66 the following types should be mentioned: PA666, PA46, PA11, PA12, and PA612.

About a decade ago, aromatic "high performance" polyamides were introduced, usually known as PPA, which stands for polyphthalamide. The latest trend is "bio-polyamides" made from long-chain monomers, e.g. PA410, PA610, PA1010, PA10, PA11, and PA612.

> **Chemical facts:**
>
> Polyamide is available in a number of variations, labeled alpha-numerically, e.g. PA66, indicating the number of carbon atoms in the molecules that make up the monomer. PA6 is the most common type of polyamide and has the simplest structure:
>
> $$\left[NH - \overset{O}{\underset{\|}{C}} - (CH_2)_5 \right]_N$$
>
> PA66 has a monomer that consists of two different molecules wherein each molecule has six carbon atoms, as illustrated below:
>
> $$\left[\underset{\text{Amide group}}{NH - (CH_2)_6 - NH} - \underset{\text{Acid group}}{\overset{O}{\underset{\|}{C}} - (CH_2)_4 - \overset{O}{\underset{\|}{C}}} \right]_N$$

Chapter 3 — Engineering Polymers

Figure 3.1 Polyamide has an excellent combination of good electrical properties, high operating temperatures, and flame-retardant capability (up to UL V-0 classification). The material is therefore used for electrical components such as fuses, circuit breakers, transformer housing, etc.

Table 3.1 This table shows the mechanical properties of a standard quality of PA66 in a DAM, Dry As Molded, (unconditioned) state and after the material has absorbed 2.5% humidity at 23 °C and 50% rel. humidity (conditioned state). The stiffness decreases by 65% and tensile strength by 35%, while toughness (elongation) increases five-fold.
The impact strength at room temperature increases three-fold but drops by 33% at low temperature. [Source: DuPont]

Mechanical Properties	DAM	Cond.	Unit
Stiffness (tensile modulus)	3,100	1,400	MPa
Tensile stress (at yield)	82	53	MPa
Elongation at yield at +23 °C	4.5	25	%
Charpy notched impact strength at +30 °C	5.5	15	kJ/m^2
Charpy notched impact strength at −30 °C	4.5	3	kJ/m^2

3.1.2 Properties of Polyamide

+ Stiffness at high temperatures (glass fiber reinforced PA)
+ High service temperatures: 120 °C constantly and a short-term peak temperature of 180 °C
+ Good electrical properties
+ Food-approved grades are available
+ Can be made flame-retardant
− Absorbs excess moisture from the air, which alters the mechanical properties and dimensional stability
− Brittleness at low temperatures if not impact modified

3.1.3 Recycling

Material recycling is preferable for PA, although incineration for energy extraction is also an option. The basic recycling code is > PA <, sometimes with additional information shown. For example, PA66 containing 30% glass fibers would be > PA66 GF30 <.

3.1.4 Application Areas

Polyamide can be processed by injection molding, extrusion, and blow molding.

Figure 3.2 The end pieces of a car radiator, molded in a special hydrolysis-stabilized grade of polyamide 66.

Figure 3.3 Covers and gasoline tanks for chain saws, manufactured in impact-modified polyamide that can withstand gasoline, oils, and rough handling at low temperatures.

Figure 3.4 The automotive industry is a major user of polyamide. Engine parts such as intake manifolds and cylinder head covers are often made of glass reinforced polyamide. There are also hot oil resistant grades.

Figure 3.5 Electrical hand tools almost always have covers made from impact-modified polyamide, since the material withstands rough treatment and has good electrical insulation properties.

Figure 3.6 The metal in car pedals has been replaced with impact-resistant glass fiber reinforced polyamide with great weight savings as a result.

Figure 3.7 Supertough polyamide is used in sport equipment such as this floorball stick.

Figure 3.8 There are special blow-molding grades of polyamide. The upper tube is made in a blow-molding grade and has a high gloss inner surface.

3.2 Acetal

Acetal is the most crystalline material of all the engineering polymers. It is denoted by POM, an abbreviation of the chemical name polyoxymethylene. There are two variants of acetal: acetal homopolymer and acetal copolymer. Acetal homopolymer was invented and introduced to the market by DuPont in the United States in 1958. Two years later, Celanese (also in the U.S.) came along with acetal copolymer. Acetal is difficult to produce, so there are only a few manufacturers world-wide, the three leading ones being Celanese with the brand name Hostaform, DuPont with Delrin, and BASF with Ultraform.

3.2 Acetal

> **Chemical facts:**
>
> Acetal is composed of the monomer oxymethylene (also known as formaldehyde). There are about 1,500 molecules of formaldehyde (marked in red) in the homopolymer chain. For the acetal copolymer, about 2.5% of other monomers are present in the polymer chain (the so-called copolymer group, marked here in blue).

The homopolymer has better mechanical properties compared to the copolymer, although the copolymer has better resistance to hot water. Processing-wise and price-wise, they are about equal.

Figure 3.9 The wheels on baskets in dishwashers are produced in acetal copolymer, which has a better hot-water resistance than the acetal homopolymer.

3.2.1 Properties of Acetal

+ The stiffest non-reinforced engineering polymer
+ The mechanical properties are affected only slightly in the temperature range −40 °C to +80 °C
+ High toughness without impact modifiers
+ High fatigue resistance
+ Good creep resistance
+ Excellent spring properties
+ Does not absorb moisture and has good dimensional stability
+ Good resistance to gasoline and solvents
+ Excellent friction and wear properties
+ Food-approved grades are available
− The maximum continuous service temperature is only 80 °C, with a short-term peak temperature of 120 °C
− Squeaks against itself (can be eliminated with lubricants)
− Sensitive to stress concentration (i.e. sharp corners)

Chapter 3 – Engineering Polymers

3.2.2 Recycling

Material recycling is preferable for POM, although incineration for energy extraction is also an option. The recycling code is > POM <.

3.2.3 Application Areas

Acetal can be processed by injection molding and extrusion. In recent years, grades which can be painted and chrome plated have been developed.

Figure 3.10 Alpine ski binding.
Acetal is appropriate for alpine ski bindings since its mechanical properties are negligibly affected by the temperature range in which the bindings are used, i.e. the release force is almost constant.

Figure 3.11 Fuel filler of a diesel car.
Acetal plastic has very good resistance to gasoline, diesel, and ethanol, and is used by the automotive industry for fuel fillers, filling pipes, level gauge, fuel pumps, etc.

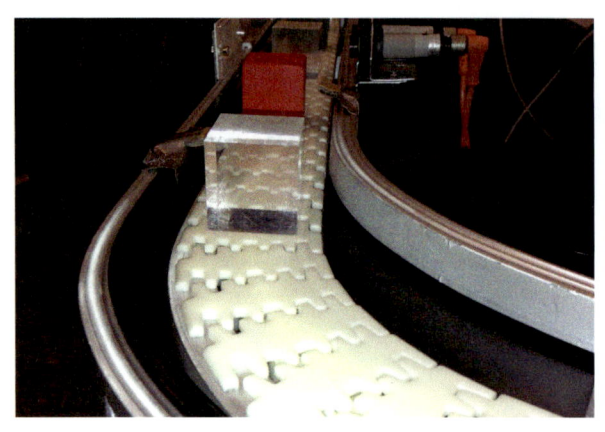

Figure 3.12 Chain links for conveyor belts.
Acetal's excellent friction and wear properties are very useful for this type of application.

Figure 3.13 Gears and cog wheels.
Acetal is the natural choice for manufacturing plastic gears, due to its excellent fatigue resistance combined with low friction and wear.

Figure 3.14 Snap fasteners.
Because acetal is the most crystalline of the engineering polymers, it has the most metal-like properties. The excellent spring characteristics are well suited to the design of various types of fasteners and clips.

Figure 3.15 Inhaler in acetal.
There are special grades of acetal used for medical applications such as inhalers and instruments. Acetal plastic can be sterilized.

3.3 Polyester

Polyester is the name of a group of plastics that can be both thermoset and thermoplastic. Thermoplastic polyester may in turn be either amorphous or crystalline. Among the crystalline types we find the engineering polymers PBT and PET, but also more advanced plastics such as LCP and PCT, which we will return to later. In this chapter, we focus on the semi-crystalline thermoplastic polyesters PBT and PET.

Polyester combines high stiffness, high temperature resistance, and good electrical properties and is used primarily in the electrical and electronics industry and in the automotive industry. However, only a very small proportion of all the polyester manufactured is used for injection molding and extrusion because its largest usage is for fibers, packaging, and film.

It was Carothers, the man who later invented Nylon, who first discovered polyester in the late 1920s at the DuPont laboratories in the U.S. However, DuPont did not launch the material onto the market, and it was not until 1940 that the German company Agfa introduced it as a synthetic fiber. The first injection-molding grade was launched in the mid-1960s by Akzo in the Netherlands, but the real breakthrough as an engineering polymer first came about in the 1980s.

Polyester PBT, an abbreviation of the chemical name polybutylene terephthalate, is the type that is easiest to process (it does not require the same extreme drying process as PET) and has therefore become the larger of the two materials in terms of volume produced. PBT was launched by Celanese in the U.S. in 1970. The leading manufacturers of PBT are Sabic with the trade name Valox, DuPont with Crastin, Lanxess with Pocan, BASF with Ultradur, and DSM with Arnite.

Polyester PET, where PET stands for polyethylene terephthalate, has better mechanical and thermal properties than PBT. However, the material is very brittle if it is not pre-dried to less than 0.02% moisture content (PBT requires 0.04% and polyamide 0.2%). This posed some problems when it was first launched, because few molders had enough knowledge and equipment to process such a difficult material to dry. Today, the drying technology has caught up, and now most molders can cope with processing PET. The leading manufacturers of injection-molding grade PET are DuPont with the trade name Rynite, DSM with Arnite, and Celanese with Impet.

> **Chemical facts:**
>
> Polyester PET has the simplest structure with the following monomer. The molecule consisting of the atoms indicated in red in the chemical formula below is known as the ester group, from which polyester derives its name.
>
> $$-\overset{O}{\underset{\|}{C}}-\bigcirc-\overset{O}{\underset{\|}{C}}-O-CH_2-CH_2-O-$$
>
> PET
>
> PBT's monomer is built in a manner similar to PET, but here includes another two carbon and four hydrogen atoms.
>
> $$-\overset{O}{\underset{\|}{C}}-\bigcirc-\overset{O}{\underset{\|}{C}}-O-CH_2-CH_2-CH_2-CH_2-O-$$
>
> PBT

Figure 3.16 Polyester is a material that can occur in many variants. The pictures here show bottles made of amorphous PET, and an iron made with glass fiber reinforced semi-crystalline PET.

3.3 Polyester

3.3.1 Properties of Polyester PBT and PET

+ Stiffness at high temperatures (when glass fiber reinforced)
+ Dimensional stability (does not absorb moisture as polyamide does)
+ High constant service temperature of 130 °C and 180 °C for PBT and PET respectively, and short-term peak temperatures of 155 °C and 200 °C for PBT and PET respectively
+ Good electrical properties
+ Good weathering resistance (UV light)
+ Can be made flame-retardant
+ High surface gloss
+ Available in beautiful colors, can be painted and metalized
− Degrades in hot water above 80 °C (hydrolysis)
− Poor resistance to strong acids and alkalis, oxidants, and alcohols

3.3.2 Recycling

Material recycling is preferable for POM, although incineration for energy extraction is also an option.

The recycling code for PET in technical moldings is > PET < and ♻1 when used in bottles or packaging. For PBT, the code is > PBT <.

3.3.3 Application Areas

Polyester can be processed by injection and blow molding, extrusion and film blowing. Sheets in amorphous PET can be vacuum formed.

Figure 3.17 Oven handles are made of a special type of color-stabilized polyester PET that will not yellow when exposed to high temperatures.
PBT is the most commonly used material for oven knobs.

Figure 3.18 Because glass fiber reinforced PBT can withstand high temperatures, is dimensionally stable, has a high surface gloss, and can be metallized, it has become a weight-saving alternative to metal in bezels.

Chapter 3 – Engineering Polymers

Figure 3.19 Both PET and PBT are used in irons because they can withstand high temperatures, are available in bright colors, and have good surface gloss.

Figure 3.20 The orange handles on Fiskars scissors are manufactured from pre-colored PBT. They are dishwasher-safe and solvent-proof.

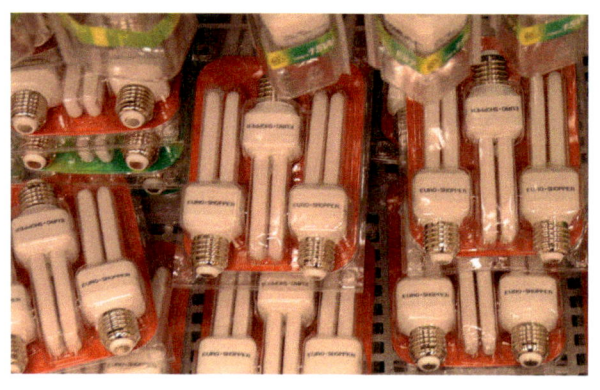

Figure 3.21 The electronic circuits in low-energy light bulbs are encapsulated in a housing of PBT. The packaging for the light bulbs in the picture is made of vacuum-formed sheets of transparent amorphous PET.

Figure 3.22 PBT has very good resistance to the sun's UV rays and is therefore suitable for outdoor applications. High-gloss exterior mirrors for cars are often made in this material.

Figure 3.23 The combination of high stiffness and good UV resistance enables glass fiber reinforced PET to be used for windscreen wipers on cars.

Figure 3.24 Electrical components in the automotive and electrical industries are often made in PBT. Both PBT and PET flow very easily in the molding tools, so many thin-walled goods can be filled without problems.

3.4 Polycarbonate

Polycarbonate is a crystal-clear very impact resistant amorphous engineering plastic, denoted by the abbreviation PC. The material was invented in the 1950s and launched onto the market by Bayer in Germany in 1958 under the trade name Makrolon. Other major manufacturers of polycarbonate are Sabic with Lexan, Dow with Calibre, and DSM with Xantar.

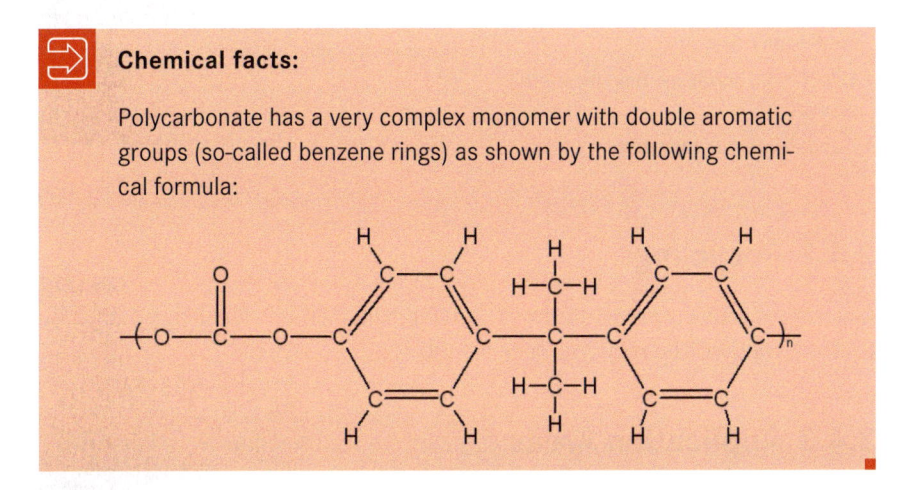

> **Chemical facts:**
>
> Polycarbonate has a very complex monomer with double aromatic groups (so-called benzene rings) as shown by the following chemical formula:

Figure 3.25 Approximately one-third of the polycarbonate manufactured is used for CD, DVD, and Blu-ray discs.

The material has grown substantially in recent years thanks to the revolutionary development of CD and DVD discs, impact-resistant plastic glass in the construction industry, and in car headlights. Approximately 10–15% of all polycarbonate is used in alloys with other plastics to get a good combination of features and price. Examples are: PC-ABS as used in mobile phone shells, PC-PBT, which copes better with the oven curing process for painting in the automotive production line, and PC-ASA, which has better UV resistance than pure polycarbonate.

3.4.1 Properties of Polycarbonate

+ Crystal clear (light permeability 89%)
+ Very high impact strength (at low temperatures down to −40 °C)
+ High operating temperature (120 °C constant, and 145 °C short-term peak load)
+ Negligible moisture absorption and good dimensional stability
+ Lower mold shrinkage than most other plastics
+ Good electrical properties
+ Self-extinguishing V-2 and can be V-0 with additives
+ Food-approved grades available
− High tendency to stress-crack under constant load
− Solvent triggers cracking
− Degrades in water hotter than 60 °C, but can be machine washed

3.4.2 Recycling

Material recycling is preferable for PC, although incineration for energy extraction is also an option. The recycling code is > PC <.

3.4.3 Application Areas

Polycarbonate can be processed by injection molding and extrusion, both with and without glass fiber. PC sheets can be vacuum formed.

Figure 3.26 Polycarbonate has a poor chemical resistance, as can be seen from the cracks caused by vinegar in the salad bowl here.

3.4 Polycarbonate

Figure 3.27 Extruded tubes of glass fiber reinforced polycarbonate are both stiff and strong and can withstand tough impact, as the paddle in this picture shows.

Figure 3.28 The glass for car headlamps is made of polycarbonate and coated with a thin layer of siloxane to improve scratch resistance, UV protection, and protection against solvents.

Figure 3.29 Polycarbonate is both incredibly impact resistant and suitable for painting, making it an excellent material for motorcycle helmets. The visor is also produced in polycarbonate.

35

CHAPTER 4
Thermoplastic Elastomers

Thermoplastic elastomers (TPE) are soft thermoplastics with a low E-modulus and high toughness. Also called thermoplastic rubbers, their toughness is sometimes indicated by Shore A or Shore D to characterize them, as with rubber. Their chemical structure consists of both thermoplastic hard segments and elastic soft segments. The crucial difference to traditional rubber is the lack of, or at least very slight, cross-linking between the molecular chains. Most of the various TPEs offer a cost-effective alternative to rubber in a variety of applications, thanks to its suitability for different processes such as injection molding, extrusion, film, and blow molding. Feature-wise, however, rubber has the advantage of higher elasticity and lower compression under constant load. All the thermoplastic elastomers are ideal for material recycling, although incineration for energy extraction is also an option.

TPEs can generally be divided into the following groups:

- TPE-O, olefin-based elastomers
- TPE-S, styrene-based elastomers
- TPE-V, olefin-based elastomers with vulcanized rubber particles
- TPE-U, polyurethane-based elastomers
- TPE-E, polyester-based elastomers
- TPE-A, polyamide-based elastomers

4.1 TPE-O

TPE-O (or TPO) thermoplastic elastomers, where the "O" stands for "olefin", are a blend of polypropylene and EPDM uncured rubber particles. Because it has a PP matrix, TPO takes on a semi-crystalline structure. TPO-based elastomers are among the largest and most cost-effective TPEs available. They have been on the market since 1970, and leading manufacturers are Elasto, Elastron, Exxon Mobile, So.F.teR, and Teknor Apex.

By mixing the levels of EPDM in PP at concentrations from 10 to 65%, a great range of properties can be achieved. With mixture concentrations below 20% we usually call the materials impact modified PP, while levels above 60% give the more rubber-like properties.

The recycling code for TPE-O is > PP + EPDM <.

4.1.1 Properties of TPE-O

+ Cost-effective substitute for rubber
+ High stretch factor
+ Good tear resistance
+ Flexible at low temperatures
+ Good surface finish
+ Good chemical resistance
+ Can be UV stabilized
+ Easy to process
+ Can be colored
+ Paintable (primer required)
− Deformation properties (i.e. setting characteristics) not as good as rubber

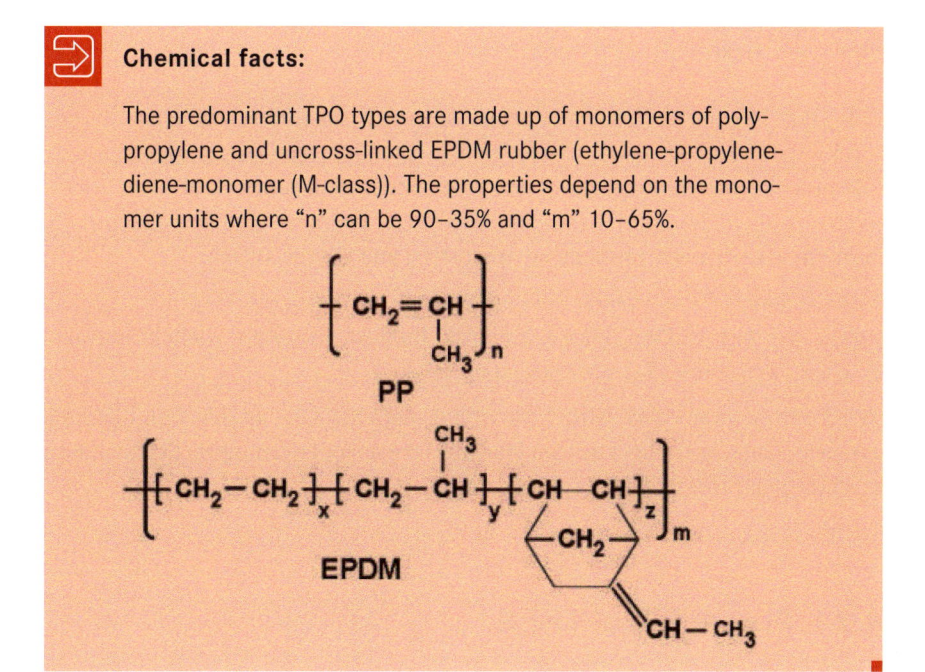

> **Chemical facts:**
>
> The predominant TPO types are made up of monomers of polypropylene and uncross-linked EPDM rubber (ethylene-propylene-diene-monomer (M-class)). The properties depend on the monomer units where "n" can be 90–35% and "m" 10–65%.
>
> $$\left[CH_2 = CH - \underset{CH_3}{|} \right]_n$$
> **PP**
>
> $$\left[[CH_2-CH_2]_x [CH_2 - \underset{CH_3}{\overset{|}{CH}}]_y [CH - CH]_z \right]_m$$
> **EPDM**

4.1.2 Application Areas

TPO-based elastomers can be used in a variety of applications in the automotive, construction, and engineering industries, also in household products, footwear, and sportswear.

Figure 4.1 The automotive industry is the largest market for TPE-O where it is commonly used in bumpers, spoilers, and interior panels. TPE-O is sufficiently rigid and has good impact resistance even at low temperatures. It can also be painted to the same finish as the car's sheet metal parts.

Figure 4.2 TPE-O is often used to make sports equipment such as fins, masks, snorkels, and other accessories for scuba diving. TPE-O is also used for sports shoes, ski boots, skates, helmets, protective gear, and the soft grips on rods, rackets, clubs, etc.

4.2 TPE-S

TPE-S (or TPS) thermoplastic elastomers, where "S" is an abbreviation for "styrene block copolymer," are usually based on SBS or SEBS (see Chemical facts below). SBS probably has the largest market and is used in applications where resistance to chemicals and aging is less critical. SEBS is characterized by substantially better heat resistance, mechanical properties, and UV resistance.

TPS elastomers have been on the market since the 1960s, and leading manufacturers are: API, ChiMei, Elasto, Elastron, Enplast, Kraiburg, Radichi, Ravago, So.F.teR, Styrolution, Teknor Apex, and Uteksol.

TPE-S can be processed using a variety of methods such as injection molding, extrusion, blow molding, and film blowing. One major advantage is that standard machinery for thermoplastics can be used.

The recycling codes are > SBS < or > SEBS <.

4.2.1 Properties of TPE-S

+ The hardness can be controlled in a wide range
+ Good abrasion resistance
+ Flexible at low temperatures
+ Can be made transparent
+ Good gas and moisture permeability
+ Can be UV stabilized

+ Easy to process
+ Easier to color than TPE-O
+ Good adhesion (for over-molding) with a number of thermoplastic resins such as PP, PS, ABS, and PA
− Less chemical resistance than TPE-O

> **Chemical facts:**
>
> SBS and SEBS are based on a styrene block copolymer with hard and soft segments. In SBS, the styrene end blocks give the thermoplastic properties, and the center blocks of butadiene give the rubber-like properties. In SEBS, it is the ethylene-co-butane molecules that provide the elastic properties.
>
> $$\text{-[CH}_2\text{-CH]}_x\text{-[(CH}_2\text{-CH=CH-CH}_2)_m\text{(CH}_2\text{-CH)}_n\text{]}_y\text{(CH}_2\text{-CH)}_x\text{-}$$
>
> SBS
>
> Poly(styrene-block-butadiene-block-styrene)
>
> $$\text{-[CH}_2\text{-CH]}_x\text{-[(CH}_2\text{-CH}_2\text{-CH}_2\text{-CH}_2)_m\text{(CH}_2\text{-CH)}_n\text{]}_y\text{(CH}_2\text{-CH)}_x\text{-}$$
>
> SEBS
>
> Poly(styrene-block-ethylene-co-butane-block-styrene)

4.2.2 Application Areas

TPS-based elastomers are used in a variety of applications across a broad range of industries: automotive, consumer products, construction, healthcare, footwear, sportswear, electrical cable, and engineering. In some of these products TPE-O can also be an option.

Figure 4.3 "Crocs," slippers, flip-flops, and rubber boots are often made of SEBS. Other footwear industry components, such as soles, insoles, and heels, are also often in SBS or SEBS.

Figure 4.4 Soft handles with high friction for tools, pens, knives, and other grips are often made of TPE-S. Many TPS qualities have good adhesion to other thermoplastics and are therefore suitable for multi-component injection molding and coextrusion.

4.3 TPE-V

TPE-V (or TPV) thermoplastic elastomers, where the "V" stands for "vulcanized," are blends of polypropylene and dynamically vulcanized (cross-linked) EPDM rubber particles. If the rubber particles in the mixture are uncross-linked, the thermoplastic elastomer is TPE-O or TPO (where the "O" stands for "olefin"). Since the TPV elastomer consists of a PP matrix, it takes on a semi-crystalline structure.

The technology behind TPE-V was patented in 1962 but was developed further in the 1970s and 1980s. The leading manufacturers of TPE-V are Elasto, Elastron, Enplast, ExxonMobil Chemicals, So.F.te.R., Teknor Apex, and Zeon Chemicals.

Compared with TPE-O, TPE-V has better mechanical properties, chemical resistance, and higher service temperature.

The recycling code is > **PP + EPDM** <.

4.3.1 Properties of TPE-V

+ Available in a hardness range of 20 Shore A to 65 Shore D
+ Good abrasion resistance
+ Good tear resistance
+ Wide service temperature range (−50 °C to +125 °C)
+ Good chemical resistance
+ Excellent ozone, UV, and weathering resistance
+ Easy to process. Good adhesion to other thermoplastics
+ Can be colored and painted (with primer)
+ Better fatigue resistance than TPO and TPS, although inferior to rubber

> **Chemical facts:**
>
> TPV materials consist of monomers of polypropylene and dynamically vulcanized EPDM rubber (ethylene-propylene-diene monomer (M-class). The hardness depends on the mixing ratio and may vary between 20 Shore A and 65 Shore D.
>
> For formulas see TPE-O!

4.3.2 Application Areas

TPE-V is widely used in the automotive industry for sealing strips in doors, bellows, air ducts, and in the electrical and electronics industry for fittings to outdoor cables, connectors, and solar panels. In engineering, construction, and appliance industries, TPE-V is used for all sorts of seals, e.g. in refrigerator and washing machine doors.

Figure 4.5 Door and window seals in cars are often made of TPE-V thanks to the material's superior resistance to wear and chemicals, its outdoor durability, and excellent sealing capacity at a wide range of ambient temperatures.

4.4 TPE-U

Thermoplastic polyurethane, known as TPE-U or TPU, is partially crystalline. It occurs as two different variants, the first based on polyester and the second on polyether.

Polyurethane was first launched by Bayer in 1940 in textile fibers known as Perlon U. Today, Covestro markets the material under the name Desmopan. Other leading producers are BASF Elastollan and Merquinsa with Pearlthane and Pearlcoat.

TPU based on polyester has the best mechanical properties and the best resistance to heat and mineral oils, while the polyether type has the best low-temperature flexibility and resistance to hydrolysis and microbiological attack.

In contrast to thermoset polyurethane PUR, TPE-U is ideal for recycling.

The recycling code is > TPU < or > TPE-U <.

4.4.1 Properties of TPE-U

+ Can be produced from renewable raw materials
+ Excellent abrasion resistance
+ Excellent performance at low temperatures
+ High shear strength
+ High elasticity

+ High transparency
+ Good oil and grease resistance
+ Good hydrolysis resistance
− Slightly more difficult to process than other types of TPE

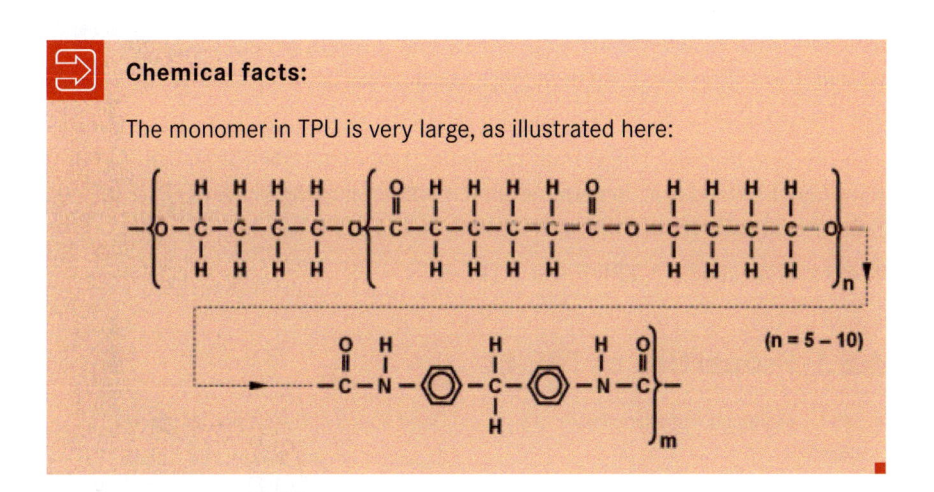

Chemical facts:

The monomer in TPU is very large, as illustrated here:

$$\left\{ O-\overset{H}{\underset{H}{C}}-\overset{H}{\underset{H}{C}}-\overset{H}{\underset{H}{C}}-\overset{H}{\underset{H}{C}}-O\left[\overset{O}{\overset{\parallel}{C}}-\overset{H}{\underset{H}{C}}-\overset{H}{\underset{H}{C}}-\overset{H}{\underset{H}{C}}-\overset{H}{\underset{H}{C}}-\overset{O}{\overset{\parallel}{C}}-O-\overset{H}{\underset{H}{C}}-\overset{H}{\underset{H}{C}}-\overset{H}{\underset{H}{C}}-\overset{H}{\underset{H}{C}}-O\right]_{} \right\}_n$$

$$\left[-\overset{O}{\overset{\parallel}{C}}-\overset{H}{\underset{}{N}}-\bigcirc-\overset{H}{\underset{H}{C}}-\bigcirc-\overset{H}{\underset{}{N}}-\overset{O}{\overset{\parallel}{C}}- \right]_m$$

(n = 5 – 10)

4.4.2 Application Areas

TPE-U is widely used in shoe soles and other footwear components because of its excellent adhesion to other materials. It is also used for the treads on wheels and for mechanical components due to its good abrasion resistance and damping capacity. The material is also used for extruded tubing to replace rubber, e.g. in fire hoses.

Figure 4.6 TPU is the predominant material in compact treads for castors and can be injection molded directly onto other plastics or metal wheels.

4.5 TPE-E

The abbreviation for polyester-based thermoplastic elastomers is usually TPE-E or TPC-ET, but sometimes TPE-ET and TEEE are also used. They can all be classified as semi-crystalline and have the following characteristics: excellent toughness and elasticity, high resistance to creep, impact and flex-fatigue, and flexibility at low temperatures. They also keep their mechanical properties at elevated temperatures.

TPE-E was launched in 1972 by DuPont under the trade name Hytrel. DuPont was also the first to make TPE-E from renewable raw materials, the so-called Hytrel RS, where RS stands for "Renewable Source." Other manufacturers of TPE-E include Celanese with Riteflex, DSM with Arnitel, and LG with Keyflex.

The recycling code is > TPE-E < or > TPC-ET <.

4.5.1 Properties of TPE-E

+ Can be produced from renewable raw material
+ High impact strength at low temperatures
+ Excellent fatigue resistance
+ Flexible at low temperatures (i.e. service temperature range of −40 °C to +120 °C)
+ Varying temperature has a negligible effect on stiffness
+ Good sound and vibration damping
+ Excellent oil and solvent resistance
+ Easy to process, even in complex geometries
+ Easy to process thanks to very good thermal stability

> **Chemical facts:**
>
> TPE-E or TCP-ET are thermoplastic polyether-ester block copolymers consisting of a hard (crystalline) segment based on polybutylene terephthalate and a soft (amorphous) segment based on long chained polyether glycols. The material properties are determined by the ratio between the soft and hard segments.
>
> $$\left\{ O-[CH_2]_X-O-\overset{O}{\underset{\|}{C}}-\bigcirc-\overset{O}{\underset{\|}{C}} \right\}$$
> $$X = 2, 4$$
> **Hard segment = PBT (4GT)**
>
> $$\left\{ \left[O-\overset{R}{\underset{R}{\overset{|}{C}}}-[CH_2]_Y \right]_Z O-\overset{O}{\underset{\|}{C}}-\bigcirc-\overset{O}{\underset{\|}{C}} \right\}$$
> $$R = H, CH_3 \quad Y = 1, 3 \quad Z = 8 - 50$$
> **Soft segment = PTMEGT (PO4T)**
>
> The structure of TPC-ET is:
>
> $-(4GT)_x-$ PO4T $-(4GT)_y-$ PO4T $-(4GT)_z-$ PO4T
> Hard   Soft   Hard   Soft   Hard   Soft

4.5.2 Application Areas

TPE-E is widely used by the automotive industry for bellows, air ducts, and air-bag covers. The electrical and electronics industry use it for cables and connectors, and it is also used in sports equipment and ski boots.

Figure 4.7 Ski boots are often manufactured in TPE-E. There are also bio-based grades. These materials can consist of almost a third renewable raw materials, but still have the same characteristics as if they were made from petroleum-based raw materials.

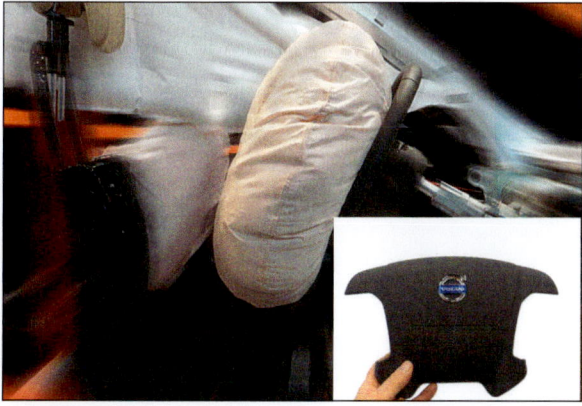

Figure 4.8 Airbag covers can be manufactured in TPE-E. This material is characterized by outstanding flexibility and high impact resistance at both low (−40 °C) and high (110 °C) temperatures. The TPE material is co-molded as a topcoat on a stiffer material that has slits to break more easily.

4.6 TPE-A

Amide-based thermoplastic elastomer is abbreviated to TPE-A or PEBA (where PEBA stands for polyether block amide). The material is similar in structure to the polyester-based TPE-E, with hard and soft segments in the polymer chain (see Chemical facts below). Compared to other advanced thermoplastic elastomers, TPE-A has a lower density, better mechanical properties, higher service temperatures, and better chemical resistance. The material can be made both transparent and with permanent antistatic properties. Due to its high service temperature (> 135 °C), TPE-A sometimes replaces silicone rubber and fluoroelastomers.

TPE-A was launched on the market in 1981 by Atochem under the trade name Pebax. Today, Pebax is manufactured by Arkema, who also has launched the first biobased grade under the trade name Pebax Rnew. Other manufacturers are Evonik with Vestamid E and EMS with Grilamid.

The recycling code is > **TPE-A** < or > **PEBA** <.

4.6.1 Properties of TPE-A

+ Can be produced from renewable raw materials
+ Can be made transparent
+ Good chemical resistance
+ Excellent strength and toughness
+ Good elastic recovery

+ Very high damping capacity
+ Excellent at low temperature
+ High service temperature
− Cost of material

> **Chemical facts:**
>
> TPE-A-materials are composed of an amorphous polyether alternately coupled to a crystalline amide segment. The polyether segments can be based on polyethylene glycol (PEG), polypropylene glycol (PPG), or polytetramethylene glycol (PTGM). Polyamide segments can be based on PA6, PA66, PA11, or PA12. The ratio of polyether and polyethylene amide segments, which can vary from 80/20 to 20/80, controls the elastomer's hardness–from soft (75 Shore A) to hard (65 Shore D).
>
> This formula shows the construction of a typical TPE-A, where:
> X = polyamide, Y = polyether
>
> $$\left[\begin{matrix} O & O \\ \| & \| \\ C-X-C-O-Y-O \end{matrix} \right]_n$$

4.6.2 Application Areas

The products manufactured in TPE-A include sport shoes, ski boots, ski goggles, flexible hoses, bellows, "breathable" film, shock-absorbing products, "silent gear wheels," conveyor belts, and medical products such as catheters.

Figure 4.9 Soft keys with a "rubber feeling" are produced in TPE-A, often in bright colors. The material is excellent to color via a number of methods, e.g. masterbatch, powder pigments, and liquid coloring. You can also print it in a variety of ways, including laser printing and "inmold decoration."

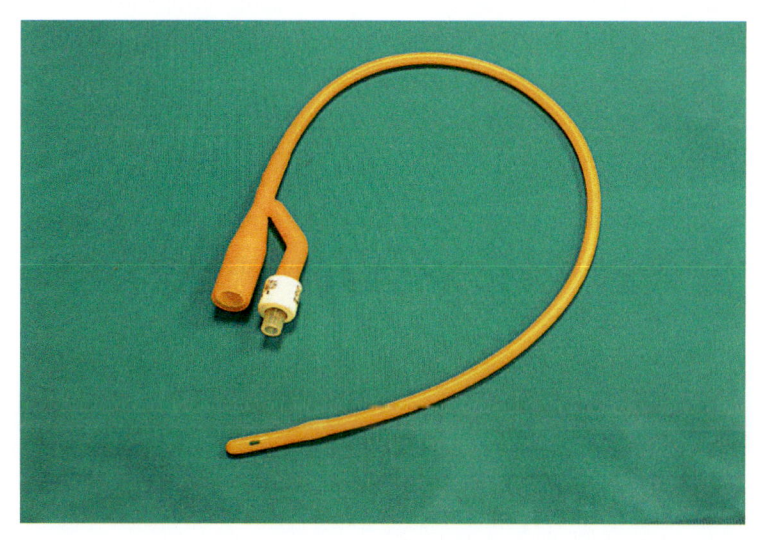

Figure 4.10 TPE-A is often used in medical applications. There are biocompatible grades and it can be sterilized. The figure shows a catheter made from low-friction grade TPE-A. The material is ideal for extrusion into tubing that can then be welded using a variety of methods such as ultrasonic welding, mirror welding, induction welding, and high-frequency welding.

CHAPTER 5
High-Performance Polymers

5.1 Advanced Thermoplastics

In everyday speech we describe this type of material as "high performance," which means plastics with the best properties. What kind of qualities do we have in mind when developing materials to belong to this category?

Below we can see the wish list the researchers may have had when they set out to improve the properties of engineering plastics:

- Improved ability to replace metals
- Improved mechanical properties such as stiffness, tensile strength, and impact strength
- Increased service temperature
- Reduced influence of ambient temperature and humidity on the mechanical properties
- Less tendency to creep under load
- Improved chemical resistance (especially considering the fluids used in cars, i. e. fuel, oil, antifreeze, detergents)
- Improved flame-retardant properties
- Improved electrical insulation properties
- Less friction and wear
- Improved barrier properties (primarily to fuel and oxygen)

In addition, any new material would be required to:

- Have a reasonable price in relation to the properties it offers
- Be easy to process using conventional machinery
- Be simple to recycle

Advanced reinforcement systems with carbon and aramid fibers or coating with so-called nano-metals can also be used in combination with advanced polymers to achieve the above goals.

Plastics designed to replace metals are sometimes called "structural materials" and clearly have a great role to play in the future, especially since to date it is estimated that only 4% of the potential applications have been converted.

This section gives an overview of the following semi-crystalline advanced polymers:

1. Fluoropolymer (PTFE)
2. High-performance aromatic polyamide (PPA)
3. "Liquid crystal polymer" (LCP)
4. Polyphenylene sulfide (PPS)
5. Polyether ether ketone (PEEK)

And the following amorphous polymers:

6. Polyetherimide (PEI)
7. Polysulfone (PSU)
8. Polyphenylsulfone (PPSU)

5.1.1 Recycling

All materials in this group can be recycled, and they have the material abbreviation in angle brackets (e.g. > **PTFE** <) as their recycling code.

5.2 Fluoropolymers

There are a number of variants in this group of semi-crystalline plastics, of which PTFE (polytetrafluoroethylene) is the largest in terms of volume in today's market. This material was discovered by chance in 1938 in DuPont's laboratories in the United States, while experimenting with different refrigerants. Launched under the trade name Teflon, this material has the lowest known friction coefficient, the best chemical resistance, and the best electrical insulation properties of all plastics, and can be used in temperatures up to 260 °C.

Figure 5.1 Teflon pans are a great example of a product that takes advantage of PTFE's superior resistance to temperature and chemicals in combination with low friction ("non-stick").

Other materials in the fluoropolymer family include PCTF, PVDF, PVF, and copolymers FEP, PFA, E/CTFE, E/TFE, THF, and so on.

5.2.1 Properties of PTFE

+ Exceptional chemical resistance
+ Extremely low friction coefficient
+ Excellent electrical properties
+ Withstands extremely low and high service temperatures (from −200 °C to +260 °C)
+ UV resistant
+ Flameproof
+ Compatible with human tissue
− Poor abrasion resistance
− Bad creep and cold flow properties
− High density (up to 2.3 g/cm^3)
− Cannot be processed by conventional methods

> **Chemical facts:**
>
> PTFE has a simple structure, consisting of only carbon and fluorine atoms in the monomer:
>
> $$-\underset{\underset{F}{|}}{\overset{\overset{F}{|}}{C}}-\underset{\underset{F}{|}}{\overset{\overset{F}{|}}{C}}-$$

5.2.2 Application Areas

PTFE cannot be injection or blow molded. It is processed by compression molding, extrusion, and then sintering. PTFE can also be dispersed and used for surface coating and film manufacture. Tapes can be produced by calendering.

The copolymer FEP, however, can be processed by injection molding, extrusion, and blow molding in special machines.

The unique properties of PTFE and other fluoropolymers make them ideal for use in pipes, coatings, and seals in extremely corrosive environments (chemical industry). It is a good material when you want low friction, as with self-lubricating bearings and low-friction surfaces, or non-stick utensils such as pans and baking tins. The electronics industry takes advantage of its excellent electrical insulation properties for use in cable insulation and other insulators. PTFE ages insignificantly and is therefore much used by the medical industry, particularly for implants.

5.3 "High-Performance" Nylon – PPA

PPA is the chemical abbreviation of polyphthalamide and consists of a group of semi-crystalline aromatic polyamides with clearly improved mechanical, chemical, and temperature characteristics compared to polyamide 6 and 66. PPA also absorbs much less moisture than PA6 and PA66 and is more dimensionally stable. Materials in this group are also denoted in certain databases as PA6T/6, PA6T/66, PA6T/PI, and PA6T/XT.

5.3 "High-Performance" Nylon – PPA

The leading manufacturers are DuPont with the trade name Zytel HTN, EMS with Grivory, and Solvay with Amodel.

Table 5.1 This table compares some of the mechanical properties of PA66, PA6, and PPA with 35% glass fiber reinforcement. A big difference in the properties of PA66 and PA6 can be seen, depending on whether they have absorbed moisture (saturated at 50% relative humidity) or if they are unconditioned (directly from the injection-molding machine). PPA is not affected at all (stiffness) or only marginally (tensile strength and elongation).

	Unit	Zytel PA66 35% Glass Fiber Reinforced		Zytel PA6 35% Glass Fiber Reinforced		Zytel HTN51 35% Glass Fiber Reinforced	
		Uncond.	Conditioned	Uncond.	Conditioned	Uncond.	Conditioned
Stiffness	MPa	11,200	8,300	11,100	7,500	12,500	12,500
Tensile strength	MPa	210	140	190	130	220	210
Elongation	%	3.2	4.6	3	5	2.4	2.1

5.3.1 Properties of PPA

+ Stiffer, stronger, and higher creep resistance than most other plastics makes PPA a suitable substitute for metal
+ High service temperature (150 °C continuous, 200 °C short-term peak temperature)
+ Much lower water absorption than PA6, PA66, and PA46 and thus better dimensional stability
+ Better chemical resistance than PA6 and PA66
+ Good electrical properties
+ Can be made flame retardant with halogen-free additives
− Less impact resistance compared to PA6 and PA66

Chemical facts:

In PPA the acid group in PA66 has been replaced with an aromatic group, which is why the monomer has the following structure:

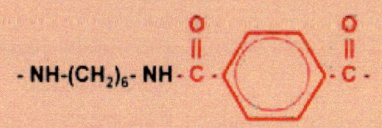

$$-NH-(CH_2)_6-NH-\overset{O}{\underset{\|}{C}}-\underset{}{\bigcirc}-\overset{O}{\underset{\|}{C}}-$$

The aromatic group is shown in red.

5.3.2 Application Areas

Most grades of PPA are glass fiber reinforced and used only for injection molding. They are used in the automotive industry (for engine parts) and in the electrical and electronics industry (for connectors and so-called backbones in mobile phones).

Figure 5.2 AD-Plast, a Swedish molder, won the Plastovationer Design Competition 2008 with these PPA cams for use in furniture assembly. The cams, previously made of zinc, maintained the dimensions while being 1.5 times stronger and weighing 7 times less than the originals.

5.4 "Liquid Crystal Polymer" – LCP

LCP materials are semi-crystalline and have self-reinforcing characteristics (build a fibrous structure similar to wood in the flow direction when they are injection molded). The materials can either be aromatic polyamides known as aramids (such as DuPont's Kevlar fiber) or aromatic polyesters.

The aromatic polyester types are thermotropic (with a transitional phase with liquid crystals in the melt).

The leading manufacturers of injection-molding-grade LCP are Celanese with the trade names Vectra and Zenite, and Solvay with Xydar.

5.4.1 Properties of LCP

+ Good balance of stiffness, strength, toughness, and superior creep resistance makes LCP an excellent replacement for metal
+ Extremely high service temperature (240 °C continuous) allows for lead-free soldering of electronics
+ Retains properties at elevated temperatures

+ Very fluid (easy to fill thin-wall material)
+ Excellent chemical resistance
+ Flame retardant (V-0 rated without additives)
+ Excellent dielectric properties
− Weak weld lines
− Poor surface gloss

> **Chemical facts:**
>
> All LCP materials are monomers consisting of several aromatic groups. Aramid (e.g. DuPont's Kevlar) has the simplest structure, with two aromatic groups:

5.4.2 Application Areas

Most grades of molding LCP are glass fiber reinforced. LCP is generally used by the electrical and electronics industry for smaller components such as coils and contacts—in lighting, mobile phones, ignition systems and sensors for automobiles, fiber optic connectors, etc.—where high service temperature, good electrical properties, and flame retardancy are required.

Food-grade LCP can be obtained and is used for pans for both microwave and conventional oven.

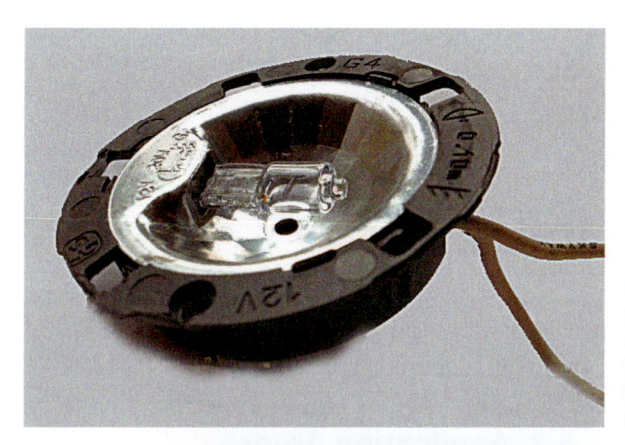

Figure 5.3 LCP has high service temperature and is self-extinguishing. It is used in lamp holders for halogen lamps in both lighting and the automotive industry, where it sometimes also replaces ceramics.

Figure 5.4 There are food-approved LCP grades which can resist very high temperatures (above 250 °C). Casseroles in LCP can be used in both standard and microwave ovens.

5.5 Polyphenylene Sulfide – PPS

Glass fiber reinforced polyphenylene sulfide is an extremely stiff material with an E-modulus of over 20 GPa. The material can withstand high temperatures and has a chemical resistance bettered only by PTFE.

The material is easily recognized because it has a metallic sound when you tap on it. The PPS launched by Chevron Phillips in the U.S. in 1968 under the trade name Ryton is the most well-known. In addition to Ryton, there are other brands of PPS on the market, such as Primef and Radel from Solvay, Fortron from Celanese, and Tedur from Albis.

5.5.1 Properties of PPS

+ Exceptionally high stiffness (with up to 65% glass fiber) and high creep resistance
+ Very high service temperature (240 °C continuous, 260 °C short-term)
+ Extremely good chemical resistance (up to 200 °C)
+ Negligible moisture absorption and good hot-water resistance
+ Excellent dimensional stability and negligible shrinkage
+ Flame retardant (UL V-0) without additives
+ Good electrical properties
− Low elongation (< 2%)

> **Chemical facts:**
>
> Polyphenylene sulfide has a simple structure with one aromatic group bound to the sulfur:
>
> -⌬- S -

5.5.2 Application Areas

Most grades of PPS are glass fiber/mineral reinforced and used for injection molding, but there are also non-reinforced grades used for extrusion.

The majority of PPS is used by the automotive industry in fuel and ignition systems, and by the electrical and electronics industry for products requiring high service temperatures and flame resistance, as well as for components that require good mechanical properties combined with high chemical and corrosion resistance.

Figure 5.5 PPS is used in demanding hot-water applications such as this black BMW thermostat housing or the brown radiator end. It can also be used in the manufacture of high-precision components that must withstand high temperatures (e.g. the diode bridge shown on the right) or chemical influences (e.g. the small impellers of fuel pumps, as shown in the middle of the picture).

Figure 5.6 PPS is approved for drinking water applications, which is a requirement in the water meter shown here. The housing in PPS has extremely good chemical and temperature resistance combined with high stiffness and creep resistance. PPS is also used in water taps as it does not absorb moisture, nor is degraded by hot water.

5.6 Polyether Ether Ketone – PEEK

Polyether ether ketone, or PEEK as this polymer is usually called, is semi-crystalline and has the best mechanical properties and temperature resistance of all the thermoplastics, combined with excellent chemical resistance, maintained even at extremely high service temperatures. The material was developed by ICI in England in the 1970s and introduced to the market in 1980 under the trade name Victrex. In 1993 a separate company was formed, called Victrex PLC, which today still has about 90% of the world market for PEEK. Other manufacturers of PEEK are Evonik, Sinof Hi-Tech Material, and Solvay.

5.6.1 Properties of PEEK

+ Very high stiffness, strength, and fatigue resistance
+ Extremely high service and softening temperatures (260 °C continuous, short-term peaks above 300 °C)
+ Excellent chemical resistance even at high temp.
+ Exceptionally good friction and wear properties
+ Flame retardant (V-0 rated without additives)
+ Good electrical properties
+ Food and medical grades are available
− Very high price

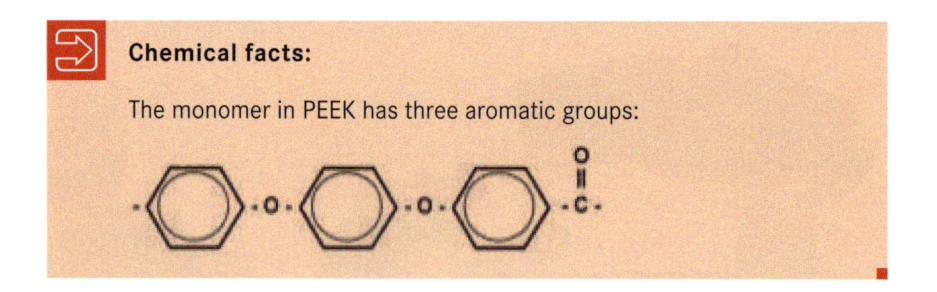

> **Chemical facts:**
>
> The monomer in PEEK has three aromatic groups:

5.6.2 Application Areas

PEEK can be processed by injection molding, extrusion, and blow molding, and is suitable for products with demanding requirements such as ball bearing rings, pistons, and valves. It is used in such industries as chemical processing, automotive and aerospace, electrical and electronics industry, and in the medical industry where it is used for implants and other sensitive applications.

Figure 5.7 Manufactured in PEEK: Bearing rings, pump parts, gears, and even hoses where high precision is required in combination with good mechanical properties, low wear, good chemical resistance, and extremely high service temperature.

5.7 Polyetherimide – PEI

Polyetherimide is an amorphous thermoplastic that combines excellent temperature with good mechanical properties, chemical resistance, and flame retardancy. Among all the amorphous plastics, PEI has the highest resistance to stress-cracking. The material was first introduced to the market in 1982 by General Electric (now Sabic) under the trade name Ultem.

5.7.1 Properties of PEI

+ High service and softening temperatures (180 °C continuous, softening temp 217 °C)
+ Good chemical resistance
+ Excellent resistance to stress-cracking
+ Flame retardant (V-0 rated without additives)
+ Very good dimensional stability
+ Good resistance to UV and microwave radiation
+ Food grade available and PEI can be sterilized
+ Attractive price level
− Low creep resistance (compared to other semi-crystalline polymers)

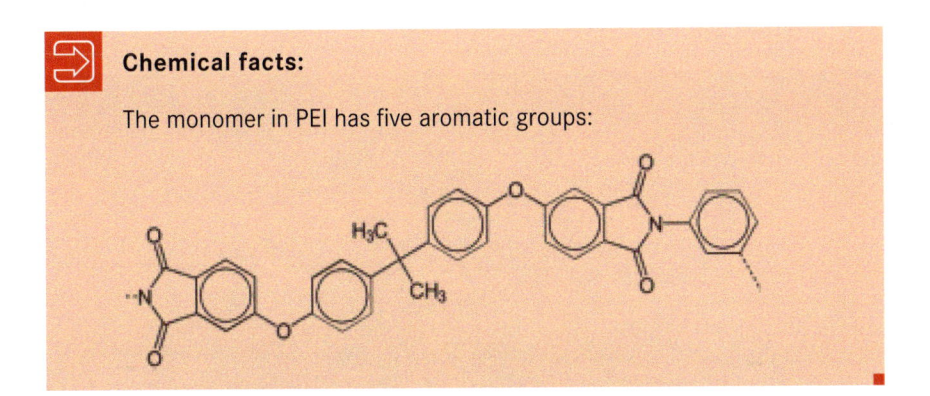

> **Chemical facts:**
>
> The monomer in PEI has five aromatic groups:

5.7.2 Application Areas

PEI is easy to work with and can be processed using injection molding, extrusion, blow molding, and vacuum molding. It is used for products in the electrical, electronics, automotive, and engineering industries where the highest standards of temperature resistance, dimensional stability, and flame retardancy are required.

Figure 5.8 The image shows an automotive windscreen camera which helps improve driving safety by lane control warning, smart braking support, distance recognition, and auto headlamp beam control. The lens holders and other plastic parts in the camera are typically manufactured in PEI as the camera housing is made in magnesium, which has the same coefficient of expansion as PEI.

5.8 Polysulfone – PSU

Polysulfone makes up the largest market share of all the amorphous sulfone plastics.

The material was first launched in 1965 by Union Carbide in the U.S. under the trade name Bakelite, which was later changed to Udel.

Union Carbide sold the rights to manufacture polysulfone to Amoco Performance Products. Today, PSU is manufactured by BASF with the trade name Ultrason and Solvay with Udel.

Polysulfone has good mechanical properties over a wide temperature range.

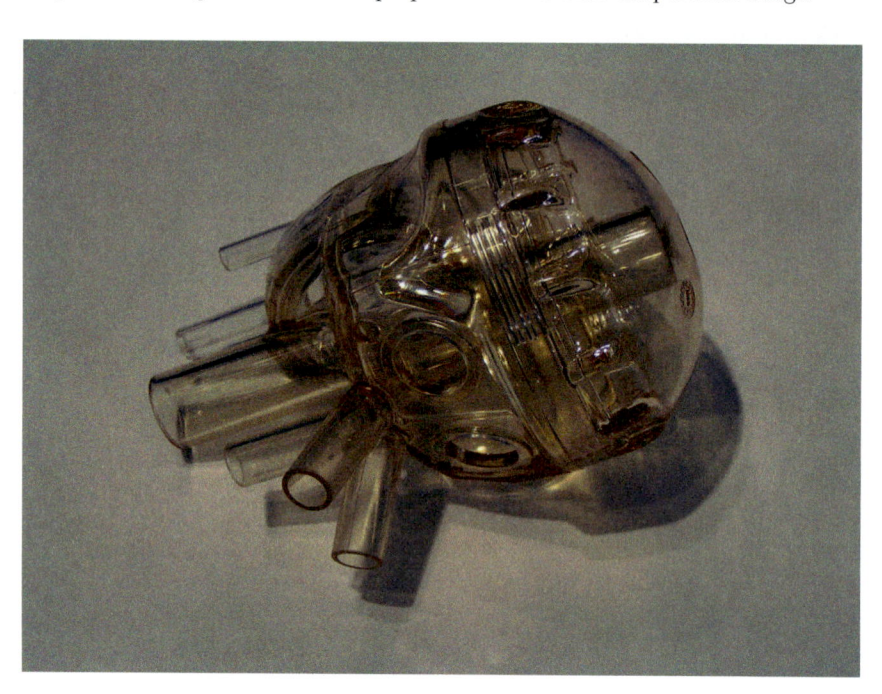

Figure 5.9 Food-grade PSU is available.
One example of use is for parts in milking machines, which need to be transparent and withstand repeated steam sterilization.

PSU can be modified with other plastics such as ABS in order to reduce the risk of stress-cracking and providing a more favorable price. Compared to pure PSU, the modified material may have reduced mechanical properties at high temperature. A common brand name is Mindel from Solvay.

5.8.1 Properties of PSU

+ Wide range of service temperatures (from −100 °C to 160 °C continuous)
+ Negligible change in stiffness across the whole service temperature range
+ Excellent hydrolysis properties (hot water)
+ Food and medical grades available
+ Can be sterilized using steam, ethylene oxide, and gamma radiation

+ Flame retardant grades are available (V-0)
− Sensitive to stress-cracking in certain solvents
− Poor UV resistance (can be improved with additives)
− Exceptionally high processing temperatures (up to 390 °C in injection molding)

> **Chemical facts:**
>
> The monomer in PSU has four aromatic groups. It is this structure that contributes to its good temperature characteristics:
>
> [Structure: –⌬–S(=O)₂–⌬–O–⌬–C(CH₃)₂–⌬–O–]

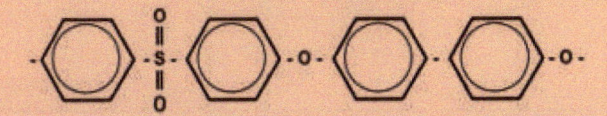

5.8.2 Application Areas

PSU can be processed with most methods used for thermoplastics. It is available in semi-finished form as bars, tubes, profiles, sheets, and films.

It is used when there is a need for a transparent product that can withstand high temperatures and hot water, especially for food and/or medical grade applications. Examples include medical devices in which PSU replaces glass, hot drink appliances (e.g. coffee makers), parts and cookware for microwave ovens, hair dryers, and automotive components.

5.9 Polyphenylsulfone – PPSU

The properties of polyphenylsulfone are much the same as PSU but PPSU has significantly better impact strength (10-fold) and carries less risk of stress corrosion. The leading manufacturer is Solvay Advanced Polymers with the trade name Radel.

5.9.1 Properties of PPSU

+ High service temperature (190 °C)
+ Exceptionally high impact strength (even at high temperatures and after heat aging)
+ Very high fireproof rating (clears FAA requirements for the interior of aircraft)
+ Good resistance to stress-cracking
+ Highly suitable for sterilizing (autoclaving)
− Must be UV stabilized for outdoor use

> **Chemical facts:**
>
> The monomer in PPSU is very similar to PSU. The key difference is that the –C(CH₃)₂– group is missing:
>
> [Structure: –⌬–S(=O)₂–⌬–O–⌬–⌬–O–]

5.9.2 Application Areas

As with PSU, PPSU can be processed with most methods suitable for thermoplastics.

The majority of PPSU is used in healthcare and medical products. Some grades can be sterilized more than 1,500 times, making it an economical alternative to disposable products.

PPSU is also used in electrical and electronic components and pipe connections where high temperature resistance combined with good chemical resistance and superior flame retardancy is required. Products for the aviation and aerospace industry are also often manufactured from PPSU.

Figure 5.10 Among all the thermoplastics, PPSU has extremely good flame resistance and is therefore used, e.g. for interior aircraft panels.

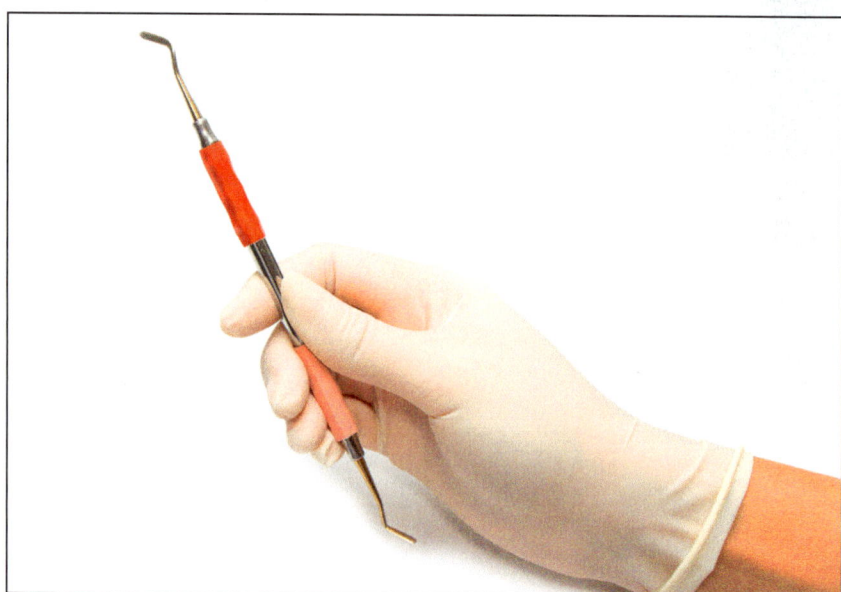

Figure 5.11 The grips of dental instruments are often made in PPSU, as they will have lower weight and be cheaper to produce. PPSU has excellent chemical resistance and can be autoclaved.

CHAPTER 6
Bioplastics and Biocomposites

6.1 Definition

If you ask a professional "what is a bioplastic?", you would get one of three different answers:

1. It is a plastic manufactured from biologically based raw materials.
2. It is a plastic that is biodegradable, i.e. can be degraded by microorganisms or enzymes.
3. It is a plastic that contains natural fibers.

Since biobased plastics are not necessarily biodegradable and biodegradable plastics do not have to be biobased, it is important to be clear about what you really mean. The percentage of renewable ingredients necessary for a plastic to be considered "bio" has not been established, although leading bioplastic suppliers deem that it should be at least 20%.

Plastics containing natural fibers are also called "biocomposites" and are mostly traditional plastics that have been reinforced or blended with natural fibers such as wood, flax, hemp, or cellulose.

In addition to the commodities PE and PP, there are also biopolyesters, such as PLA.

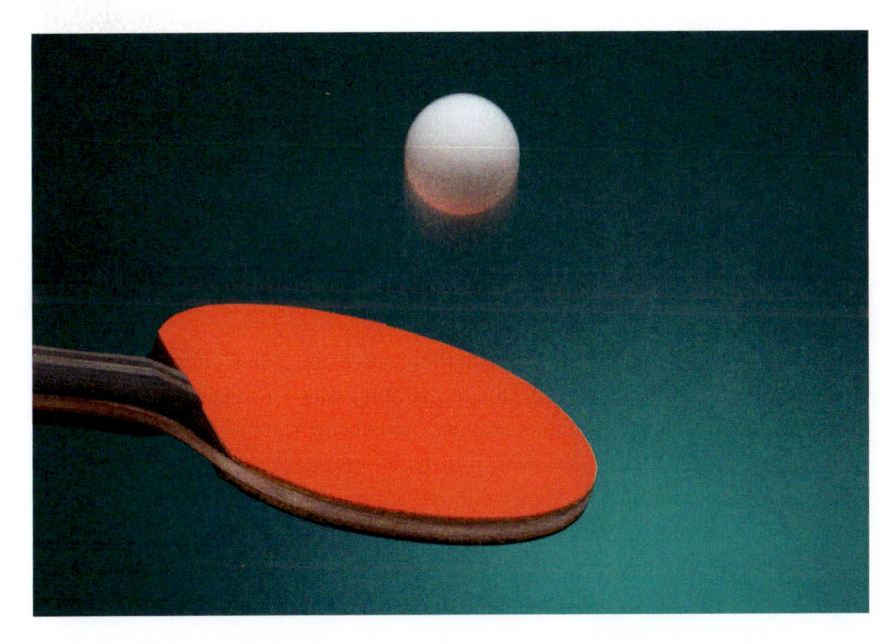

Figure 6.1 Some of the first thermoplastics were manufactured from cellulose, but they currently have little commercial significance, except for viscose fiber. Ping pong balls were originally made of celluloid and are still produced from cellulosic materials.

6.1.1 What Do We Mean by Bioplastic?

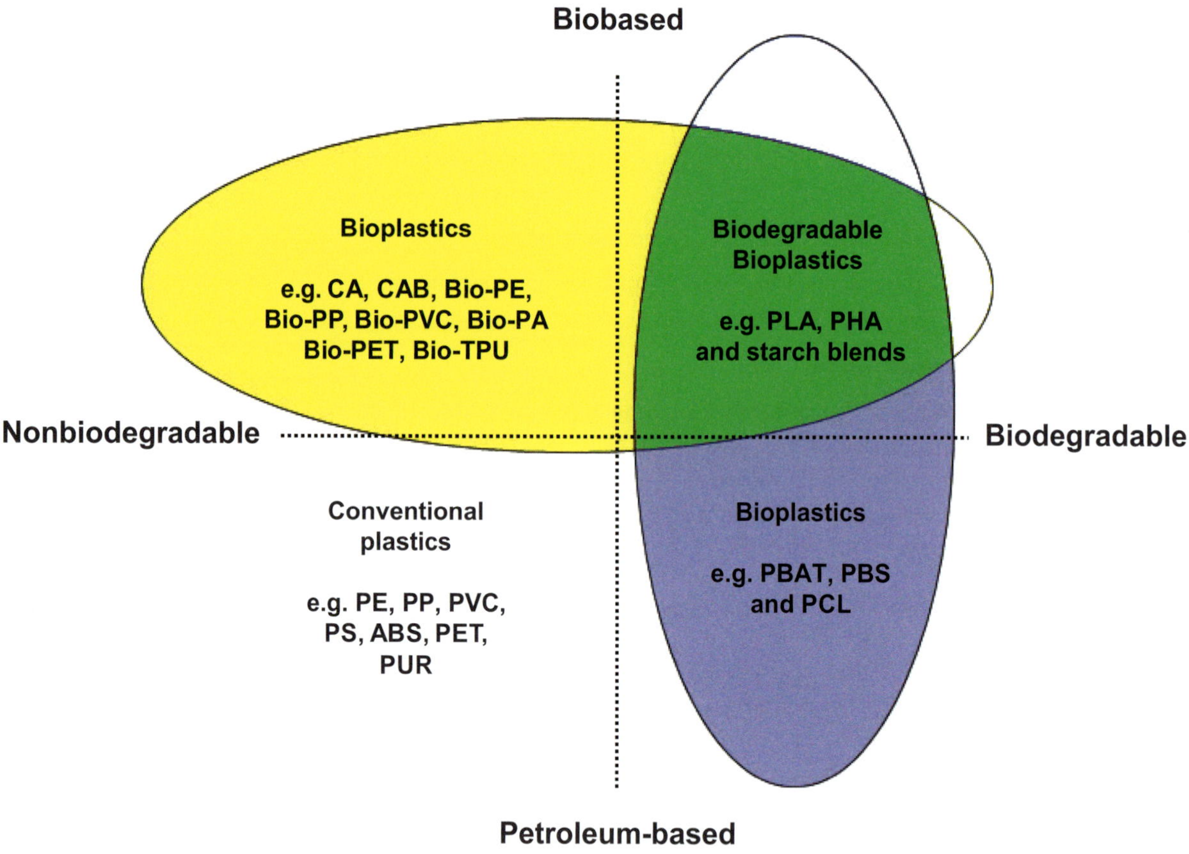

Figure 6.2 The illustration shows how to divide thermoplastics into conventional petroleum-based plastics and different types of bioplastics. [Source: IfBB Institute for Bioplastics and Biocomposites]

6.2 The Market

According to the organization European Bioplastics, the global annual production of bioplastics (both biobased and biodegradable) was 1.2 million tons in 2011. Compared to the 250 million tons of plastic produced, the market for bioplastics is still exceedingly small. However, growth is strong, and by 2017 production exceeded 2 million tons (see Figure 6.3 below). Most of the world's major manufacturers of plastic raw material are developing bioplastics. The goal is to replace 5–10% of traditional plastics (not counting PE and PVC based on bioethanol from sugar cane) with bioplastics. Production of PE based on bioethanol started in 2009 (in Braskem, Brazil). It is believed that >50% of biomaterial will be made of sugar in the near future.

6.2 The Market

Global production capacities of bioplastics 2017 (by material type)

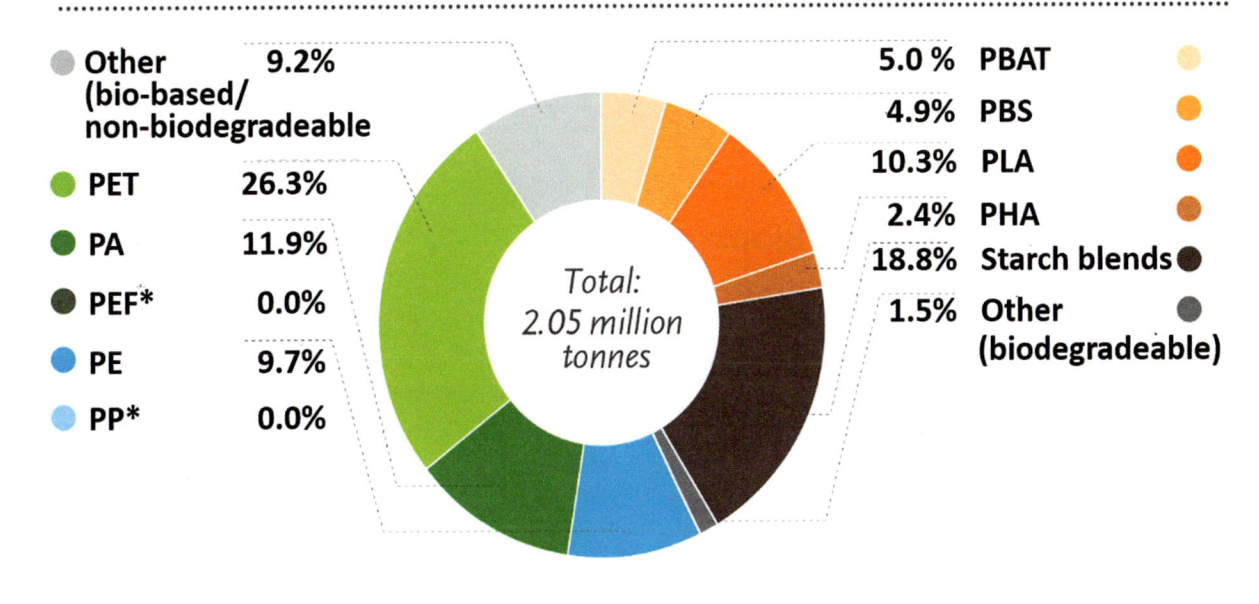

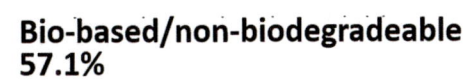

Bio-based/non-biodegradeable		Biodegradable	
Other (bio-based/non-biodegradeable)	9.2%	PBAT	5.0%
PET	26.3%	PBS	4.9%
PA	11.9%	PLA	10.3%
PEF*	0.0%	PHA	2.4%
PE	9.7%	Starch blends	18.8%
PP*	0.0%	Other (biodegradeable)	1.5%

Total: 2.05 million tonnes

Bio-based/non-biodegradeable 57.1%

Biodegradable 42.9%

*Bio-based PP and PEF are currently in development and predicted to be available in commercial scale in 2020.

Figure 6.3 The picture shows how the global production of bioplastics is divided among various polymers 2017. [Source: European Bioplastics, Nova Institute (2017)]

What are the reasons for using biobased plastics in Europe? The answers could be:

1. *Strategic:* as supplies of fossil fuels decrease, their price will increase.

2. *Regional politics:* providing support for EU farmers to find new and more profitable outlets. The sale of agricultural products for industrial purposes is not limited by World Trade Organization regulations.

3. *Environmental:* the use of biobased raw material for plastic production can help reduce carbon dioxide emissions. Whether biobased plastics are better from a carbon standpoint than fossil resins depends on a number of factors, such as:

- Energy consumption from planting through to harvesting, and plastics manufacture, as well as the source of this energy
- If the properties are sufficiently good that you don't need more material to achieve the same function as a non-bioplastic
- Waste management

6.3 Bioplastics

There are different ways to manufacture bioplastics:

- **Biopolymers** are naturally occurring polymers, most of which are starch and cellulose. Every year, around six thousand billion tons of biomass are produced, of which we humans use about 3.5%–two-thirds as food and one-third for energy, paper, furniture, and clothing. At present, biobased plastics do not account for even one-thousandth of this (i.e. 0.0006%).
- **Biobased polymers**, where the monomer has been produced by fermentation. One example is lactic acid, which is then used to produce polylactic acid (PLA).
- **Microorganisms** that produce biobased polymers, such as polyhydroxyalkanoate (PHA).
- **Bioethanol or biomethanol** produced by fermentation or gasification of biomass. These are then converted to the monomers ethylene and propylene, which are then used in the manufacture of commodities such as PE, PP, PVC, and PS.

When manufacturing biocomposites, you can use:

- **Plant fibers**, such as cotton, flax, and hemp
- **Wood waste**, such as sawdust and wood powder

6.4 Biopolymers

Figure 6.4 Corn is high in starch, which is a natural biopolymer. Fermenting corn produces bioethanol, which in turn can be used to produce biobased polymers.

Figure 6.5 The strawberry plants in this image are protected from weeds by a film made of PLA. After the harvest, the film can be plowed into the soil as it is fully compostable.

Cellulose has long been used to produce plastics such as cellulose acetates CA, CAB, and CAP. Because pure cellulose is difficult to process, these plastics consist of chemically modified cellulose. Cellulose plastics are crystal clear and very tough. Cellulose can be extracted from wood, among other things. Adding wood powder, flax, cotton, or hemp to a conventional petroleum-based plastic means the new material can be classified as bioplastic, since a part of the raw material used is renewable. There are several smaller companies that develop these blended bio-composites, while the large forestry companies are investing in research into completely cellulose-based plastics.

Starch can be extracted from corn, potatoes, seeds, sugar beets, sugar cane, grain, or other crops that are also used for food or biofuels. This is obviously a source of conflict, and there are numerous research projects underway to try to use the straw or the tops of such sources instead of the nutritious parts.

Pure starches are also difficult to process and must therefore be modified with chemicals or additives to be formed into plastic products.

6.5 Biobased Polymers: Biopolyester

Biopolyester is made from monomers produced e.g. by fermenting starch or sugar. Examples are Bio-PET, PLA, and PTT. Even petroleum-based polyesters that are biodegradable or contain a certain proportion of renewable raw material can also be called biopolyester.

Bio-PET is commercially the largest of the biopolyesters. Petroleum-based PET has been around for more than 70 years. Today's Bio-PET is made from 30% renewable materials, usually sugar cane.

Chapter 6 — Bioplastics and Biocomposites

Figure 6.6 More and more manufacturers are using Bio-PET in their bottles.

PLA (polylactic acid or polylactide) is a polyester made of a lactic acid monomer. It is produced by fermentation of simple sugars (monosaccharides), of which there is plenty in e.g. sugar beets, potatoes, corn, and wheat. The lactic acid is subsequently polymerized.

By controlling the polymerization process, both amorphous and semi-crystalline PLA can be produced.

PLA is made from 100% biobased raw materials and is also biodegradable. It is water resistant and has good barrier properties, but it is not as heat resistant as PET (max 55°C).

> **Chemical Facts:**
>
> The monomer in PLA is produced from lactic acid, which has the following structure:
>
> $$-\overset{O}{\underset{\|}{C}} - \overset{CH_3}{\underset{|}{CH}} - O -$$

Figure 6.7 DuraPulp is a biodegradable material from the forestry group Södra in Sweden. It consists of 70% cellulosic fibers and 30% PLA. [Photo: Södra]

Figure 6.8 The picture to the left shows a crematorium urn injection molded in Bio-Flex, renewable and biodegradable PLA from the German material supplier FkuR. After just a few months in the ground, the plastic will be completely degraded. [Source: Tojosplast]

PTT (polytrimethylene terephthalate) is a semi-crystalline polyester that can be prepared from renewable raw materials. Petroleum-based PTT has been around since 1941, but in 2000 DuPont launched Sorona, a fiber made through the fermentation of corn syrup. After further development, DuPont created the thermoplastic Sorona EP, which consists of between 20 and 37% renewable materials and exhibits performance and processing characteristics similar to polyester PBT.

6.6 Biobased Polymers: Biopolyamides

LCPAs (long-chain polyamides), also called biopolyamides or green polyamides, have been introduced to the market over recent years, including PA410, PA610, PA1010, PA10, PA11, PA612, and PA1012. These offer an alternative to the petroleum-based PA12.

LCPAs typically consist of renewable raw material extracted from castor oil derived from castor bean plants grown in the tropics.

Leading manufacturers include DuPont with Zytel Long Chain and Zytel RS, BASF with Ultramid Balance, EMS with Grilamid, DSM with EcoPaXX, Arkema with Rilsan, Solvay with Technyl Exten, and Evonik with Vestamid Terra.

Compared to standard polyamides such as PA6 and PA66, these materials have better dimensional stability, lower water absorption, and better chemical resistance.

Chapter 6 — Bioplastics and Biocomposites

Figure 6.9 Castor bean plant

Figure 6.10 Gas pipes and fittings for gas pipes can be made of PA11. This material is the first high-performance polyamide that has been approved to be used for pipes up to 100 mm (4 inches) in diameter at operating pressures up to 14 bars. This material may be biobased and contain 100% renewably sourced ingredients by weight.

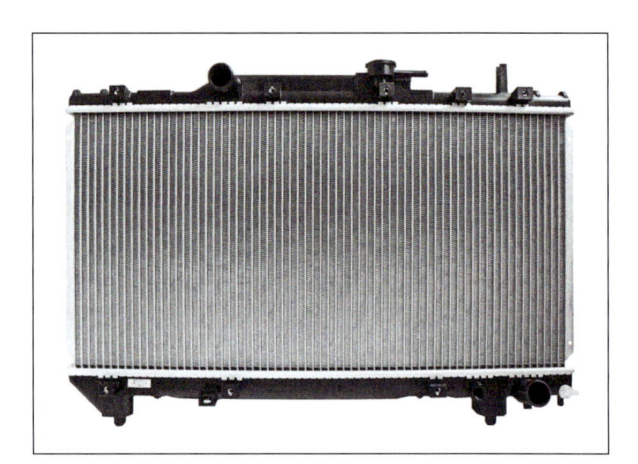

Figure 6.11 Automotive radiator end tanks can be produced from PA610 bio-polyamide and resist the hot, chemically aggressive underhood environment. PA610 also has low water absorption. Some PA610 contains more than 40% renewably sourced ingredients by weight.

6.7 Biobased Polymers from Microorganisms

PHA (polyhydroxyalkanoate) is a linear semi-crystalline polyester produced by the bacterial fermentation of sugar, glucose, or lipids, i.e. a group of substances consisting of fats and fat-like substances. The material was developed by ICI in the 1980s, and there are very few producers in the market. The material has good weathering properties and low water permeability. Overall, it has properties similar to PP.

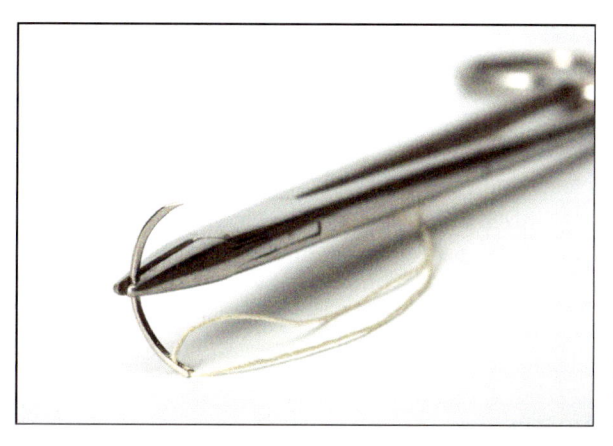

Figure 6.12 PHA has many medical applications. PHA fibers can be used to suture wounds.

6.8 Bioethanol or Biomethanol

PE is a commodity that has begun to be produced again of renewable biobased raw materials.

In the 1970s, a substantial proportion of India's ethanol was used for the manufacture of PE, PVC, and PS. In the 1980s, companies in Brazil began to manufacture biobased PE and PVC. However, when oil prices dropped in the early 1990s, production stopped. Twenty years later, production is beginning to build up again.

Today, the Brazilian company Braskem is a world leader in biobased PE. Commercial production started in September 2010, using sugar cane as a raw material to prepare bioethanol, which is then converted into ethylene, used in the production of PE. Total production is currently around 200,000 tons and represents 17% of the market for bioplastics.

Bio-PE is nonbiodegradable.

Other commodities that can come from renewable resources are PP and PVC.

Figure 6.13 A plastic shopping bag produced in green Bio-PE.

6.9 Biocomposites

Conventional petroleum-based plastics such as PE and PP containing at least 20% **plant fiber** from cotton, flax, or hemp are counted as bioplastics. The choice of the plastic matrix is limited by the processing temperature, because the fibers are carbonized if the temperature is too high. Instead of plant fiber, sawdust or wood powder can be used to produce a bioplastic from a conventional petroleum-based plastic.

Figure 6.14 The picture shows a floor decking that is made of 30% HDPE, 10% pigment and UV additives, as well 60% wood fibers.

Figure 6.15 These toiletries from German company FkuR are all made from bioplastics. The manicure instruments at the bottom of the picture are made of Bioflex, a mixture of PLA and wood fibers. [Source: Polymerfront AB]

Figure 6.16 An extruded profile made from a 50% HDPE/50% wood blend. The product shown here is manufactured in a natural color, but the raw material can also be colored with pigment. [Source: Talent Plastics AB]

6.10 More Information about Bioplastics

Additional information can be found in the "Plastic Guide," a free app for iPhone and Android: http://www.plasticguide.se/eng or by European Bioplastics: www.european-bioplastics.org.

CHAPTER 7
Plastic and the Environment

At first glance, the title of this chapter may seem ambiguous. Do we mean how plastic affects our environment? Or how various environmental factors affect plastic? We will consider both aspects.

The use of plastics is constantly increasing. One important reason is the fact that plastics contribute to increased resource management–for example, saving energy and reducing emissions. Plastics also contribute to technological development.

The plastic industry wants to contribute to a sustainable society. That's why they invest considerable resources in the production of environmentally friendly materials and resource-efficient processes.

Figure 7.1 The use of plastics reduces our climate impact by saving energy and reducing CO_2 emissions.

7.1 Plastic is Climate-Friendly and Saves Energy

That plastic can slow climate change by saving energy and reducing our emission of greenhouse gases is not something we may instantly think of. The recent study "Plastics' Contribution to Climate Protection" concluded that the use of plastic in all EU member states plus Norway and Switzerland contributes to the following environmental benefits:

- Plastic products enable energy savings equivalent to 50 million tons of crude oil—that's 194 very large oil tankers.
- Plastic prevents the emission of 120 million tons of greenhouse gas emissions per year, which is equivalent to 38% of the EU's Kyoto target.
- The average consumer causes around 14 tons of carbon dioxide emission. Only 1.3% of that, about 170 kg, is derived from plastic.

7.1 Plastic is Climate-Friendly and Saves Energy

In the automotive and aerospace industries, the use of plastic saves weight and thus reduces fuel costs. In the construction industry, plastic is increasingly used as a superior insulation material that provides a good indoor environment and reduces energy consumption.

Figure 7.2 Plastic accounts for approx. 12–15% of a modern car's weight, which in Europe alone results in annual savings of 12 million tons of oil and a 30 ton reduction in CO_2 emissions.
The body of this sports car is made of carbon fiber reinforced plastic and has an even higher proportion of plastic than an ordinary car.

Without plastic, transportation costs for the retail industry would increase by 50%. On average, plastic packaging accounts for between 1 and 4% by weight of all products packaged in plastic. For example, a film that weighs 2 g is used to pack 200 g of cheese, and a plastic bottle weighing 35 g packs 1.5 liters of drink. If you also include containers and shipping material, then plastic packaging increases its share to 3.6% on average.

Figure 7.3 A 330 ml glass Coca-Cola bottle weighs 784 g when full and 430 g empty (including the lid), i.e. 55% of the product weight is in the packaging. By comparison, a 500 ml bottle in PET is 554 g when full and just 24 g when empty (incl. lid), i.e. only 4% of the weight is packaging.

Plastic also has many uses in climate-friendly energy production. For example, the wings of wind turbines are made of vinyl ester with internal PVC foam; pipes in solar collectors are made from polyphenylsulfone; and the casings for fuel cells are manufactured out of polyetherimide.

Chapter 7 – Plastic and the Environment

7.2 Environmental Effects on Plastic

Like all materials, plastics are affected to some degree by the environment they are used in. Over time, they are broken down by various environmental factors, such as exposure to:

1. The sun's UV rays
2. Oxygen in the air
3. Water or steam
4. Temperature changes
5. Micro-organisms (e.g. fungi and bacteria)
6. Different chemical solutions

Figure 7.4 The sun's rays affect and degrade many plastics, sometimes over just a few months. But for some plastics it is believed that it takes many thousands or perhaps even millions of years.
Therefore, most plastics intended for outdoor use must be UV-stabilized.

Figure 7.5 Ozone cracking in a bicycle tire.
Rubber, a synthetic polymer used in boots, tires, and so forth, is prone to cracking after prolonged exposure to the oxygen in the air, where ozone degrades the material over time.

Figure 7.6 Some plastic materials are dissolved by water. Others absorb water and then crack due to the temperature dropping below freezing.
Plastics that end up in the ocean can also disintegrate into smaller pieces (sometimes down to nano-size) due to wave erosion.

Figure 7.7 There are a number of plastics that are biodegradable, which means they must be able to be broken down completely in just a few months by soil micro-organisms in a given environment defined by EU standard 13432.
This picture shows a Rigello bottle that has been in nature for 40 years. The material was launched in the 1960s as degradable in nature, but today we know better.

7.3 Recycling Plastic

The interest in using recycled plastic in new products is increasing in line with our greater environmental awareness. Regarding the situation today (2019), the European Commission has given a preliminary assessment which shows that EU industry is significantly committed to recycling plastics: at least 10 million tons of recycled plastics could be supplied by 2025 if the pledges are fully delivered. However, on the demand side, only 5 million tons are expected so far, demonstrating that more will be needed to achieve the objective of a well-functioning EU market of recycled plastics.

Most companies also consider how their products can be recycled when they wear out–for example, ensuring that different materials are easily separated from each other.

Marking materials in a widely understood way can therefore be important (see the recycling symbols in Figure 7.12).

In Sweden, the home country of the author of this book, the recycling of plastic packaging accounts for a significant part of the recycling as Sweden has one of the best collection systems in the world. Some 28% of all plastic packaging is collected and recycled, a figure that rises to 84% for PET bottles. To improve the collection rates even more, the producers run regular information campaigns about the benefits of recycling. Many people do not know that each piece of plastic packaging returned is a big win for the environment. One kilogram can reduce carbon emissions by a factor of 2, since the packaging is ground, washed, and used as raw material in place of oil, which is a limited natural resource.

Figure 7.8 Households and other consumers are required to separate plastic packaging from other waste and return them to collection sites or leave them for collection by the local authorities. Collected plastic is then sorted and made into new products.

7.3.1 Plastic Recycling in the EU

Since many plastic products have a long life, e.g. as construction products or automotive components, we typically use twice the amount of plastic per year compared to what becomes waste. Within the EU, around 40% of the waste in landfills is plastic, which is a waste of the earth's resources. Therefore the plastics industry is pushing for a landfill ban in the EU by 2020. Many countries already have such a ban, e.g. Sweden. The good news is that there is already an upward trend and more and more plastic is being recycled. Throughout the EU (and Norway and Switzerland) around 25% of plastics are recycled into new products (material recovery) and 34% is used to produce energy and heat (energy extraction). There is a big difference between the member countries, particularly in terms of energy extraction. Plastic makes up about 9% of the waste for an average EU household, but this plastic in turn accounts for 30% of the energy used.

Figure 7.9 This chair is made from recycled ketchup bottles!

7.3 Recycling Plastic

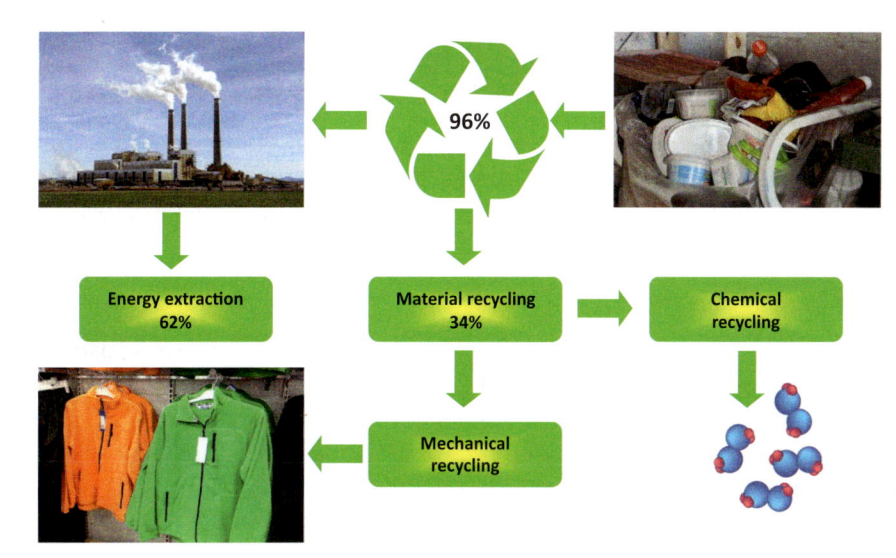

Figure 7.10 There are many different possibilities for recycled plastics. We can utilize the intrinsic energy of plastic for heat (energy extraction). Or we can make new products (material recovery); e.g. discarded PET bottles can become fleece jackets. One method is to chemically convert the product to its original monomers, which can then be re-used as plastic material.

Figure 7.11 Almost all plastic packaging can be taken advantage of.
All of this plastic packaging can be recycled, depending on where you reside:
- Bag-in-box (wine packaging)
- Bags
- Bottle caps
- Bottles
- Bubble wrap
- Buckets for jam
- Canisters
- Cans
- Caps
- Freezer bags
- Plastic sheet and plastic film
- Refill packs for detergents, etc.
- Styrofoam
- Tubes
- Vacuum pack trays (for meat, fish, etc.)

A little effort from you is often a great contribution to the climate!

„Figure 7.12" Recycling symbols for plastic packaging and the plastics they correspond to.

CHAPTER 8
Modification of Polymers

This chapter describes the polymerization of thermoplastics and how to control their properties by using various additives.

Figure 8.1 95% of all the plastics produced are based on natural gas and oil. The remaining 5% comes from renewable sources, i.e. plants.
In 2010 plastics accounted for about 4% of the total oil consumption, as follows:
- Heating 35%
- Transport 29%
- Energy 22%
- Plastic materials 4%
- Rubber materials 2%
- Chemicals and medicine 1%
- Other 7%

8.1 Polymerization

The polymerization of monomers obtained by cracking of oil or natural gas creates polymers (synthetic materials) that can be either plastic or rubber. The type of monomer determines which type of material you get, while the polymerization process itself can create different variations of the molecular chains, such as linear or branched as shown below.

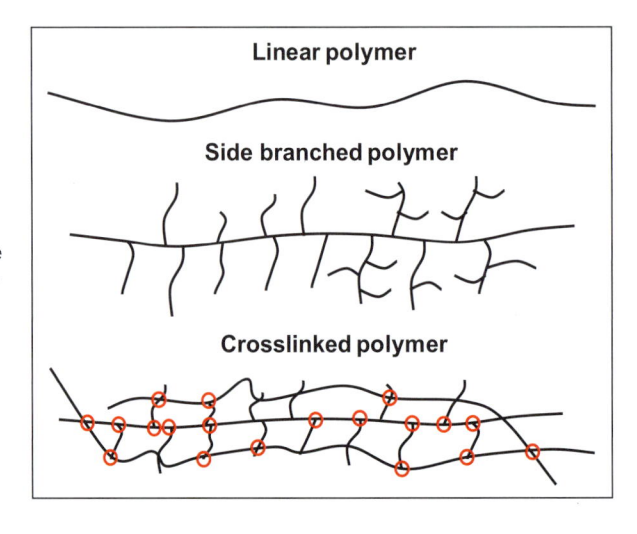

Figure 8.2 Polymerization of ethylene can produce different variants of polyethylene. LLDPE is made up of linear chains like the one at the top of the figure. LDPE has a branched chain structure, as shown in the middle. And PEX has cross-linked chains, i.e. where there are molecular bonds between the chains, as shown at the bottom.

If a polymer is made up of a single monomer it is called a homopolymer. If there are more monomers in the chain it is called a copolymer. Acetal and polypropylene are resins that can occur in both these variations. The copolymer group (the second monomer) is mainly located after the main monomer in the chain. In the case of acetal there are about 40 main monomers between every copolymer group. The copolymer may also occur as a side branch in the main chain, in which case it is known as a graft copolymer.

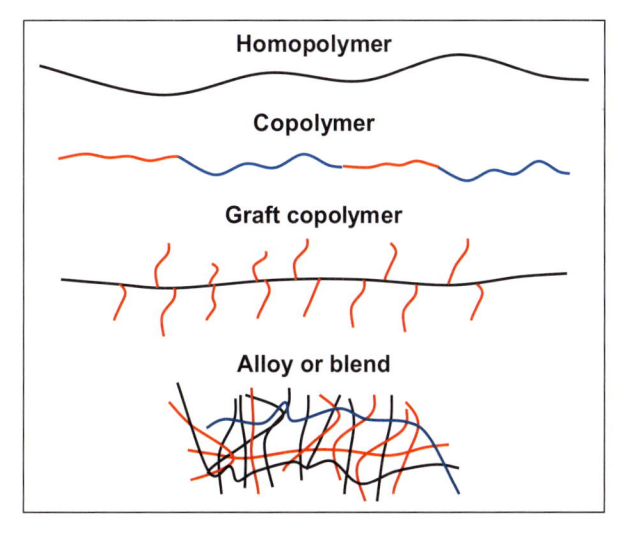

Figure 8.3 At the top we can see the linear chain of a pure polymer, such as polypropylene. By adding ethylene you get a polypropylene copolymer with a block structure according to the second chain from the top. This material has much better impact resistance than normal polypropylene.
By adding EPDM (rubber monomer) you get a graft polymer with a chain structure and a material with extremely high impact strength.
You can also create a copolymer by mixing the granules from different polymers. In this case, the material is known as an alloy or blend. ABS + PC is an example of this type of copolymer.

An additional way to modify the polymer is to control where the different molecules end up in the chain (see next).

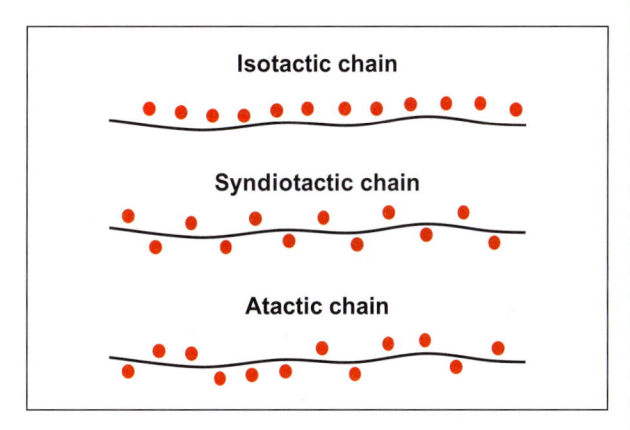

Figure 8.4 To a certain extent, we can control the properties of a polymer by influencing how a particular molecule in the chain is oriented. The red circles in the top two chains symbolize the $-CH_3$ group in polypropylene. If all the $-CH_3$ groups are oriented in the same direction, it is called isotactic.
In polypropylene, with the help of a so-called metallocene catalyst, you can orient the groups so that they are evenly distributed in different directions. In this case the chain is called syndiotactic.
In a material such as polystyrene, there is an aromatic molecule with 6 carbon atoms in a ring (symbolized by the red circle in the lower chain). This molecule ends up completely random both in orientation and distribution in the chain. Such a chain is called atactic.

Chapter 8 – Modification of Polymers

8.2 Additives

Polymeric materials are never used without being modified with different additives. For example, thermoplastic resins used in injection molding are first modified with a heat stabilizer so that they will not degrade thermally when they are molten in the cylinder of the injection-molding machine. A mold release agent (i. e. lubricant) is also added so that the finished part will be easy to eject from the mold.

In addition to improving the processing parameters, different additives are used to tailor the materials in terms of:

- Physical properties
- Chemical properties
- Electrical properties
- Thermal Properties

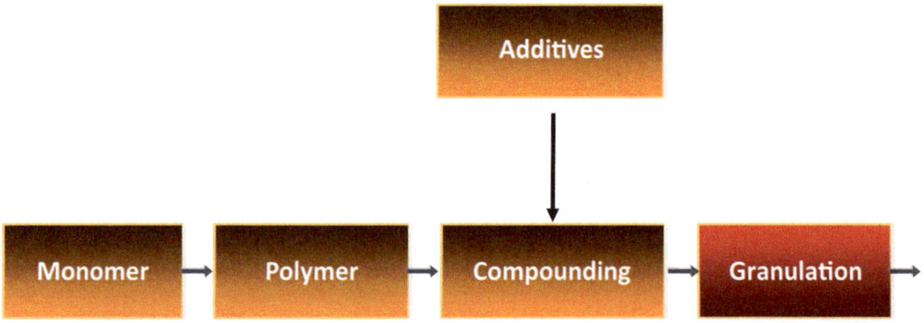

Figure 8.5 Additives are added after polymerization, in a manufacturing step called compounding.
In most cases, after the additives have been dispersed, plastic granules are produced that can then be used in the manufacture of plastic parts, profiles, or film.

Figure 8.6 These two different granulation methods are normally used.
If the material has a relatively low processing temperature (e. g. polyethylene or polypropylene) a rotating knife nozzle is used (see the picture on the left). The method is called *melt cut*, and the granules will in this case be lens-shaped and air-cooled as they fall down into the container.
If the material has high processing temperature (such as polyamide) a method called *strand cut* is used. The extruded strands are cooled in water, after which they will be cut into cylindrical-shaped granules (see middle picture). Although polyamide is sensitive to moisture, the cooling is so quick that there is not enough time to take up a significant amount.
The picture on the right shows black granules made by the strand cut and white granules made by the melt cut method.

- Mechanical Properties

  The term mechanical properties usually means:
  - Stiffness
  - Tensile strength
  - Surface hardness
  - Wear resistance
  - Toughness (elongation or impact strength)

8.2.1 Stiffness and Tensile Strength

To increase stiffness and strength, plastics are reinforced with different fibers, the most common and cheapest being glass fiber. Carbon fiber is the best and most expensive.

If you need both high stiffness and wear resistance, aramid fiber (e.g. Kevlar) is a good choice, falling between glass and carbon fiber in terms of price.

Figure 8.7 Kevlar aramid fiber is extremely strong and offers good reinforcement effects in both polyamide and acetal. The gears on the left are made of Kevlar reinforced Delrin acetal resin. The natural color of the Kevlar fiber is light yellow as seen on the image. The advantage of choosing the more expensive Kevlar fiber instead of glass fiber is that it saves weight and has very good abrasion properties.

8.2.2 Surface Hardness

Reinforcement additives will improve surface hardness and scratch resistance. Glass fiber reinforcement carries a risk of warping, which can be avoided by using glass beads or mineral reinforcement (e.g. aluminum silicate particles).

8.2.3 Wear Resistance

In addition to Kevlar fibers you can also improve wear resistance by adding different types of surface lubricants such as molybdenum disulfide, silicone oil, or fluoroplastics (i.e. Teflon).

8.2.4 Toughness

When we talk about a material's toughness, we mean either the elongation at yield/break or its impact strength. When you reinforce a material its elongation decreases while the impact strength (the energy required to break a test bar) may increase.

Figure 8.8 Many materials (such as polypropylene and polyamide) become brittle when the temperature drops below zero. By modifying them with different impact modifiers such as EPDM, used in automotive body parts, the impact strength will be increased significantly.

8.3 Physical Properties

The term physical properties usually means:

- Appearance
- Crystallinity
- Weather resistance
- Friction
- Density

8.3.1 Appearance

Appearance usually refers to the color and surface structure.

Figure 8.9 Different types of color pigments are the most common additives used to improve the appearance of a material. These can be added early at the compounding stage or afterward in the processing in the form of so-called masterbatch.
[Photo: Clariant]

8.3.2 Crystallinity

You can influence the speed of crystallinity of semi-crystalline plastics. Cable ties are manufactured in the millions, and an extremely short cycle time is required in order to be competitive. To get the polyamide to solidify quickly, nucleation agents are added.

It is not unusual to produce cable ties in large multi-cavity molds with a total cycle time of less than 4 seconds in nucleated PA66.

8.3.3 Weather Resistance

Numerous plastics are degraded by the UV rays in sunshine. This is first noticed as a color change, and then the strength of the material decreases. Some pigments (e.g. carbon black) have a UV protective effect. There are also transparent UV stabilizers.

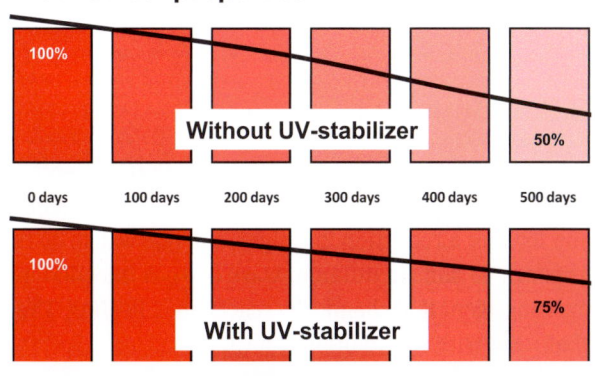

Figure 8.10 The picture shows the color change and strength decline of a red plastic material exposed outdoors for 500 days. The strength is decreased by 50%. When you add a UV stabilizer this will not eliminate the negative impact of the sun's UV light, but it will delay both the color change and the decline in strength—a loss of 25% after 500 days outdoors.
The best protection against the sun's UV rays is normally by using black pigments containing carbon black.

Figure 8.11 Many plastic parts, such as this leaf collector attached to a drainpipe, will last for several years outdoors. To avoid a reduction in strength the manufacturers use special UV-stabilized grades.

8.3.4 Friction

Fluoropolymer materials (e.g. Teflon) have the lowest friction and can be added to other plastics in small quantities in order to improve the friction characteristics.

However, fluoropolymers are very expensive lubricants, and sometimes other cheaper but less effective alternatives can be an option.

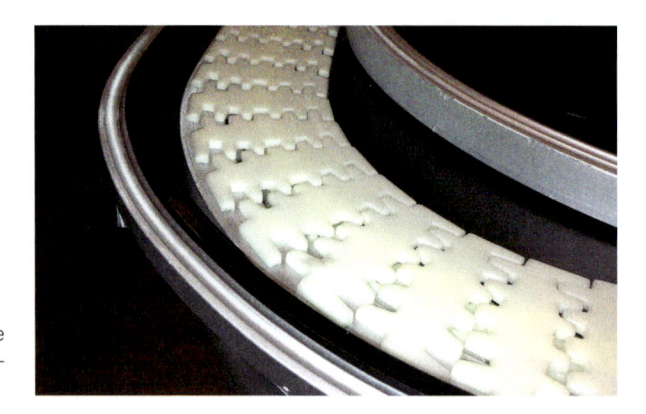

Figure 8.12 Conveyor links are produced almost exclusively in acetal. To reduce friction, which in turn reduces both wear and the power output of the drive motors, silicone oil or fluoropolymers are often used.

8.3.5 Density

It is extremely rare to add something with the intention of increasing a product's weight, although it is of course possible (with the help of metal particles). Usually you look for a reduction in weight, and one way to do this is to foam the material. This can be done in different ways like gas injection into the melt or by adding a foaming agent that chemically reacts upon heating. There are also foaming agents that react with the addition of a catalyst.

Figure 8.13 Styrofoam is foamed polystyrene.
By adding different foaming agents you can chemically control the density of thermoplastics.

8.4 Chemical Properties

The term chemical properties usually means:
- Permeability (barrier properties)
- Oxidation
- Hydrolysis

8.4.1 Permeability

Good permeability is often required by law, e.g. to prevent leakage of environmentally hazardous substances, and by food manufacturers who need reliable packaging, e.g. for carbonated beverages.

Figure 8.14 Authorities around the world, led by California, are increasing the requirements for reduced emissions of gasoline vapors from automotive plastic fuel tanks. One way to improve the permeability of blow-molded polyethylene tanks is to add a special type of polyamide that creates tight layers inside the polyethylene wall.
Another solution is the Akulon Fuel Lock FL40-HP, a PA6-based blow-molding grade from DSM that can be used in HDPE tooling (as shown to the left). [Source: DSM Engineering polymers]

8.4.2 Oxidation Resistance

Some thermoplastics (e.g. polyamide) are sensitive to contact with oxygen in the air at elevated temperatures. In order to prevent rapid degradation, antioxidants can be added.

Figure 8.15 The figure shows two railway insulators made in polyamide 66. Polyamide must be dry when processed (moisture content below 0.2%) and therefore pre-dried 2–4 hours at 80°C.
The material in the left insulator has been dried properly. In the right, which has turned yellow because of oxidation, the material has been dried too long or at too high temperature.

8.4.3 Hydrolysis Resistance

Some thermoplastics (e.g. polyesters) are sensitive to contact with water or water vapor at elevated temperatures. A chemical reaction degrades the material.

Figure 8.16 Polyamide is excellent for maintaining chemical resistance in hot water. If an extremely good hydrolysis resistance is required, as in the radiator panels for cars, some special hydrolysis stabilizers must be added.

8.5 Electrical Properties

The term electrical property usually means:

- Electrical insulation (volume resistivity)
- Resistance to creep currents on the surface (surface resistivity)
- Antistatic resistance
- Electrical conductivity

Volume and surface resistivity are dependent on the polymer chosen. In the case of electrical conductivity and static uploading, some additives can influence these properties (e.g. carbon black pigment).

Figure 8.17 The wheels in audio or video cassettes, used before CDs, were made of conductive (antistatic) acetal to reduce the risk of demagnetization of the tape.

8.6 Thermal Properties

The term thermal property usually means:

- Heat stabilization of the melt
- Service temperature
- Heat deflection temperature (HDT)
- Flame retardant classification

8.6.1 Heat Stabilization

With most thermoplastics some type of heat stabilizer is used to prevent thermal degradation in the heated machine cylinder. In addition to a higher melt temperature, the heat-stabilized material can withstand even longer residence time in the molten state before it begins to degrade.

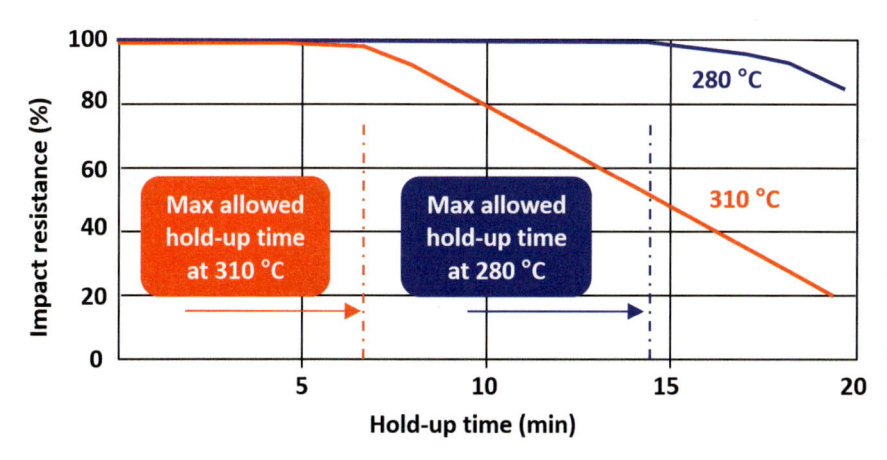

Figure 8.18 Toughened polyamide 66 is sensitive to a hold-up time longer than 15 minutes in the cylinder at the recommended processing temperature of 280 °C. At elevated temperatures (310 °C) it begins to degrade thermally and loses impact strength after just 7 minutes in the cylinder.
[Source: DuPont]

It is possible to maintain the mechanical or physical properties of a thermoplastic at elevated service temperatures by using a special thermal or color stabilizer.

Figure 8.19 Thermoplastic polyester (PET) is used in irons and oven handles due to its high service temperature. Over time, however, the material will turn yellow unless a special color stabilizer has been added.

Chapter 8 — Modification of Polymers

8.6.2 Heat Deflection Temperature

Heat deflection temperature (HDT) refers to the temperature at which a plastic sample deforms under a specific load (normally 0.45 or 1.8 MPa).

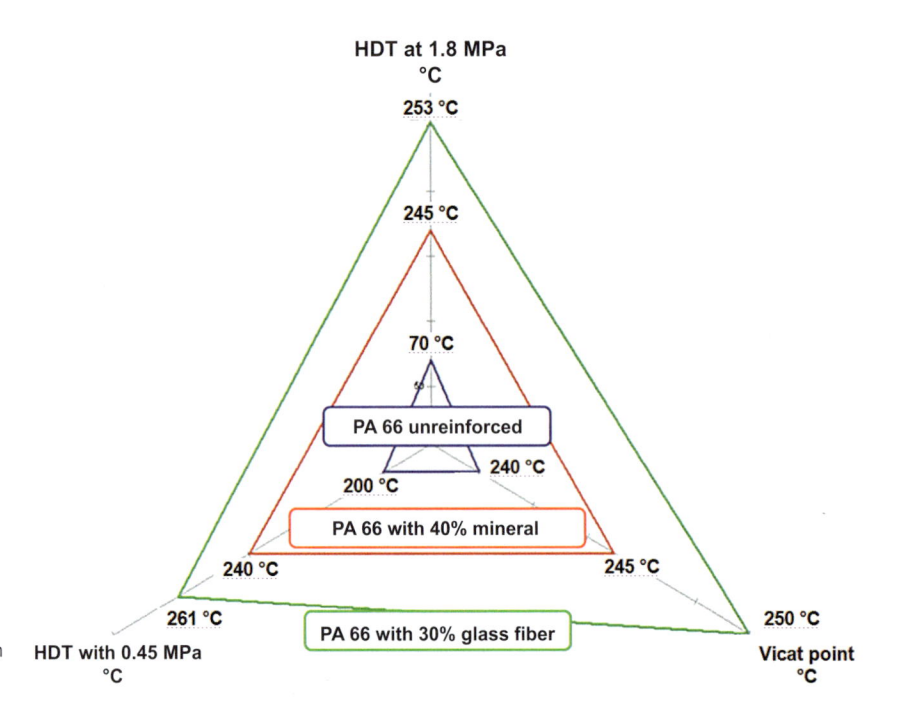

Figure 8.20 A polar chart taken from the CAMPUS material database showing the effect of glass fiber or mineral content in PA66 on the dimensional stability at elevated temperatures. HDT at 1.8 MPa increases from 70 °C to 245 °C when adding 40% mineral (aluminum silicate) and to 253 °C when adding 30% glass fiber.

8.6.3 Flame Retardant Classification

Many electrical and electronic products in our daily lives are modified with flame retardants. Unfortunately these are normally hazardous compounds containing halogens (e.g. bromine) or phosphorus, although the trend is slowly moving toward more environmentally friendly (halogen-free) alternatives, which are more expensive. The only way to speed up the transition to the environmentally friendly alternatives is probably by international law, as was the case with the EU cadmium ban in 2004.

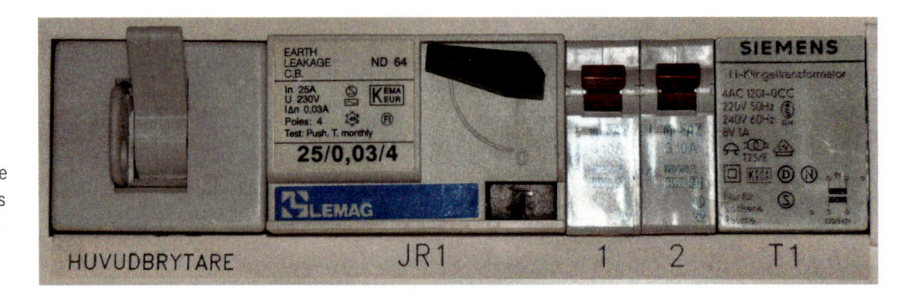

Figure 8.21 Many of the products we use daily are modified to meet various safety requirements. Flame retardant classification is one of these requirements.

8.7 Material Price

The cost of material in a plastic product can be influenced by the additives used (listed below), which may affect the mechanical or physical properties of the virgin material:

- Regrind (runners, scrapped parts)
- Foaming agents

Figure 8.22 Most molders can grind runners, unfilled parts etc. The normal recommendation is to mix up to 30% regrind into virgin resin since this affects the mechanical properties only marginally. If you have several different colors in the same material you usually add 1–2% black masterbatch.

CHAPTER 9
Material Data and Measurements

In this chapter we will go through those properties of thermoplastics that are often requested by designers and product developers when they are looking for a material in a new product or when they must meet different industry or regulatory requirements, such as electrical or fire classification.

When plastic producers develop a new plastic grade they usually also publish a data sheet of material properties. Sometimes this is made as a "preliminary data sheet" with only a few properties. If then the product will be a standard grade, a more complete data sheet will be published. Many suppliers publish their material grades in the CAMPUS or Prospector materials databases on the Internet, which can be used to some extent free of charge (see next chapter).

Figure 9.1 What are the different requirements from authorities on a so unremarkable product as an electrical outlet that must be fulfilled to be sold on the market?

CAMPUS is very comprehensive and can describe a material with over 60 different data types, and at the same time you can get graphs (e.g. stress-strain curves) and chemical resistance to many chemicals.

The most requested data when it comes to thermoplastics and that are usually in the "preliminary data sheet" are:

- Tensile or flexural modulus
- Tensile strength
- Elongation
- Impact strength
- Maximum service temperature
- Flame resistant classification
- Electrical properties
- Rheology (flow properties)
- Shrinkage
- Density

9.1 Tensile Strength and Stiffness

Stiffness, tensile strength, and toughness in terms of elongation can be obtained by the curves in tensile testing of test bars.

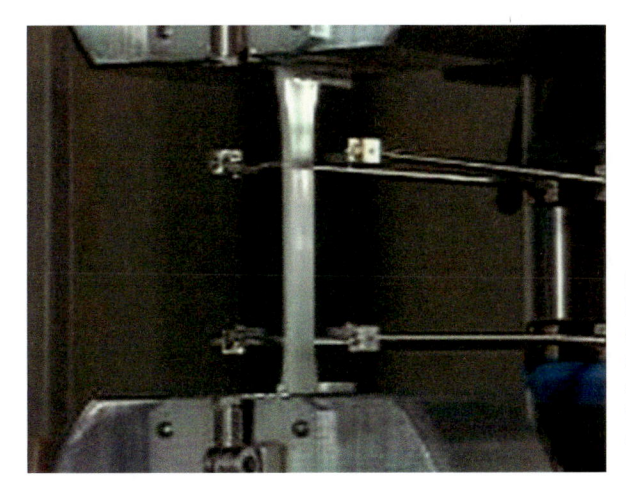

Figure 9.2 The picture shows a test bar in a tensile tester. All plastic producers measure the mechanical properties on specimens manufactured according to various ISO standards, which makes it possible to compare data between different manufacturers. [Photo: DuPont]

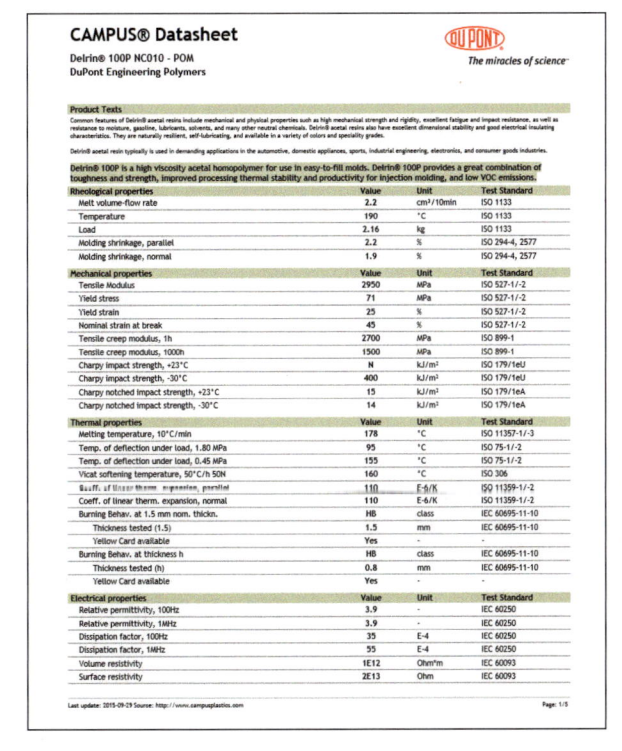

Figure 9.3 In the CAMPUS materials database on the Internet, you can pick up thousands of datasheets. Depending on the type of plastic, different characteristics are shown. Often the following properties are displayed:
- Mechanical
- Thermal
- Rheological
- Electrical
- Fire classification
- Other (e.g., chemical)

Chapter 9 – Material Data and Measurements

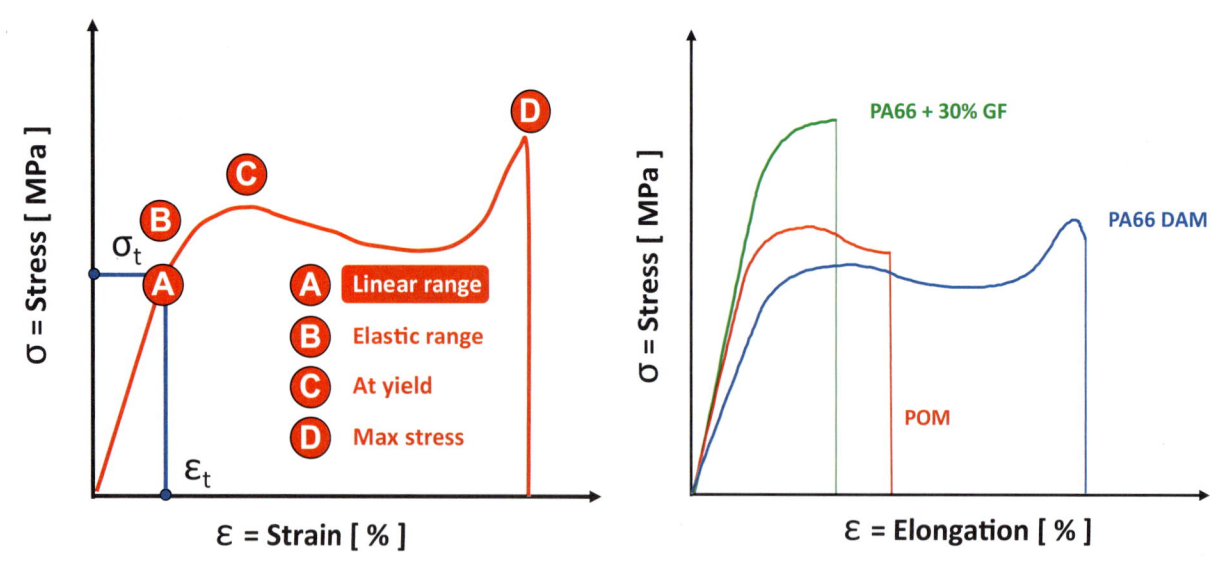

Figure 9.4 The picture shows a typical stress-strain curve for unconditioned PA66 obtained by tensile testing. The curve can be divided into the following segments:
(A) Linear range
(B) Elastic range
(C) At yield
(D) Maximal stress/strain

Figure 9.5 In this figure you can see tensile curves for different plastics. Note that stress at break for the acetal curve (red) is lower than the maximum stress. This is due to necking. The reason why the unconditioned PA66 gets increased stress at the end of the curve depends on molecule orientation, which gives a hardening effect. If you compare the green curve (PA66 with 30% glass fiber) and the blue (unreinforced PA66) you can see the effect of glass fiber reinforcements on stress and strain. You can get significantly stronger but more brittle materials. Higher elongation at break means tougher material.

In the linear region, it's easy to make a strength calculation because you can use Hooke's law, i.e. σ = Force/Area (MPa).

Figure 9.6 As long as you are in the range of (A) or (B) in the graph in Figure 9.4, a test bar regains its original shape after unloading (picture on the left here). If you exceed the so-called elongation at yield at (C), necking occurs (middle image) that will extend until the breakage at (D) (the test bar on the right).

From the stress-strain curve (Figure 9.4) you can get the following mechanical properties:

1. Stress at yield σ_y i.e. (C) in the curve

2. Elongation at yield ε_y i.e. (C) in the curve

3. Stress at break σ_B i.e. (D) in the curve

4. Elongation at break ε_B i.e. (D) in the curve

5. The stiffness of the material E_t is specified as the tensile modulus and can be calculated in the linear region with the following equation:
$E_t = \Sigma_t / \varepsilon_t$

If the tensile curve is nonlinear you need an approximation and calculate the tangent or secant modulus.

9.1 Tensile Strength and Stiffness

Tension is measured in units of MPa (megapascals) and elongation in percentage. In Figure 9.9 the relationship between the various units is presented.

In addition to the specification of stiffness as tensile modulus E_t you can also specify it as flexural modulus E_S. At present, the tensile modulus is much more common than the flexural modulus in the plastic raw material suppliers' data sheets. Figure 9.7 and Figure 9.8 show the curve for bending stress-strain.

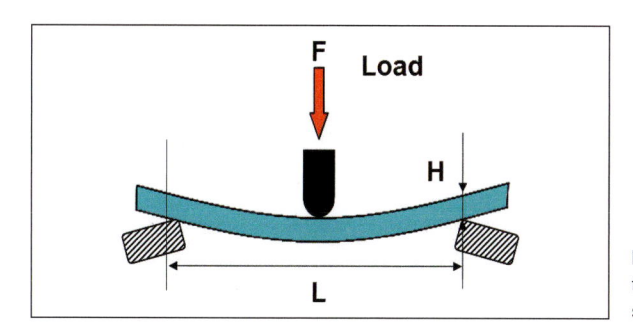

Figure 9.7 This picture shows that the test bar is fixed horizontally on two supports and loaded in the middle.

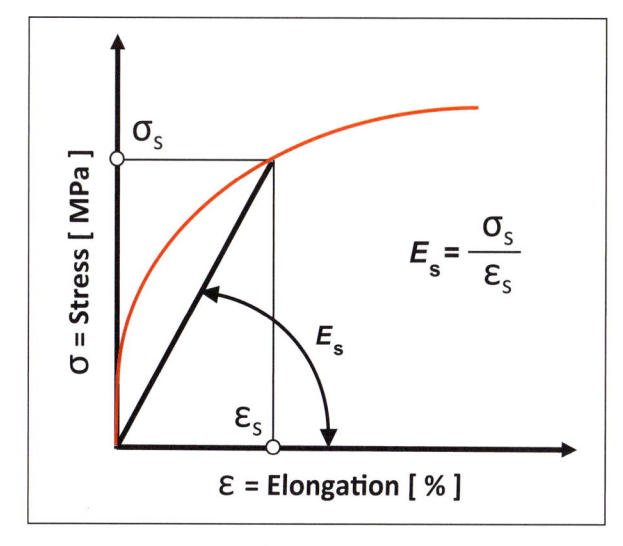

Figure 9.8 While making a bending load the curve becomes nonlinear. You must approximate and use a secant (diagonal) to calculate the flexural modulus $E_S = \sigma_s/\varepsilon_s$.

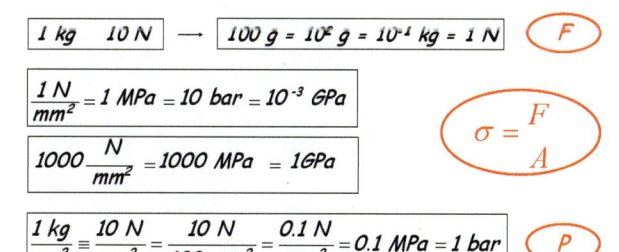

Figure 9.9 If you hang a weight of 1 kg on a string, the string will be loaded by the force F = 10 N (Newton). The stress in the string depends on area A and will be σ = F/A, i.e. if the string area is 1 mm² the stress will be 10 N/mm² = 10 MPa.
The pressure P is also specified in MPa. Earlier it was specified in bar.
The locking force of injection-molding machines is often specified in tons. The correct unit is however MPa and 1 ton = 10 MPa.

Chapter 9 — Material Data and Measurements

9.2 Impact Strength

The impact test that is dominant today is the "Charpy." You fix the test bar at both ends in a horizontal position and allow the pendulum to hit it in the middle. The unit of the Charpy test is kJ/m^2. In the past another impact test according to "Izod" was commonly used. Here you need to fix the lower half of the test bar in a vertical position and hit the upper half of it. The unit for the Izod test is J/m, and there is no factor that allows you to convert the values between the two test methods.

Figure 9.10 This picture shows a test bar with a milled notch to be used in impact tests.

Figure 9.11 The picture shows a pendulum impact tester.
The pendulum is locked in its starting position. When it passes and hits the test bar it loses energy. The energy loss value indicates the impact resistance. Usually you measure the impact at 23 °C or at −30 °C with or without notch.

The impact test is a sensitive and good quality control method. Many molders are using a self-built drop weight tester. You can e.g. drill holes with about 5 cm distance in a 50 mm plastic drain pipe. A cylindrical weight with a ball-shaped bottom is then hoisted up in the pipe to a certain height and fixed with a pin. When the pin is pulled out the weight falls and hits the part that has been fixed under the pipe. If the part passes the required height with no damage the impact strength is OK. If it breaks something is wrong with the material or the process.

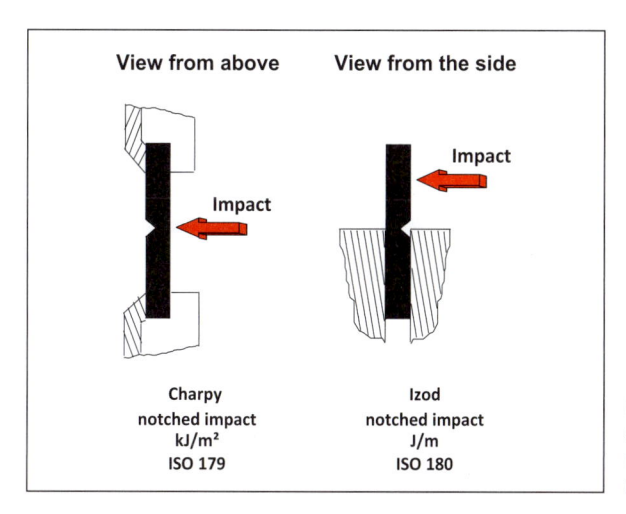

Figure 9.12 The picture shows the attachment of the test bars according to Charpy to the left and Izod to the right.

9.3 Maximum Service Temperature

9.3.1 UL Service Temperature

It is easy to get confused when trying to determine a material's maximum service temperature as this can be specified differently. A leading international test institute called Underwriters Laboratories has developed the specification of the maximum continuous service temperature and calls it "UL service temperature." To do this you have to put test bars in ovens at different temperatures and wait 60,000 hours (I.e. almost 7 years). Then you bring out the test bars and test them. The temperature that has affected the test bars so much that they have lost 50% of their initial values is specified as the maximum continuous service temperature (UL service temperature). It is specified both for mechanical and electrical properties.

9.3.2 Heat Deflection Temperature

In most plastic data sheets you will find values of the material's heat deflection temperature at different loads. Heat deflection temperature is abbreviated to HDT.

Chapter 9 — Material Data and Measurements

Heat deflection temperature

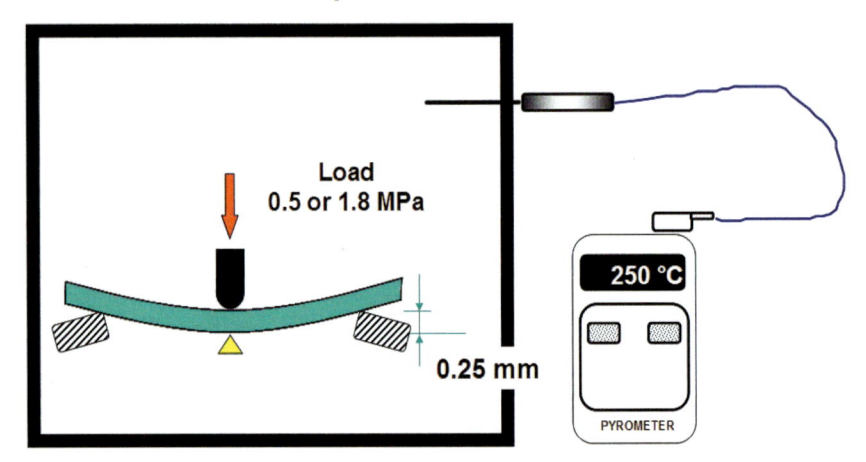

Figure 9.13 When measuring HDT you have to fix a test bar horizontally at both ends. Then you put it into an oven and load it in the middle with either 0.45 or 1.8 MPa.
You let the oven temperature rise by 2 °C per minute and record the temperature at which the sample bar has bent down 0.25 mm as the HDT.

In the tables below with values from the CAMPUS materials database (see next chapter) you will find HDT for a number of thermoplastics. NOTE! Some deviation from the values below may occur depending on the viscosity and additives of the materials.

Table 9.1 Table with common plastics heat deflection temperatures

Type of polymer	HDT at 0.45 MPa	HDT at 1.8 MPa	Melting point
ABS	100	90	–
Acetal copolymer	160	104	166
Acetal homopolymer	160	95	178
HDPE, polyethylene	75	44	130
PA 6	160	55	221
PA 6 + 30% glass fiber	220	205	220
PA 66	200	70	262
PA 66 + 30% glass fiber	250	260	263
Polyester PBT	180	60	225
PBT + 30% glass fiber	220	205	225
Polyester PET	75	70	255
PET + 30% glass fiber	245	224	252
PMMA (acrylic plastic)	120	110	–
Polycarbonate	138	125	–
Polystyrene	90	80	–
PP, polypropylene	100	55	163
PP + 30% glass fiber	160	145	163

Note: The amorphous materials have no melting point

9.4 Flammability Tests

The international testing institute Underwriters Laboratories has developed various tests to specify a material's fire resistance. You select test bars with different thickness and ignite them either horizontally or vertically. We specify this as HB (= horizontal burning) or V-2, V-1, or V-0 (V = vertical burning). For a material to be classified as fire resistant, it must be extinguished by itself within a certain distance (HB) and at a certain time. When testing a material for V-0 to V-2 you will also give attention to possible drops that ignite cotton (see below).

9.4.1 HB Rating

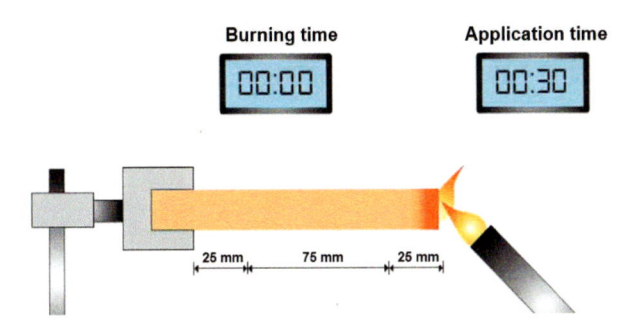

Figure 9.14 The flame is applied for 30 seconds before the ignition speed is measured. HB classification is obtained if the ignition speed measured between two points does not exceed:
1. 40 mm/min for 3–13 mm test bars
2. 75 mm/min for test bars < 3 mm
3. If the flame goes out before the first mark

9.4.2 V Rating

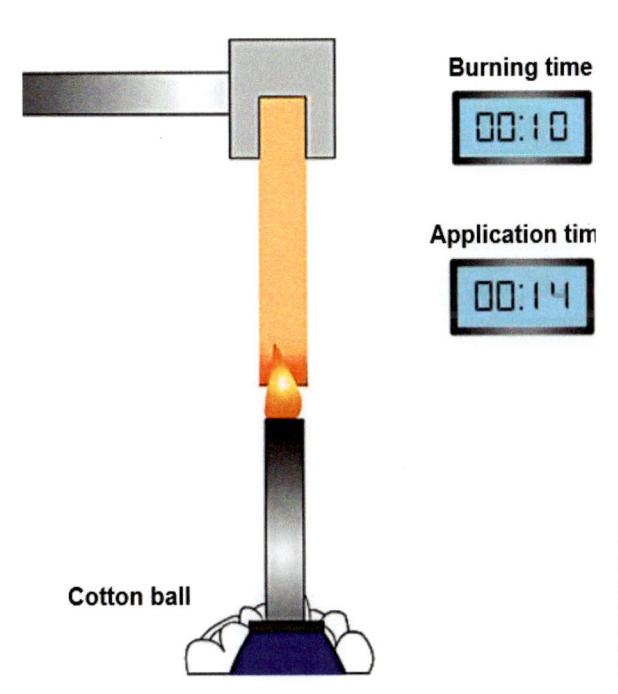

Figure 9.15 When testing a test bar in a vertical position you will apply the flame twice during each 10 seconds. The contact time of the second ignition begins immediately after extinguishing the test bar of the first flame.

Table 9.2 This specifies the times that must be met for the test to be approved. Under the test bar there is cotton, and attention is given if resultant drops will ignite it. Finally, if any afterglow occurs, the time of this will be measured. [Source: Underwriters Laboratories]

Flame application	20 mm high Tirill burner flame		
Flame application time	2 × 10 s		
The second flame application time begins as soon as the ignited specimen is extinguished or immediately if the specimen does not ignite			
Flammability rating UL 94	**V-0**	**V-1**	**V-2**
Burning time after flame application (sec)	< 10	< 30	< 30
Total burning time (s) (10 flame applications)	< 50	< 250	< 250
Burning and afterglow times of specimens after second flame application (s)	< 30	< 60	< 60
Dripping of burning specimens (ignition of cotton batting)	No	No	Yes
Specimens completely burned	No	No	No

9.5 Electrical Properties

There are a variety of test methods for electrical properties of plastics. Usually, you specify the material's insulating capability to or resistance to creep currents on the surface.

The following methods are often published in data sheets:

1. Dielectric strength
2. Volume resistivity
3. Arc resistance
4. Surface resistivity
5. "Tracking" resistance CTI (Comparative Tracking Index)

Figure 9.16 Tests according to methods (1) to (3) above are made in test equipment with the principle shown in the figure to the left, and tests according to (4) and (5) are made in test equipment with the principle shown in the figure to the right.
If you want to know more about electrical test methods for plastics we can recommend this website: www.ul.com.

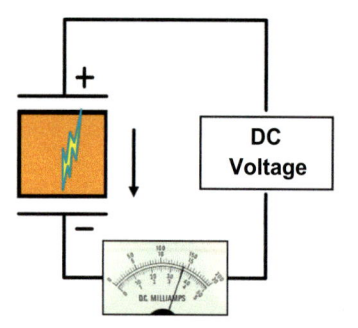

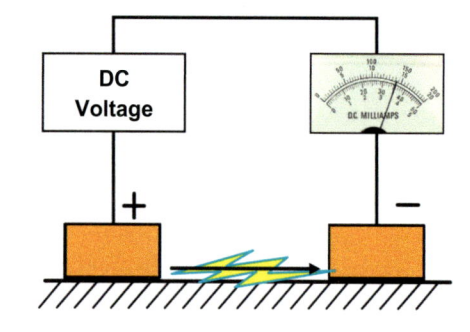

9.6 Flow Properties: Melt Index

You can measure the melt flow properties of thermoplastics by using a test method called the melt flow index, MFI. Another name for this method is melt flow rate, MFR.

Below you can see the principle for melt index.

Figure 9.17 When you are testing the flow properties of a thermoplastic melt you start by heating up the granules in a cylinder. The temperature according to the standard will vary depending on the polymer. Once the material has reached the specified temperature you put a weight (also polymer dependent) on the piston and record the time it takes for the material to flow out of the cylinder.
You specify the MFR in units of cm^3/10 minutes. MFI is specified as the weight of the material flow after 10 minutes and the unit is g/10 minutes.

9.7 Shrinkage

Mold shrinkage is the difference between the dimensions of the cavity and the dimensions of the molded part.

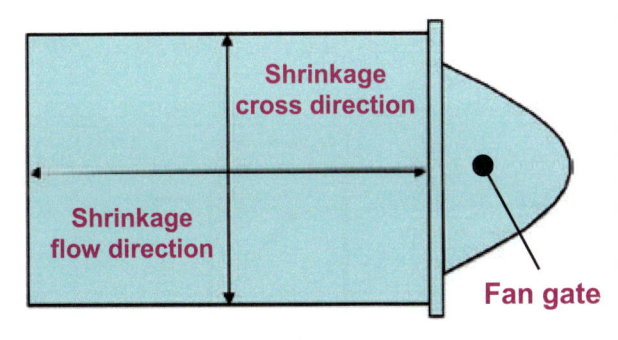

Figure 9.18 The mold shrinkage is measured after one day (at least 16 hours). Semi-crystalline materials undergo a post-crystallization that can last for months depending on the ambient temperature and type of polymer. This causes a shrinkage called post-shrinkage.
The total shrinkage = mold shrinkage + post-shrinkage.
Shrinkage is usually measured in both the flow and cross-flow directions.

CHAPTER 10
Material Databases on the Internet

A good way to find information about different plastic materials is to visit the raw material producer's websites or visit independent material databases on the Internet. In this chapter you will find three leading global databases: CAMPUS and Material Data Center from the European company M-Base and the Prospector Materials Database from the U.S. company UL IDES. The great advantage of all the databases is that you can compare the material data no matter who the producer is, as all the materials in the databases are tested in exactly the same way.

10.1 CAMPUS

About 20 large plastics raw material producers use CAMPUS to inform their customers about their products. Software for CAMPUS is offered free by the producers, and can be downloaded directly via the Internet: www.campusplastics.com.

The database is updated regularly and can be updated via the CAMPUS website.

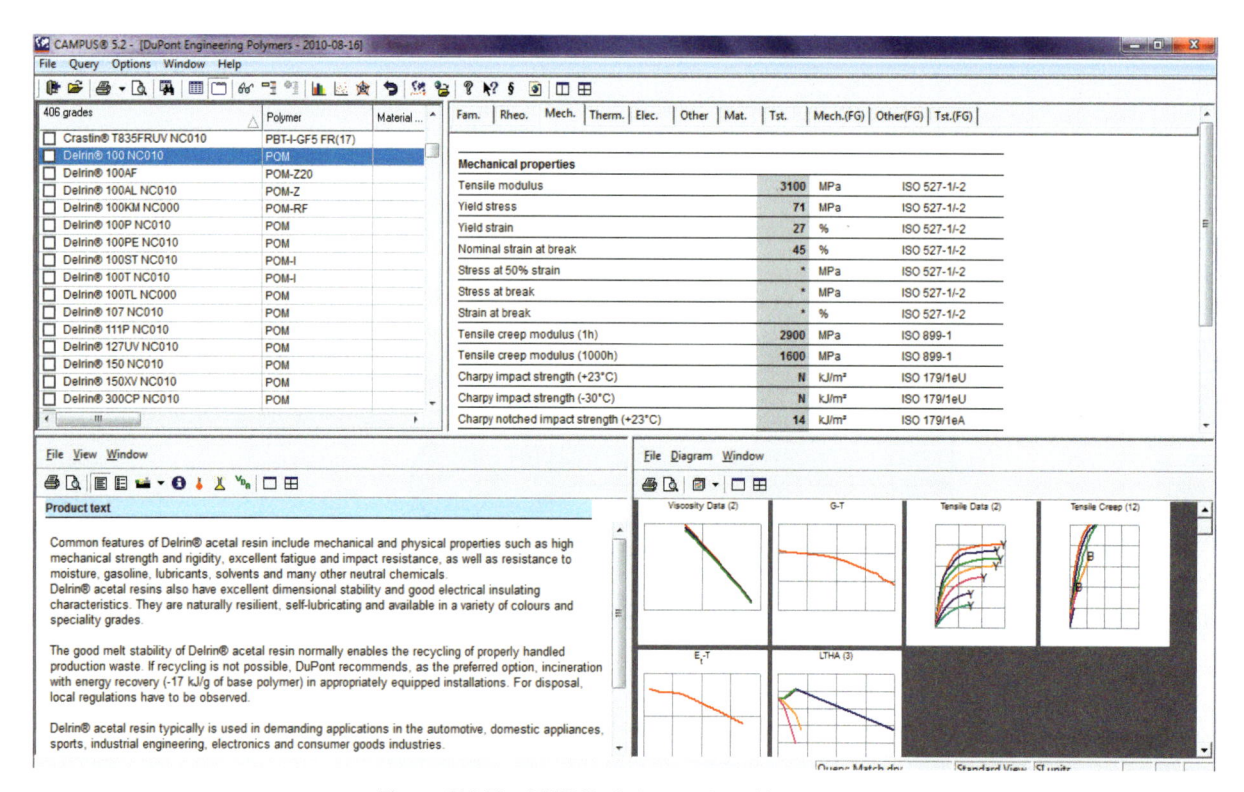

Figure 10.1 The CAMPUS window consists of four smaller windows. The top left is the list of all the materials. The top right is the properties window, which in this case shows the mechanical properties of the selected grade (Delrin 100 by DuPont). The bottom left is the information window with the information about Delrin 100, and to the right we can see the different curves for this material in the graphics window.

10.1.1 Properties of CAMPUS 5.2

+ The database is free to download from the Internet
+ You can sort the properties in tables
+ You can compare different materials in a tabular form
+ You can compare different materials graphically
+ You can get chemical resistance information for the materials
+ You can specify and print your "own" data sheet
+ You can search for materials meeting various criteria on properties
+ You can get the material process data in "curve overlay" and "polar" charts
+ You can get the material's flow properties (to be used in mold flow simulations)
− You can only compare materials from one specified material producer at a time
− The database must be updated manually

10.2 Material Data Center

This database contains more than 40,600 plastic grades from more than 330 different raw materials producers. In order to use Material Data Center you must be registered and pay an annual fee of € 390 (about US$ 447; January 2019). But you can do a 7-day free trial.

Here is the link to the website: www.materialdatacenter.com.

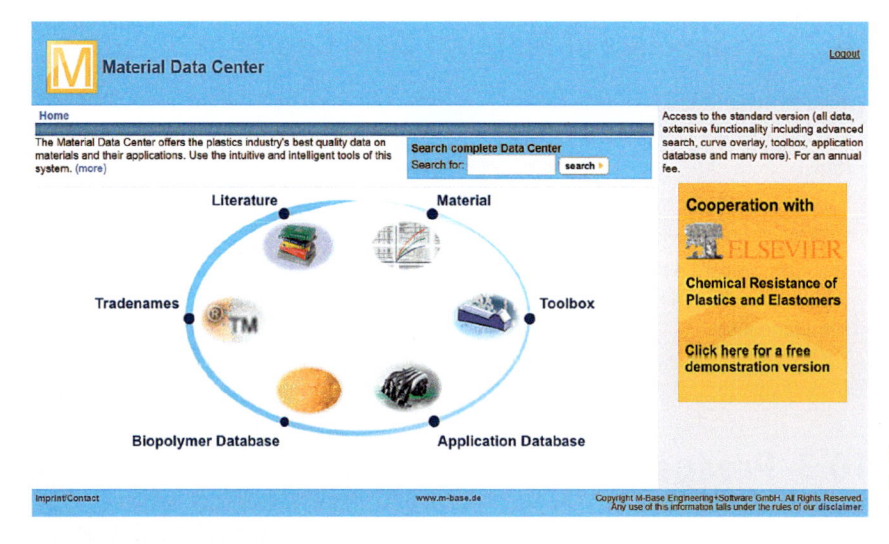

Figure 10.2 The start screen in the Material Data Center shows the different functions of the database.

Chapter 10 – Material Databases on the Internet

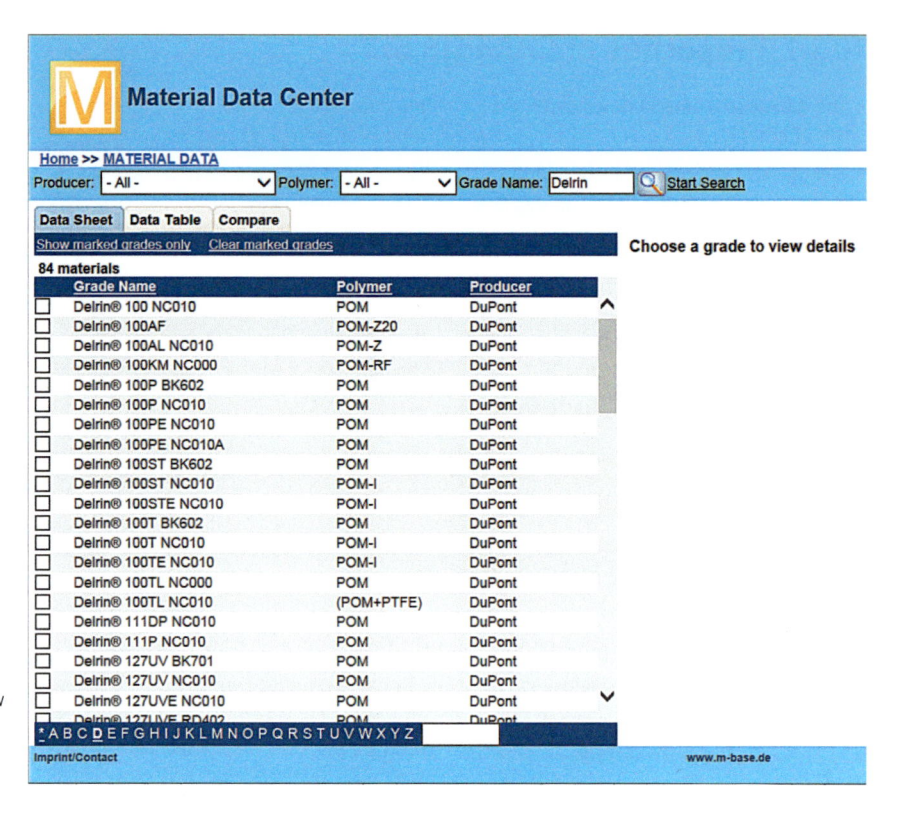

Figure 10.3 In the Data Sheet window you can search for Producer, Polymer, or Grade Name. Here you can see some Delrin grades by DuPont.

10.2.1 Properties of Material Data Center

+ You can specify and print data sheets
+ You can sort the properties in tabular form
+ You can search for materials that meet various criteria on the properties
+ You can compare different materials in tabular form from all producers simultaneously
+ You can compare different materials in the "curve overlay" chart
+ You get access to special databases with plastic literature, applications, and bioplastics

10.3 Prospector Plastics Database

This database is the largest and contains more than 120,000 material grades from about 900 different producers (February 2019). Prospector contains both free and fee-based features.

Here is the link to the website: www.ides.com/prospector/.

10.3 Prospector Plastics Database

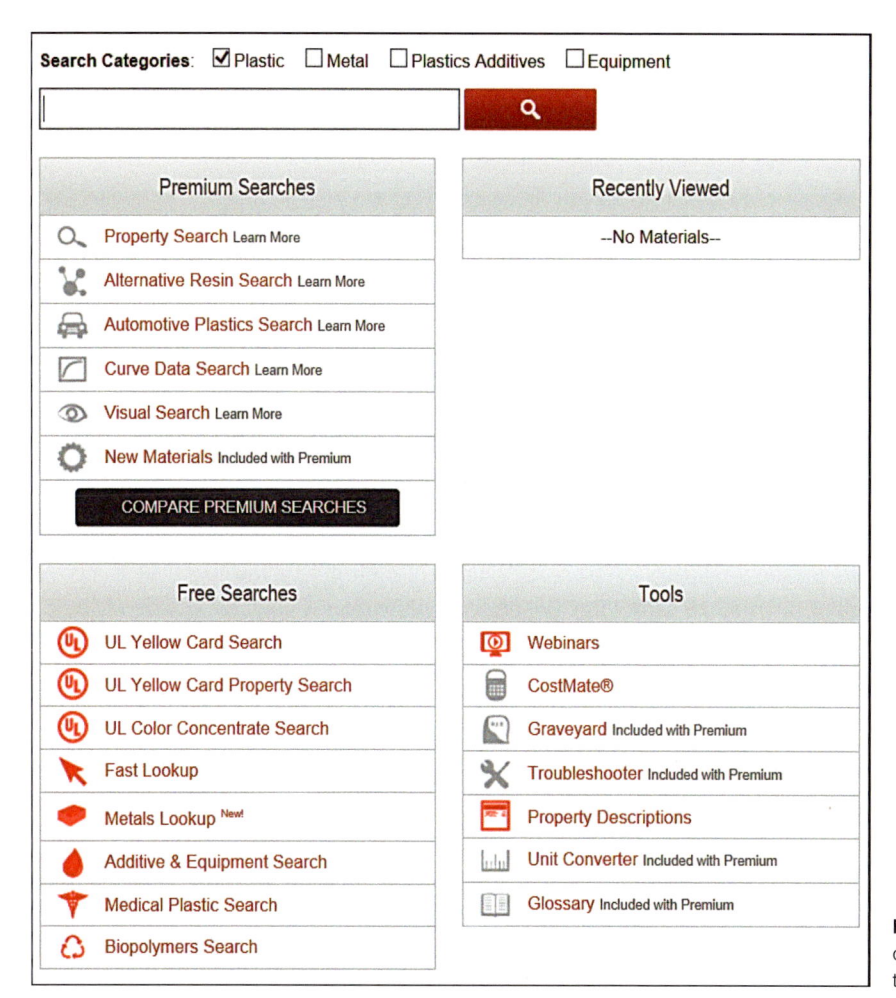

Figure 10.4 To use the free functions of Prospector you must first register on the Ides Prospector website.

Free Functions in Prospector

- Data sheets for all materials including process information
- Troubleshooting guide
- Search for keywords, trade names, polymer abbreviations, applications, bio plastics, and medical grades
- Description of various test methods
- Plastic Technical Glossary
- Video clips

Fee-Based Functions in Prospector

- Search on material properties
- Search for alternative materials and approved "automotive grades"
- Material comparisons in tabular form in "curve overlay" charts
- Cost estimate for moldings

101

CHAPTER 11
Test Methods for Plastic Raw Materials and Moldings

In this chapter we will describe the plastic raw material producer's quality control data, the various material defects that a molder may find, as well as the test methods you can use when you want to analyze these kinds of defects.

11.1 Quality Control during Raw Material Production

The plastic producers measure the quality of their plastic raw material at regular intervals (random sampling). Depending on the type of polymer and the included additives, they use different test methods during production. In general they are testing:

- Viscosity, which is dependent on the molecular chain length
- Fiber content, i.e. the ash content after complete combustion of the polymer
- Moisture content of each batch at the packing station

Figure 11.1 Here we can see the test results of 12 different batches of a 30% glass fiber reinforced grade. The aim is to be as close as possible to 30%, but as long as the result is within the green lines (30 ± 2%), the material is approved for delivery. Batches 7 and 11 are not acceptable and must be redone in order to fall within the delivery limits.

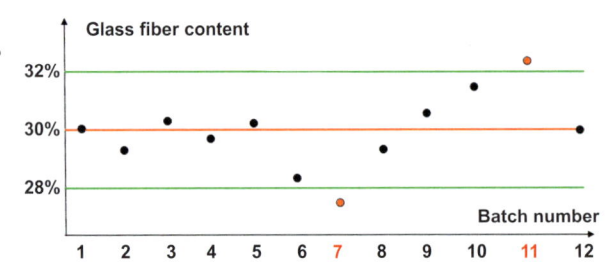

Table 11.1 In this table we can see that thermal and mechanical properties are tested at least once a year.
These values are then used for the published values in the producer's literature or in databases. It is only in exceptional cases that molders can get their material regularly tested with these types of testing.

	Test Method (Unit)	ISO	Specified
A	Moisture content (%)	15512	≤0.20
A	Ash content (%)	3451	31–35
B	Melting point (°C)	3146	250–265
B	Density (g/cm^3)	1183	1.34–1.41
B	Tensile strength (MPa)	527	≥157
B	Elongation (%)	527	≥1.8
B	Stiffness – E-modulus (MPa)	527	≥8,000
B	Impact strength Charpy (kJ/m^2)	179	≥7.8
B	HDT at 1.8 MPa (°C)	D789	≥245

A: Every batch; B: Annually

The test values that the producers receive during the random sampling of the various production batches will as a rule be attached together with the material (or invoice) in the form of a delivery certificate. In this certificate you will find the same lot number (also called batch number) that you will find stamped on the bags or octabins. It is very important to keep these certificates in case of a complaint because the production plants always want information about the batch.

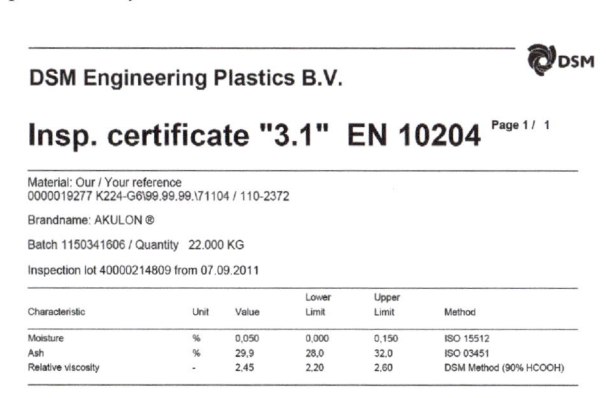

Figure 11.2 Here you can see a delivery certificate from DSM for Akulon K224-G6 (natural PA6 with 30% glass fiber).
Here they have measured:
- Moisture content of 0.050% and indicated the upper delivery limit to be 0.150%
- An ash content (glass fiber content) of 29.9% and indicated the limits of supply to be between 28.0% and 32.0%
- A relative viscosity in a solution of formic acid of 2.45, which is well within the tolerance limits.

11.2 Visual Quality Control of Plastic Granules

It is very rare that an operator discovers that something is wrong with a material just by inspecting the granules. Only when the material is in production is it normal to discover defects.

Figure 11.3 The defects that you sometimes can see by inspecting the granules are different color, wrong size, clumping of granules, black specks in bright material, or metal particles on the surface. It may also be defects such as dust, dirt, or contamination within the bag of granules that affect quality and performance of production.

11.3 Visual Inspection of Plastic Parts

If there are defects with the plastic raw materials, this is something that is discovered in most cases when the production of parts has begun. Most often you will discover defects by visual inspection, but there are also occasions when defects are discovered during measuring or mechanical testing.

Figure 11.4 Black specks
Black specks occur due to thermal degradation in either the machine cylinder or within the production of the granules. If they already are present in the granules there is no doubt that this is a defect of the raw material, and a complaint must be made.

Figure 11.5 Metal particles
Many molders use magnetic bars in the hopper to catch granules containing ferrous metal particles. If the metal particles are made of brass or stainless steel they may slip through, and they can damage both the screw and the cylinder or get stuck in the gate of the mold.

Figure 11.6 Discolorations
If the color pigment in the raw material is poorly dispersed, you can get dark "shadows" of discoloration on the surface of the part. Increased back pressure and lower screw speed may in some cases help to solve this problem.

Figure 11.7 Foreign granules
Occasionally foreign granules end up in bags when packing. If these have a higher melting point and a different color it may look like this picture, where green polyamide granules have fallen into natural-colored acetal. The nylon melts about 80 °C above the melt temperature of acetal.

If an error occurs when you are changing to a new bag or a new batch and there is an indication that the new material behaves differently it is a sign of defects in the material.

In many cases you can solve this by making an adjustment of the process parameters. But if this is not successful you should test the material in another machine before filing a complaint.

Figure 11.8 Silver streaks
Polyamides are moisture sensitive in the melt. Vapor bubbles form and appear as silver streaks on the surface. The resin should be sufficiently dry in the bag, but if the bag has been punctured this is enough to cause normal pre-drying to be insufficient.

Figure 11.9 Unfilled parts
If the plastic raw material has higher melt viscosity than normal, there can be problems with unfilled parts. Usually this can be corrected by using a higher pressure or increased melt or mold temperature.

Figure 11.10 Flash
If the plastic raw material has a lower melt viscosity than normal, there can be problems with flash. Usually this can be corrected by using a lower pressure.
In the polyesters PBT or PET, flash may occur due to insufficient drying.

Figure 11.11 Wrong size of the granules
Sometimes the plastic producer offers a so-called "cube blend". This is a masterbatch-colored resin.
As you can see in this picture some of the granules are unusually long, and these granules may cause problems during the dispersion of color within the cylinder of the molding machine.

11.4 Tests That Can Be Performed by the Molder

The eye is the best instrument for quality control!

Besides visual inspection, most molders have other methods for testing their parts in order to determine if defects are present.

Below there are some of the most common test instruments or tools used by molders:

1. Precision balance in order to determine if the weight deviates from normal
2. Measuring equipment in order to determine if the dimensions deviate
3. A saw in order to determine if the parts contain unmelted granules, foreign particles, or voids

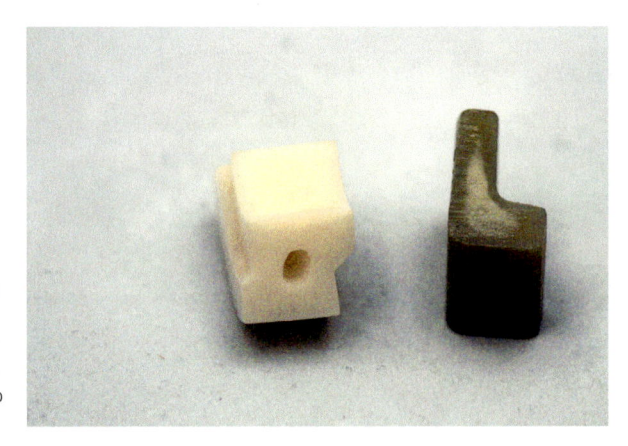

Chapter 11 — Test Methods for Plastic Raw Materials and Moldings

Figure 11.12 Sometimes there are surprises when you saw apart a plastic part. This picture shows the presence of a large void within the acetyl part to the left and micro-pores present in the glass fiber reinforced polyamide part to the right.

Less common tests or equipment found among molders:

4. Drop-weight test for tests of toughness
5. Tensile testers for mechanical testing
6. Equipment for moisture analysis
7. Equipment for measuring melt flow index
8. Combustion testing (see Table 11.2 below)
9. Color testing

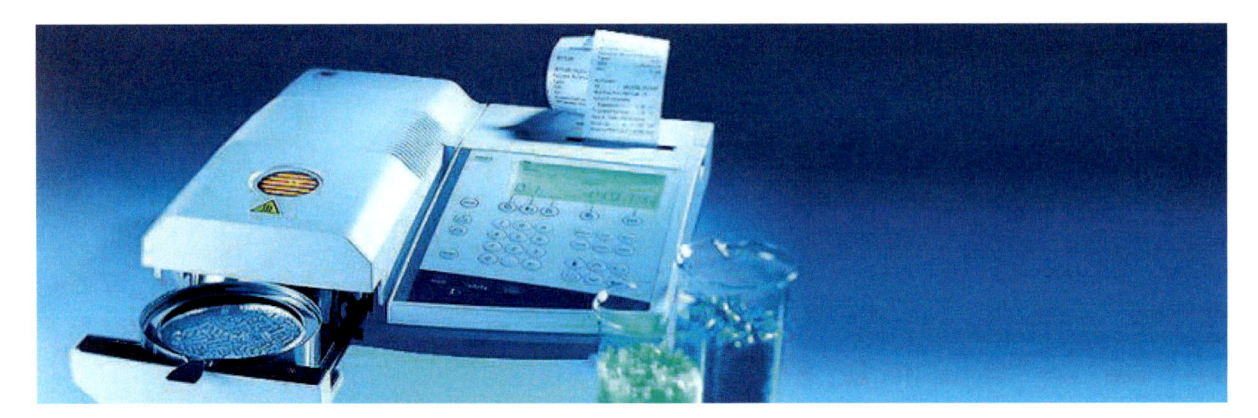

Figure 11.13 The equipment used for moisture analysis generally consists of a precision balance enclosed in an oven. You begin by weighing the granules before the oven is turned on. When all the moisture has been dried off from the granules you will weigh the granules once more. The difference in weight will show how much moisture the granules contained.
This picture shows a HR83, a modern high tech instrument used to measure levels of moisture by the use of halogen technology from Mettler Toledo. [Source: se.mt.com]

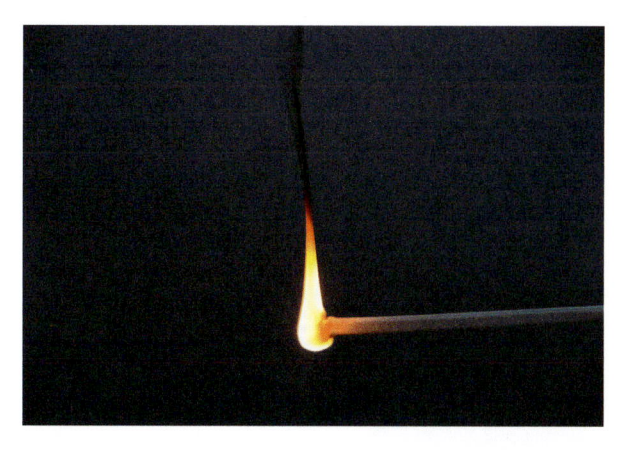

Figure 11.14 Most plastics have a very characteristic flame combined with smoke, smell, and fire, which can be used to determine the type of polymer.

Table 11.2 This table describes the flame progression and odor of some common plastics.

Material	Fire progression	Odor
ABS	Yellow dripping flame with black smoke	Very typical for ABS
PA – Polyamide	Blue flame with yellow top. The flame melts and drips with clear viscous drops	Burned wood
PC – Polycarbonate	Yellow flame with smoke. The test bar melts and carbonizes	
PE – Polyethylene	Blue flame with yellow top. The flame melts and drips with clear burning drops	Candle
POM – Acetal	Blue flame without smoke	Ammonia
PP – Polypropylene	Blue flame with yellow top. The test bar swells and drips	Sweetish
PS – Polystyrene	Yellow flame with black sooty smoke	Coal gas
PVC	Yellow flame with green edges. The test bar is softening	Hydrochloric acid
SAN	Yellow flame with a black sooty smoke	

11.5 Advanced Testing Methods

Most major raw material producers use advanced test equipment for quality control and material development. Many offer analysis to their customers in order to determine the causes of defects on plastic parts.

Some of these test methods are:

1. Accurate moisture analysis of granules or parts

2. Viscosity tests of granules or parts

3. Ash analysis after burning off the polymer to measure the level of reinforcement or filler

4. Material identification by the use of an infrared spectrophotometer (IR spectra)

5. Material identification using differential scanning calorimetry (DSC)

Chapter 11 — Test Methods for Plastic Raw Materials and Moldings

6. Error analysis by the use of scanning electron microscopy (SEM)
7. Error analysis using a microtome-cut sample in combination with light microscopy

Figure 11.15 The image shows an infrared spectrophotometer. With this equipment you can analyze organic material. The result is the type of spectrum shown below.

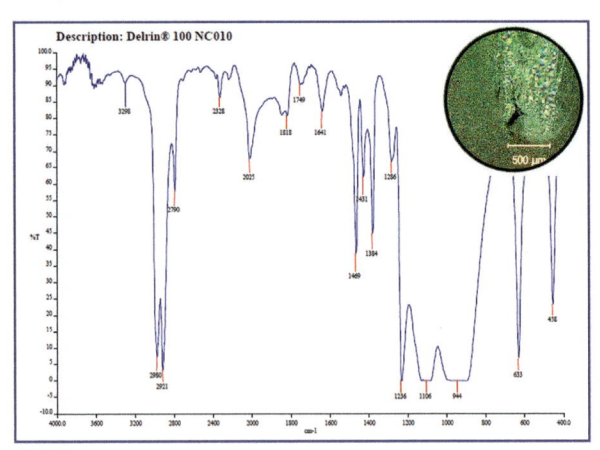

Figure 11.16 This diagram shows the IR spectrum of Delrin® 100 NC010 acetal. An IR spectrum can be compared to a fingerprint. From each IR spectrum you can directly see which polymer and organic additives are present in the sample.
When looking at a possible contamination in the surface of a plastic part using this method, it is possible to see what it consists of, unless the contamination is organic. [Source: DuPont]

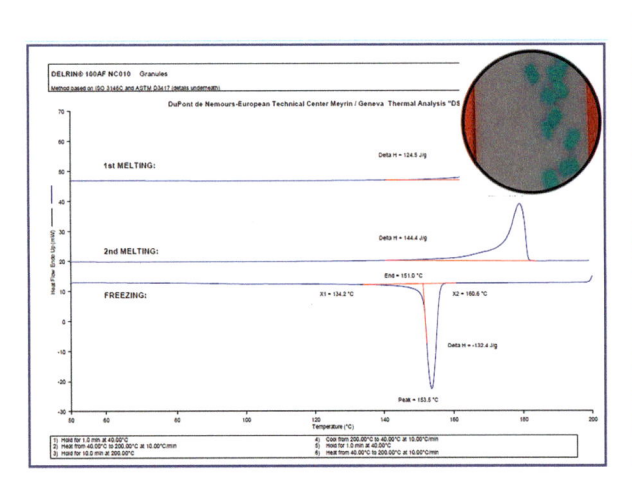

Figure 11.17 A different method where you can analyze organic material is DSC.
Here you can see what happens when the material is heated. It gives you, for example, the melting point of the contamination of the green granulates present in the circle. [Source: DuPont]

Figure 11.18 A scanning electron microscope is a very expensive equipment used for analysis of inorganic inclusions in plastic.
It also allows analysis of fractures on the plastic part surfaces or surface structure as shown in the illustration in Figure 11.19. [Photo: DuPont]

11.5 Advanced Testing Methods

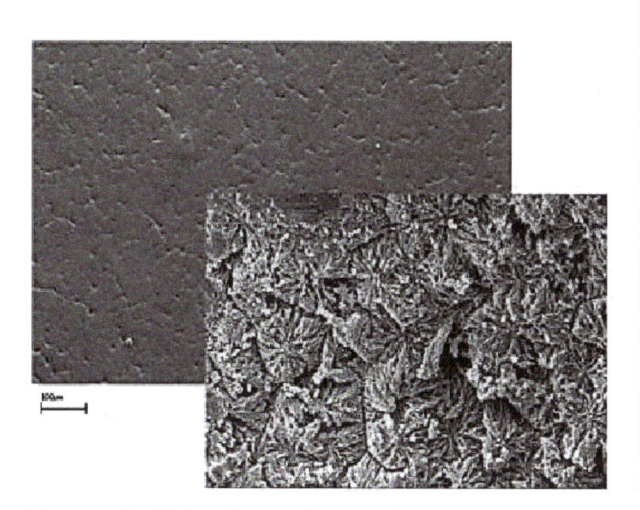

Figure 11.19 This is a high magnification of the surface of a perfectly molded part in acetal. To the right you can see how the surface has changed after exposure to sulfuric acid. In addition to the surface becoming etched, the tensile strength will be significantly lower. [Photo: DuPont]

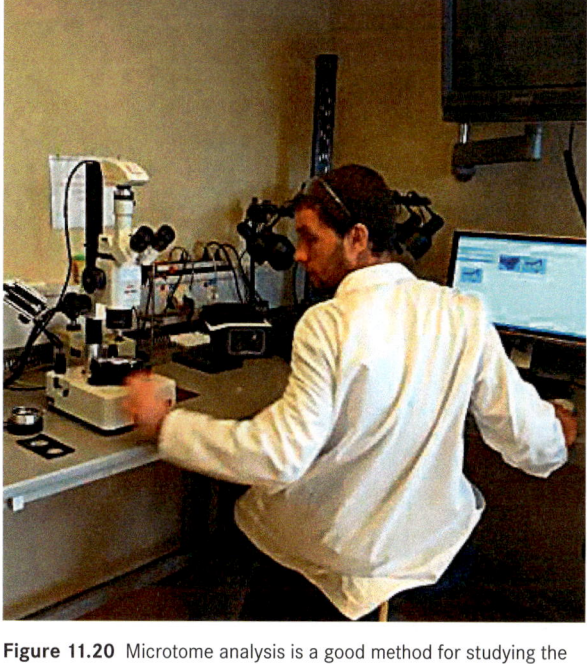

Figure 11.20 Microtome analysis is a good method for studying the structure of injection-molded parts. A very thin layer is sliced from the surface of the plastic part by using a microtome. The structure of the material is then illuminated from below by polarized light. In order to make thin slices of soft materials the material must be frozen. [Photo: DuPont]

Figure 11.21 The image at the top left shows the slicing of a frozen plastic layer. To the right there is a sample mounted between two glass plates.
At the bottom left, the sample is illuminated with polarized light. At the bottom right you can see a black enclosure of degraded material in the bottom of the part (see red arrow). [Photo: DuPont]

109

CHAPTER 12
Injection-Molding Methods

12.1 History

Injection molding is the predominant processing method for plastics, allowing the production of parts in both thermoplastics and thermosets. In this chapter, however, we focus on injection molding for thermoplastics.

The method was patented in the U.S. back in 1872 by the Hyatt brothers, who began producing billiard balls in celluloid. The first molding machines were piston machines where the plastic material was filled in a heated cylinder. Once the plastic is thus melted, it is pressed into a cavity by means of the piston. The first screw machines, i.e. the type used today, were not introduced until the 1950s.

Injection molding has become the most popular machining process for thermoplastics today because it provides such great cost advantages over conventional machining or other casting methods. The process has also undergone great development in the last fifty years and is now completely computerized.

Figure 12.1 The picture shows an old piston injection-molding machine. The locking mechanism was a knee-joint type that is commonly used even today, but the locking and opening movements required arm muscles to operate the levers.

Figure 12.2 A modern injection-molding machine manufactured by Engel. This machine has a hydraulic locking mechanism. You can also get "all-electric" machines, which are significantly quieter than the hydraulic ones. [Photo: Engel]

12.2 Properties

Injection molding is a completely automated process that often produces finished components in one go. In addition:

- + The components can be very complex in shape without any need for post-operations
- + It has a very high production rate (in extreme cases, with thin-walled packaging, a cycle time of only 3 to 4 seconds)
- + It can manufacture everything, from millimeter-size precision parts (e.g. gears in watches) to large body parts for trucks, with lengths over 2 m
- + It can fabricate really thin walls of just a few tenths of a millimeter, or thicker up to 20 mm
- + Several different plastic materials can be combined by co-injection in the same shot (e.g. a soft grip on a rigid handle)
- + Metal parts can be overmolded (Figure 12.3)
- + Components with so-called Class A surfaces (Figure 12.4), can be produced, suitable to paint or chrome-plate, as often seen on cars
- + Automated post-processing operations can easily be made, such as removing gates and runners, assembling (e.g. welding), or surface coating the components
- + Runners or rejected components can be directly recycled at the injection-molding machine

12.2.1 Limitations

If one had to identify a disadvantage with injection molding, it would be that the process requires relatively expensive equipment (machines and molds), which generally require large batch sizes (>1000 components) to be really profitable. Another problem that can arise is the shrinkage of the part compared with the cavity dimensions, which can lead to tolerance problems. The process also requires draft angles of 0.5 to 1° to facilitate ejecting the parts from the mold.

Figure 12.3 An injection-molded buckle for a car safety belt, where an engineering polymer is molded over stamped and chrome-plated steel.

Figure 12.4 An injection-molded bumper, made from PP with EPDM. It has a so-called Class A surface and can therefore be painted with the same paint as the bodywork in steel plate.

Chapter 12 — Injection-Molding Methods

12.3 The Injection-Molding Machine

An injection mold consists of two parts: the injection unit (where the plastic material is filled and melted), and the clamping unit (where the two halves of the mold are assembled). With the help of the clamping unit, the mold can close and open. On the machine is a computerized control panel where the actual machining process is set and controlled.

12.3.1 The Injection Unit

The injection unit consists of a hopper or dosing unit at the back of the cylinder heated by heating bands. Inside the cylinder is a screw that either rotates and doses the plastic material for the next shot or performs linear movement acting like a piston when the material is injected into the mold. At the end of the screw is a tip with a back-flow valve that prevents the material from being pushed backward when the mold is being filled. The cylinder enters the nozzle butting the mold, and the material flows into the mold cavities through the nozzle during the injection phase.

Figure 12.5 Plastic granules usually come in the form of rice-sized grains. The granules are fed into the top of the cylinder using a metering device that sucks up the material from a container or drying unit.
The dosing unit replaces the hopper used in earlier machines.

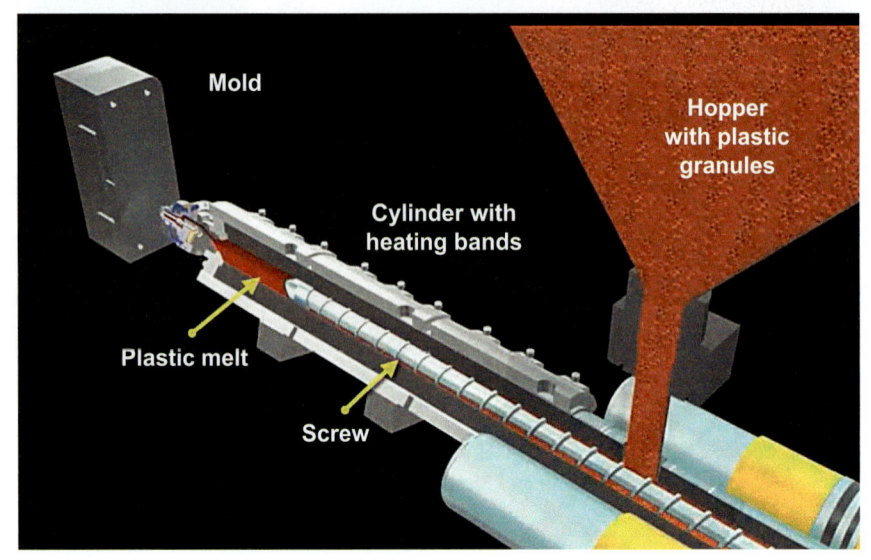

Figure 12.6 This figure shows an exposed view of a cylinder. The hopper sits at the back and the nozzle at the front. Inside the cylinder is a screw, and on the outside are heating bands that, together with the frictional heat from the screw, melt the plastic granules. [Image: Engel]

12.3.2 Locking Unit

The mold usually consists of two halves (or occasionally three in special stack molds). In the so-called fixed mold half, the gate normally is mounted to the machine's platen. The moving mold half, where the ejectors are, is mounted on the machine's moving platen. The locking unit is normally a hydraulic cylinder with a knee-joint mechanism or a more powerful hydraulic cylinder with a direct locking piston. It provides the machine's moving platen with a back and forth movement, thereby opening or closing the mold.

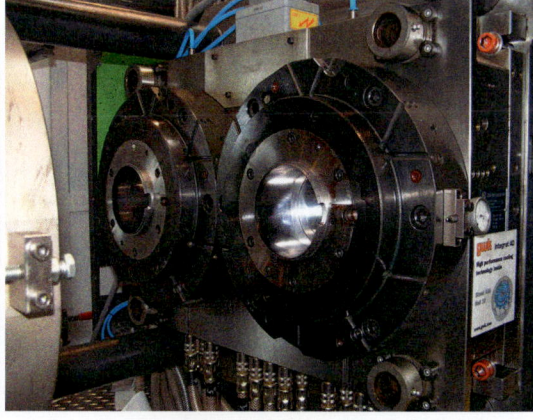

Figure 12.7 In the left-hand picture, we see plastic jars in PP. To the right we see the solid tool half, where the jars are produced. This is a two-cavity mold, which means that two jars are produced simultaneously in each shot. The gate is centered on the bottom of each jar.

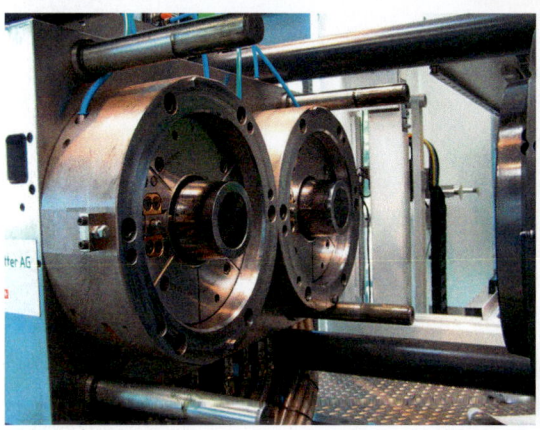

Figure 12.8 In the left-hand picture, we see the moving mold half for the plastic jars in Figure 12.7. Usually, the machine has ejector pins to release the products from the mold cavities. In this case we have an ejector ring. The picture on the right shows the knee-joint mechanism that will open and close the mold.

12.3.3 Injection-Molding Cycle

The injection-molding cycle starts with the mold closing, at which point the molten material can be injected through the nozzle into the cavities (known as the injection phase). The screw does not rotate during this phase but is controlled by hydraulic pressure, resulting in a linear motion so that it acts as a piston.

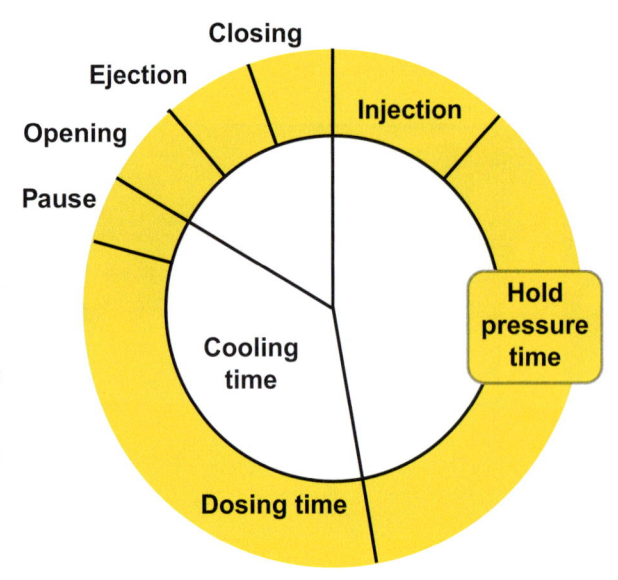

Figure 12.9 This image illustrates the injection-molding cycle. During the injection and hold pressure phases the screw does not rotate but moves forward like a piston, with the back-flow valve closed. During the dosing phase the screw rotates and is simultaneously pushed backward by the pressure buildup in front of the screw tip caused by the molten plastic. In this phase, the back-flow valve is in the open position. The screw is stationary and to the rear during the pause and opening, closing, and ejecting phase.

During the hold pressure phase, the screw does not rotate but is kept under high pressure (50–100 MPa). It moves slowly forward a few millimeters when the material in the cavity is packed, which is known as shrink compensation.

A hold pressure phase is needed due to the considerable difference (up to 20%) between the specific volumes of the molten versus the solidified material (especially when it comes to semi-crystalline plastics). One must therefore compensate for the volume difference by packing more material into the mold cavity during solidification to avoid sink marks or voids being formed.

During the dosing phase of the cooling time, the screw rotates and feeds the plastic melt forward in front of the screw tip. The screw speed must be adapted to the material's melt viscosity to avoid too high a shear between the flanks of the screw and the cylinder wall. The shearing produces friction heat, and if this becomes too high the material will degrade.

In principle, when injection molding semi-crystalline plastics, the mold can be opened directly when the dosing phase is complete. However, it is safer to add a small margin (about 0.5–1 sec) to compensate for variations in dosing times (i.e. pause time). For amorphous materials, a greater pause time is needed after the dosing phase so that the material has time to become stiff enough not to deform on ejection. The opening, closing, and ejection phase usually takes a few seconds even if you allow a robot to pick up the parts from the mold.

12.4 Alternative Injection-Molding Methods

In recent decades the injection-molding process has been developed to enable different materials to be combined in the same part and to make thick-walled components hollow or to add foam material inside the wall. You can also insert foils into the cavity to get finished parts with printing, a textile surface, or wood imitation.

12.4.1 Multi-component Injection Molding

Normally, only two different materials are combined in the same part, even if more than two materials could be molded in the same shot. Multi-component injection molding uses a special machine with a separate cylinder for each material.

Two-component molding commonly uses a rotating mold. These molds are relatively expensive but save a lot of time compared to the other multi-component method, which is known as "robot transfer".

Figure 12.10 Rotating molds. The left-hand image displays a mold with a rotating half that produces a special specimen for testing the adhesion between two materials—one light in color, one black. The lighter material is injected in two parts separated by a central core in the fixed half of the mold. When the mold is then rotated, this core is absent in the second cavity in the fixed half, and instead the black material is injected. [Photo: DuPont]

Figure 12.11 Robot transfer. To make this screwdriver, the steel rod is put into a mold in one machine and then overmolded with blue polypropylene. A robot will then move the blue screwdriver to another mold in another machine, where a black thermoplastic elastomer will be injected on top of the blue shaft.

12.4.2 Gas or Water Injection

With injection molding you can also produce hollow parts. In this case, a standard molding machine has to be equipped with a special unit that injects water or gas (usually nitrogen or carbon dioxide) at the end of the filling phase. A shut-off nozzle is also required to stop the gas or water from being pushed into the cylinder.

There are various methods to inject the gas or water, either with a special nozzle or valves, directly into the cavity. In some cases, the cavity is completely filled with plastic, and then the so-called overflow pockets are opened, letting the gas or water form a hollow space in the middle of the wall, while the equivalent plastic volume is simultaneously squeezed into the overflow pockets.

The main advantage of water injection over gas injection is that you get a significantly shorter cycle time and a better surface finish on the inside.

Gas injection reduces the locking force needed in the injection-molding machine as it replaces the high hold pressure with a much lower gas pressure, without affecting the outer surface finish.

One can also atomize the gas in a special nozzle and produce foam inside the wall. There are several different methods for this, MuCell being the most well-known.

Figure 12.12 This armrest is made with gas injection, which is a plus in terms of weight and cycle time, but it results in varying wall thickness and poor surface texture on the inside. [Photo: DuPont]

Figure 12.13 This pipe for a VW engine is manufactured by water injection, which has the benefits of short cycle time, consistent wall thickness, and a good surface texture inside the tube. [Photo: DuPont]

CHAPTER 13
Post-molding Operations

13.1 Surface Treatment of Moldings

Generally, you get completely finished parts when performing injection moldings. Parts with the correct color are ready to be used immediately or ready to be assembled with other components. However, there are opportunities to further enhance improvements of the injection-molded part by surface treatment. Usually surface treatments are made to improve the aesthetic value but may sometimes be required to meet the functional needs.

Figure 13.1 A headlamp housing made in PBT
In this picture you can see the chrome-plated reflector through the glass (which has been made in polycarbonate). The surface treatment has been made in order to get the optic properties as well as adding protection to the surface from the high-heat-generating light sources.

The various surface treatment methods used for thermoplastics that we will cover are:

- Printing/labeling
- IMD, In-mold decoration
- Laser marking
- Painting
- Chrome plating or metalizing

13.1.1 Printing

There are many different reasons for printing on plastic products. You often want to add a label or add instructions onto the surface of the product. The printing methods that we will describe in this chapter are:

- Hot stamping
- Tampon printing
- Screen printing

Chapter 13 – Post-molding Operations

Figure 13.2 Containers, cans, and bottles with labels and instructions added to the surface. On many of these an adhesive label in either paper or plastic foil has been attached. There are also various methods to print directly on the plastic surface.

13.1.2 "Hot Stamp" Printing

When performing hot stamp printing a foil is used. This foil may have a shiny metallic gloss (shown on the lens cover in Figure 13.3). The foil is pressed against the plastic product by the use of a heated engraved stamp, making the print stick to the surface.

Figure 13.3 This picture shows a few examples of products in which hot stamping has been used in order to enhance the aesthetic look of the brand name and/or to add symbols for various functions necessary for proper use of the product.

13.1.3 Tampon Printing

When performing tampon printing, a soft (rubber-like) silicone tampon is used. It is first pressed against an ink-coated printing block and is then pressed onto the surface of the product.

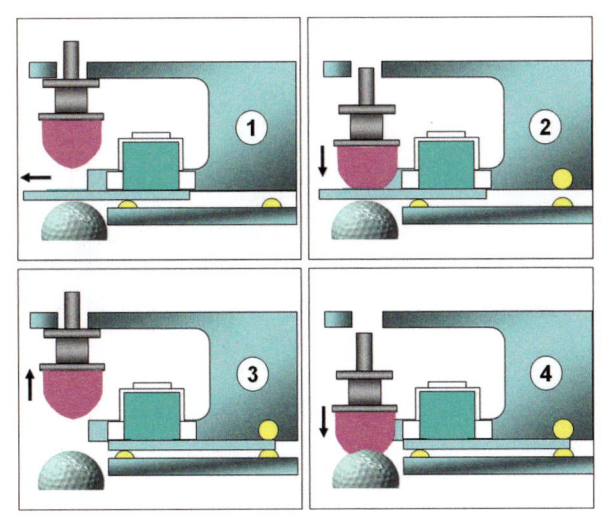

Figure 13.4 This picture illustrates how tampon printing of a golf ball is performed.
In the first frame ❶ the tampon (in pink) is in its starting position. Below the tampon is an attached printing block with a stained pattern (in turquoise).
In the second frame ❷ the pattern istransferred to the surface of the tampon.
In the third frame ❸ the tampon is elevated and the printing block is pulled back and coated with color paste from the turquoise container.
In the last frame ❹ the pattern is carried over from the tampon to the golf ball.

13.1.4 Screen Printing

Screen printing is an old printing method, invented by the Chinese 2000 years ago. This method can be used on both flat and cylindrical surfaces and is used when the method of tampon printing becomes too small for the surface area. A stencil (carrying the printing pattern) and a canvas impregnated with color paste are used so that the stencil is reproduced onto the canvas. When the canvas is then pressed against the plastic surface the color paste is transmitted onto the surface and you get a print.

Figure 13.5 This "ice cream man" has most likely been made by the use of vacuum-formed plates in an amorphous plastic that have been printed on in cheerful colors by screen printing.

Chapter 13 — Post-molding Operations

13.1.5 IMD: In-Mold Decoration

In-mold decoration (IMD) is the designated term used when a foil or fabric is placed into the mold in the injection-molding process. This is usually done by using a robot, which also carries out the part after ejection. This method enables the production of very fine and aesthetic-looking parts. See the figures below.

Figure 13.6 This panel looks very much like a wooden walnut panel and is quite common in the interior of cars. But it is an ABS panel that has been coated with a foil placed directly into the mold.

Figure 13.7 IMD is commonly used in mass production of containers. The cycle time including insertion of the foil into the mold as well as ejection of two finished containers of the kind shown above may take less than four seconds to complete.

13.1.6 Laser Marking

This is one of the latest techniques for marking of plastic surfaces. In order for this method to work, the plastic material must have a special pigment that will change color when exposed by a laser beam.

Figure 13.8 This picture shows laser marking equipment with a YAG laser of 1,064 nm.
The diameter of the laser is only 0.05 mm. It moves biaxially over the fixture that holds the plastic part and at the same time exposes its surface. [Source: Vadstena Lasermärkning]

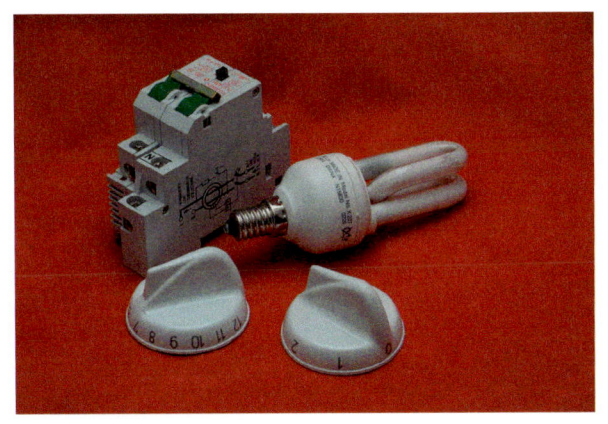

Figure 13.9 In 90% of the cases, laser marking of plastic products has been chosen for functional reasons rather than aesthetic ones. Above is a fuse where information has been printed onto the side surface. This is also the case for the low-energy bulb. However, when it comes to the oven handles, laser marking has been chosen for aesthetic reasons. The numbers on the handles cannot be worn away.

13.1.7 Painting

For most colored, injection-molded parts, a colored plastic resin has been used. The reason you need to paint a plastic part could be because the material has such a natural color that you cannot color the resin to get the required aesthetic value. If the part is to be assembled with painted metal parts (such as a door mirror on a car), it is virtually impossible to get the same color on a colored plastic part as on the painted surface. Sometimes the surface of a part is painted in order to improve the UV or scratch resistance.

Figure 13.10 Bumpers, side mirrors, etc. made in plastic must be painted using the same paint as used for the body parts in stamped steel in order to get a perfect color match.

Chapter 13 — Post-molding Operations

13.1.8 Metalizing/Chroming

If you want to increase the aesthetic value of a plastic part and give the impression that the part is made of metal, chrome plating is used as a suitable surface treatment. The material most suitable for chrome plating is ABS, but also engineering polymers such as polyamide, acetal, and thermoplastic polyester can be chrome plated. Other reasons for metalizing or chrome plating a plastic part is to improve scratch resistance or protect the plastic from UV light or heat radiation. By coating the surface with metal this can also be a shield against the electromagnetic field. The headlamp shown in Figure 13.1 is an example of using metal coating in order to reflect light and protect against heat radiation.

Figure 13.11 A shower handle in chrome-plated ABS. Only its light weight gives an indication that it is made of plastic and not metal.

Figure 13.12 Above to the right is a spinning reel made in plastic. Here the material has a thin coating of nano-particles of metal (which are much smaller than the particles in conventional metalization). This is one of the latest metalization methods for plastics. You get a substantial improvement in stiffness compared to unmodified plastics. The metal shell is just a few hundredths of a millimeter thick, so the weight increase is marginal. [Photo: DuPont]

CHAPTER 14
Different Types of Molds

In this chapter we will cover different types of molds, and in the next chapter, we will look at them in more detail. If you ask an operator within the molding business what types of molds are commonly used, their likely reply is "common molds and hot runner molds". Hot runner molds will be discussed in the next chapter. What the operator probably means by "common molds" is shown below:

- Conventional two-plate molds
- Three-plate molds
- Molds with slides
- Molds with rotating cores for parts with inner threads
- Stack molds
- Molds with ejection from the fixed half
- Family molds
- Multi-component molds
- Molds with melt cores

The list covers most of the common types of molds but does not claim to be complete. In most of the above types, you can also choose between either cold or hot runner systems for the molds.

14.1 Two-Plate Molds

Two-plate molds are the most common type of molds for injection molding.

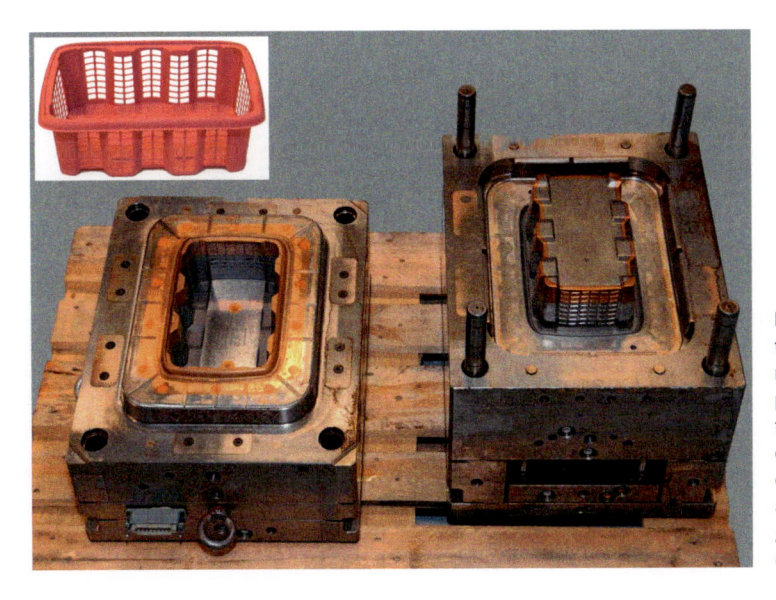

Figure 14.1 Here is an example of a two-plate mold with one cavity used to manufacture the basket shown in the picture at the top left corner. It is easy to see that the right half is the movable one, since the ejector plate (bottom) can be seen here. The left half is fixed and has a hot runner system integrated. The four columns are used to center the mold halves together.

14.2 Three-Plate Molds

In a three-plate mold the sprue, normally created between plate one and two, is separated from the product that is formed between plate two and three. The advantage is that no sorting will be necessary as the sprue and the details easily can be separated when they fall down.

Figure 14.2 This figure shows the principle for a three-plate mold.
❶ shows the fixed plate where the sprue is formed. ❷ indicates the first of the two movable plates.
To the right of these plates, the sprue is formed ❹ and to the left, the outer surface of the part ❺.
❸ shows the second movable plate with the core that forms the inside of the part. Here is the ejection ring ❻ that ejects the part once the mold is completely opened. When the mold is opened, the upper lug ❼ pushes the plate ❷ that, when reaching its end position, causes the sprue to fall. Then, the cone ❽ releases the lever, and the movable mold plate ❸ is opened and the detail is ejected.
[Source: M. Kröckel]

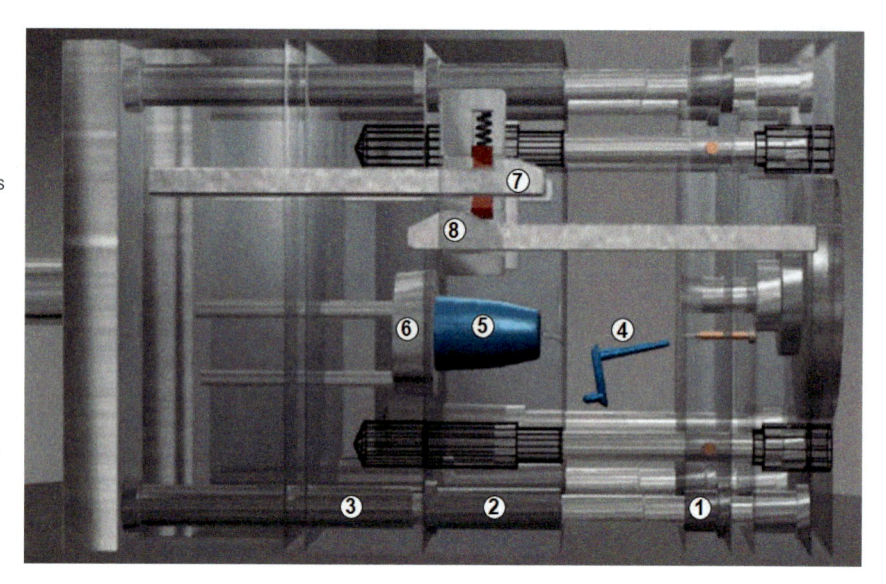

14.3 Molds with Slides

For molds with slides the movement of the cores can either be controlled by an electric, pneumatic, or hydraulic force or by using angled pillars (see Figure 14.3). The pillars cause the sliding cores to open when the mold is opened.

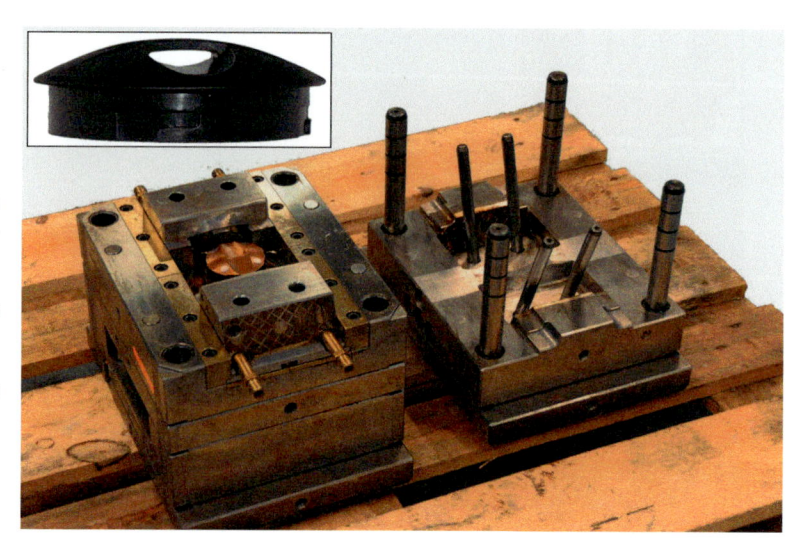

Figure 14.3 The picture shows a mold with sliding cores that is used to manufacture the lid above to the right. The two sliding cores are controlled by the angled pillars. They create both the grip and the bayonet lock. The sliding cores have a separate temperature control that is connected by the brass pipes. In order to get an efficient temperature control of the core, it is made of a beryllium copper alloy of high strength called "Moldmax" supplied by Uddeholm Tooling.

14.4 Molds with Rotating Cores

If you ever wondered how the screw cap of most plastic bottles is produced, the answer is: By using a mold with rotating cores.

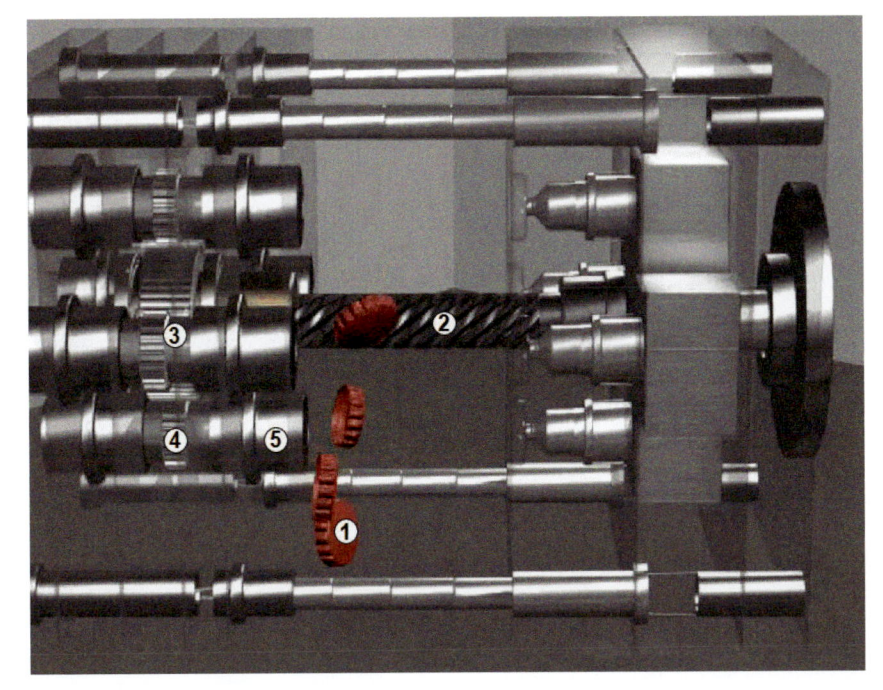

Figure 14.4 The figure shows a mold with rotating cores.
The mold has four different cavities. As the caps ❶ are produced they will fall down when the rotating cores ❺ that create the outer surface area of the cap cause them to release from the fixed core.
The fixed core creates the inner surface of the cap. The rotating core ❺ gets its rotation from the worm gear ❸ that revolves around a fixed screw ❷. The rotating cores ❺ then transmit the rotation to the gears ❹ that are placed at the back of the rotating cores.
[Source: M. Kröckel]

14.5 Stack Molds

A stack mold is to some extent similar to a three-plate mold. Three different plates are used here too. Instead of separating the sprue, there are two cavities with mirror orientation on both the right and left side of the center plate. The advantage of using a stack mold is that one achieves a high production rate without having to increase the size of the molding machine.

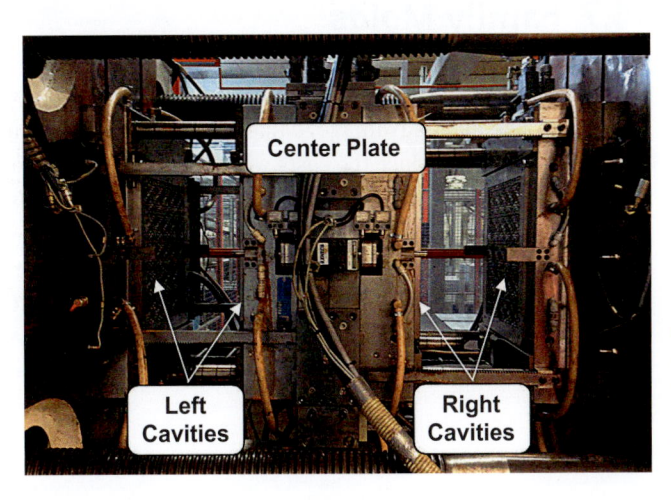

Figure 14.5 The figure shows the principle of a stack mold. The two cavities are placed on either side of the center plate. Both the fixed and the movable outer half have an ejection mechanism.

14.6 Molds with Ejection from the Fixed Half

For various reasons it can be an advantage to have the ejection from the fixed half (for example to hide the gate). For a stack mold this is a must, but this type of ejector solution can also be used on a two-plate mold. One can solve this by dragging an ejection plate that is located on the cylinder side of the mold by using various mechanical devices such as chains or gear racks, etc.

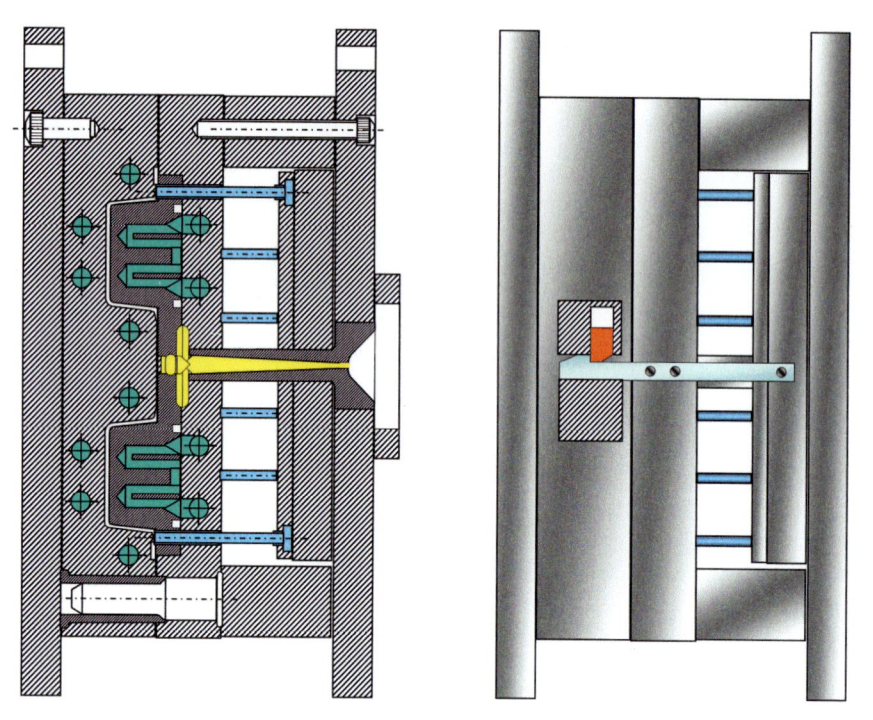

Figure 14.6 The pictures above show a mold with ejection from the fixed half. To the left is the cross section of the mold. To the right you can see the arm that pulls the ejector plate when the mold opens.
The red locking pin comes loose when the parts are ejected and the mold continues to open.

14.7 Family Molds

Family molds are multiple-cavity molds where the cavities have different geometric shapes. The main advantage is that you do not have to make a mold for each detail and can thus reduce the costs when producing small series. The disadvantage is that the cycle time required for the thickest detail also controls the cost structure of all the other details. There is also a risk that there is not enough time to pack all the cavities (shrinkage compensating) before the gates and runners freeze. Another disadvantage is that you get a lot of waste if the demand for one part is much larger than the others. Shown below are details with attached runner systems from a 1 + 1 cavity family mold.

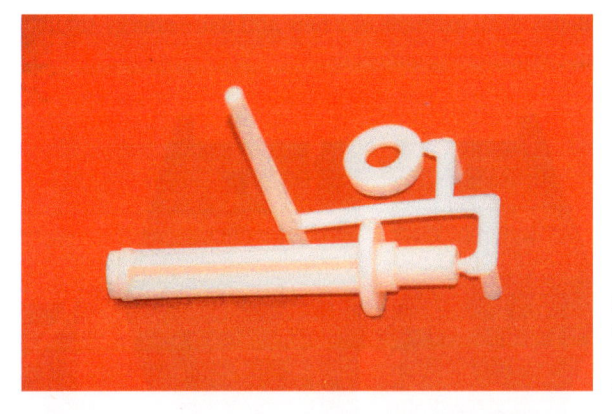

Figure 14.7 The figure shows a piston with a ring that is made of a thermoplastic elastomer. The runner system is of conventional type with a sprue.

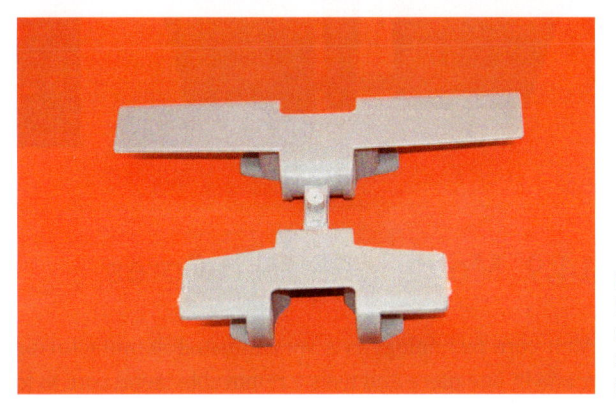

Figure 14.8 The figure displays a larger and a smaller conveyor link made of acetal. The runner system is hot runner with a sub-runner between the links.

14.8 Multi-component Molds

There are several different mold solutions if you want to use different materials simultaneously when producing a molded part. The most common approach is to combine two different materials in either different colors or to cover a hard stiff part with a soft thermoplastic elastomer skin.

Figure 14.9 The figure shows a mold with a rotating table.
The upper right-hand corner of the figure shows the part that is manufactured in this mold.
The movable half is attached to a rotating table ❶. The black material is injected into the cavity ❷ in the first step. In the second step, after a 180° rotation has been completed, a finished part is ejected from the mold cavity ❷. The part from the first step is then covered with the next material in the cavity ❸. [Source: Ferbe Tools AB]

Chapter 14 — Different Types of Molds

Below is an example of a new and more energy-saving solution to produce a part in two different materials.

Figure 14.10 The picture to the left shows the principle for a new technique called rotating inserts. The mold to the right is an example of a mold where this technology has been applied. The inserts are placed on the middle plate. The cap that is produced in this mold is shown in the upper right corner. The white plastic is injected in the first step followed by an injection of the black plastic, covering the cap, in the second step.
[Source: Ferbe Tools AB]

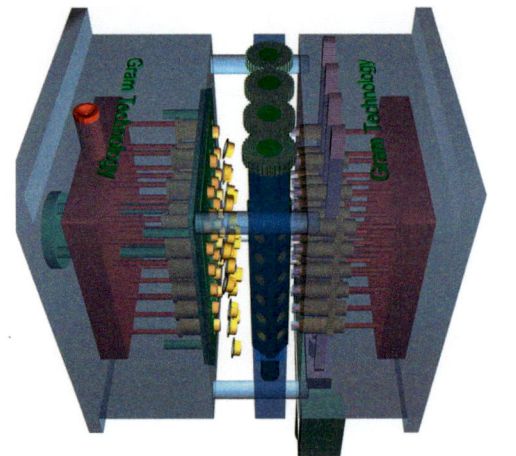

14.9 Molds with Melt Cores

Melt core technology is a very unusual manufacturing process. The product is made in one piece and uses a removable metal core to form undercuts that would be impossible to produce using conventional molds. This core is made of a metal with a very low melting point (similar to tin solder). The core is overmolded with plastic and thus becomes embedded into the product. The product is then placed in an oven with a temperature above the melting point of the metal core, but below the melting point of the plastic. Once the metal is melted and has drained from the plastic part, it can be collected and reused to cast a new melt core.

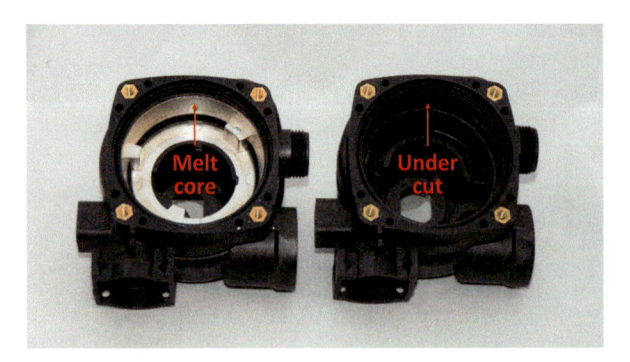

Figure 14.11 To the left we can see one of the halves of a circulation pump for hot water systems. The pump is made of glass fiber reinforced PA66 and has a geometric shape with a sharp undercut. The undercut is created by a metal core that has been covered but is visible inside the part. After the metal core has melted away in the oven and has been removed, we can see the finished part with the sharp undercut to the right.
Although this is an expensive injection-molding method, it can usually compete with a product in cast metal. The four threaded brass inserts can either be placed into the mold and be overmolded or can be pressed into the detail afterwards.

CHAPTER 15
Structure of Molds

In this chapter we will focus on how a common two-plate mold is constructed.
We will look at the following:

15.1　The function of the mold
15.2　Runner systems – cold runners
15.3　Runner systems – hot runners
15.4　Cold slug pockets/pullers
15.5　Tempering and cooling systems
15.6　Venting systems
15.7　Ejector systems
15.8　Draft angles

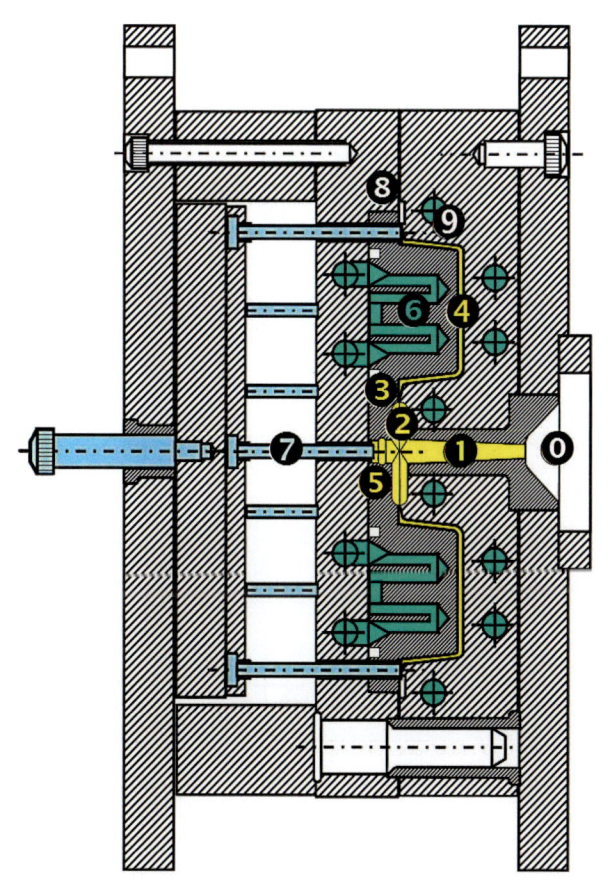

Figure 15.1 Schematic diagram of the structure of a two-plate mold.
- ❶ shows the nozzle centering of the tool.
- ❶ shows the sprue and the channel between the nozzle of the cylinder and the runner.
- ❷ shows the runner that leads the material through the gate into the cavities.
- ❸ shows the gate that leads the material into the cavity.
- ❹ shows the cavity.
- ❺ is the cold slug pocket.
- ❻ is the cooling system.
- ❼ shows the ejector system.
- ❽ shows the venting and
- ❾ shows that the part has a draft angle.

129

Chapter 15 — Structure of Molds

15.1 The Function of the Mold

There are many demands on a mold in order to obtain high-quality products:

- The dimensions have to be correct
- Filling of the cavities has to be shear free
- Good venting is necessary during the filling process
- Controlled cooling of the plastic melt in order to obtain correct structure of the material
- Warp-free ejection of the part

15.2 Runner Systems – Cold Runners

The cold slug pocket system can be divided into different parts:

1. The sprue
2. The runners
3. The gate

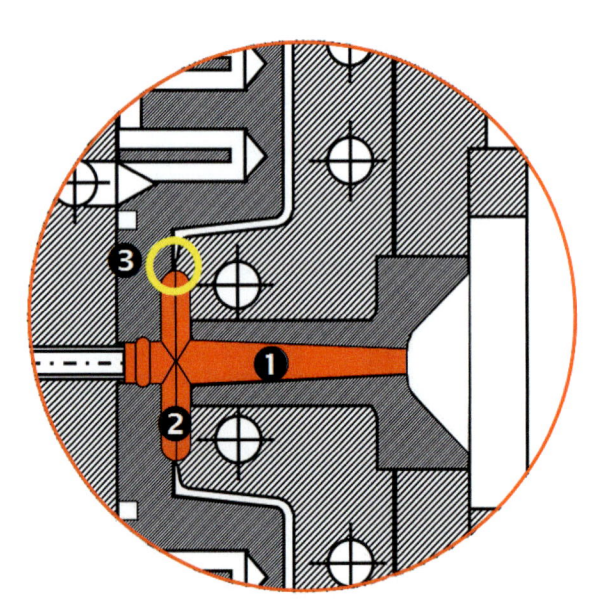

Figure 15.2 The runner system. ❶ is the sprue, ❷ is the runner and ❸ shows the location of the gate.

The sprue ❶ shown on the right is the connection between the cylinder nozzle and the runners ❷ of the mold. In most cases, it has a conical shape in order to avoid sticking in the mold when it has been packed. The sprue should be easy to pull out from the fixed half when the mold is opened at the end of the injection-molding cycle. For some semi-crystalline materials such as acetal the sprue can be cylindrical in shape. The dimensions of the nozzle should be adjusted with a diameter of about 1 mm less than the smallest diameter of the sprue.

15.2 Runner Systems – Cold Runners

The runner system leads the material from the sprue to the cavities. If you have multiple cavities, these should be balanced so they get even filling and have the same pressure drop.

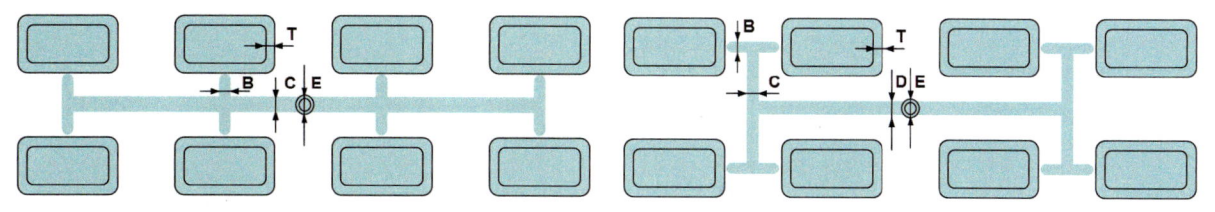

Figure 15.3 Unbalanced runner system (left) and a balanced one (right).
The balanced runner is preferred as all cavities will be filled at the same time. A rule of thumb when it comes to the dimensions of the runners is that if the wall thickness of the cavity is T, the runner B should be T + 1 mm, C should be T + 2 mm, and D should be T + 3 mm. The sprue to the left D should be T + 4 mm and the spruce to the right E should be T + 5 mm, based on the smallest diameter of the sprue if it is conical. The nozzle diameter should be 1 mm less than the diameter of the sprue.

The material flows into the cavities through the gate. It is important that the dimensions are correct so that semi-crystalline material does not freeze too early. The gate must be well rounded to avoid high shear during the injection phase.

The location of the gate is very important on a plastic part as it is, together with the weld lines, the weakest spot. The gate should also be located in the thickest wall of the part in order to make it possible to pack and compensate for shrinkage by using hold pressure. In Figure 15.4 you can see some different types of gates:

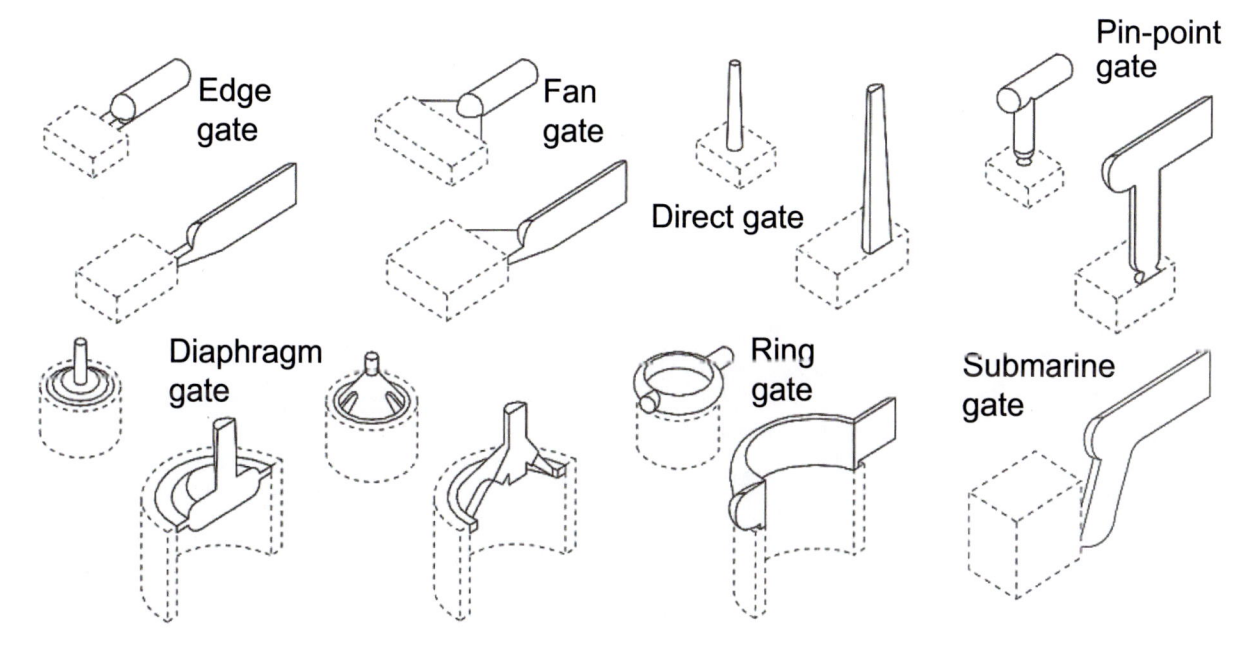

Figure 15.4 Different gate designs [Source: DuPont]

Submarine gate for unreinforced polyamide

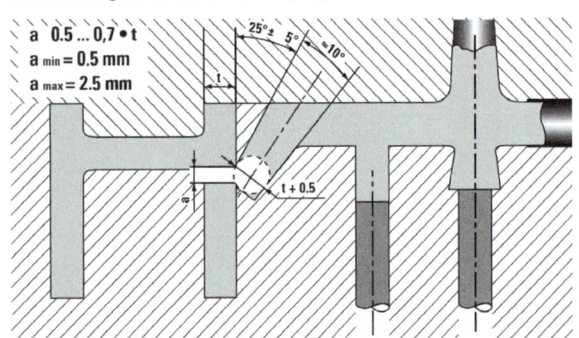

Submarine gate for glass-reinforced polyamide

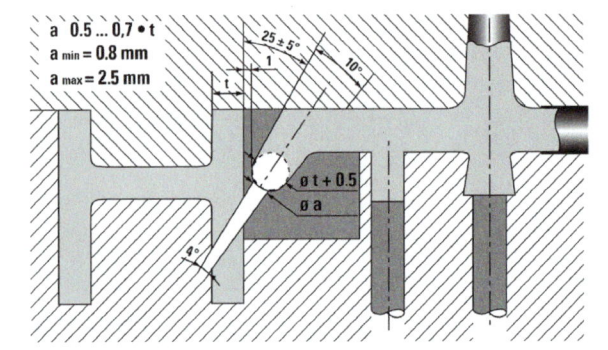

Figure 15.5 Suggestions of dimensions in correlation to the thickness of the wall (t) for unreinforced polyamide to the left and glass-reinforced polyamide to the right. [Source: DuPont]

15.3 Runner Systems – Hot Runners

Most of the molds that are described in the previous chapter can be equipped with hot runner systems. The advantage of using this type of runner is that regrinding of the runners is not necessary and 100% virgin resin can be used in the production.

The main disadvantage is the longer start-up times that are required and that all materials are not suitable for hot runner systems because longer hold times may cause degradation of the material.

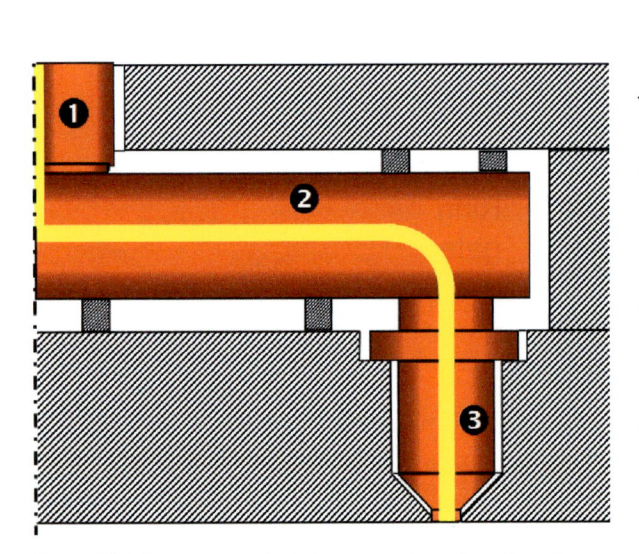

Figure 15.6 Components of a hot runner system, where the yellow channel is the plastic melt and the numbers are:
❶ The hot sprue bush
❷ The manifold
❸ The hot runner nozzle
All three are electric heated and keep the material melted at all times.
[Source: DuPont]

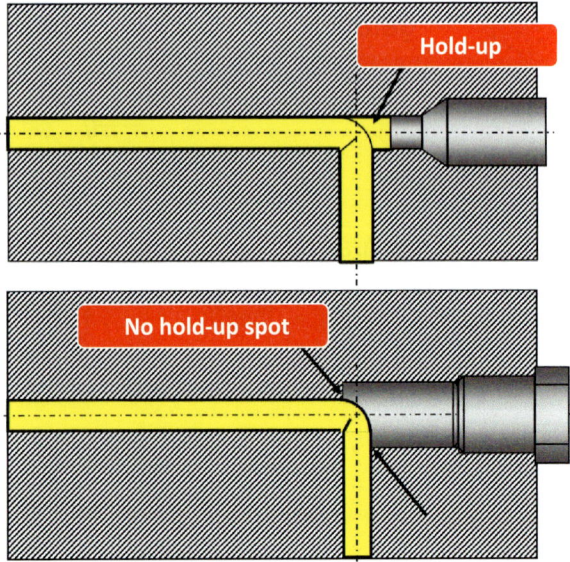

Figure 15.7 The top picture shows a hot runner angle with a disadvantageous hold-up spot.
Below is a properly designed hot runner angle without a hold-up spot.
[Source: DuPont]

15.4 Cold Slug Pockets/Pullers

It is important that the hot runner system is dimensioned in relation to the shot volume and that no hold-up spots are formed (see Figure 15.7). Material that is caught in the hold-up spot will degrade and causes black specks on the plastic parts. These specks tend to occur irregularly now and then.

The cold slug has two main functions. One is to capture any cold slugs that are formed in the nozzle during the dosing, opening, ejection, or closing phases of the injection-molding cycle, especially if semi-crystalline materials are used.

The second function is to pull the sprue when the mold is opened and the part is ejected. Various solutions for pullers are shown in Figure 15.9.

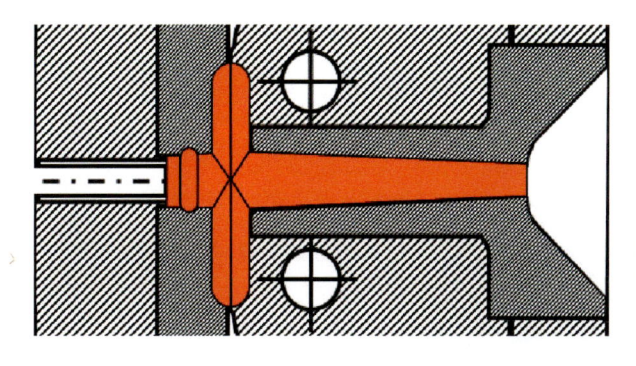

Figure 15.8 The cold slug pocket is an extension of the sprue that both captures any cold slugs and assists in the pulling of the sprue out of the fixed half of the mold. In this figure, this is accomplished by using the ring-formed undercut inside the yellow ring.

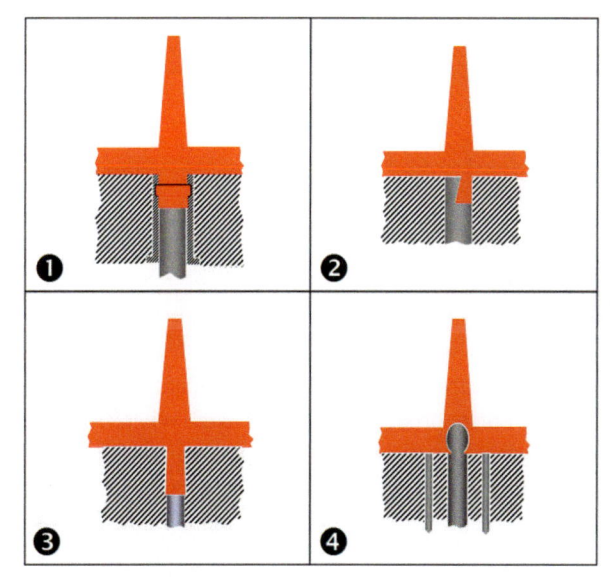

Figure 15.9 Various types of pullers:
❶ Cold slug pocket with a circular groove.
❷ Cold slug pocket with a Z-shaped puller
❸ Cold slug pocket with a cone
❹ Puller with a knob
In ❹ there is no cold slug pocket, and therefore it is not suitable for semi-crystalline materials.

Figure 15.10 Various runners with missing cold slug pockets. If more pullers are needed these should always be located opposite to the sprue in order to avoid having an eventual cold slug run into the cavity.

15.5 Tempering or Cooling Systems

Figure 15.11 shows the tempering of cavities and runner systems (colored in blue). It is important to have a correctly dimensioned tempering system in order to quickly get the correct mold temperature at the start-up or after an interruption in the production. The system should have sufficient capacity to provide a minimal temperature variation of the mold during the work cycle of the machine or the room temperature variations over the day or season of the year. The mold temperature is a very important process parameter that may affect the quality of the part in several ways:

- Surface finish
- Strength and structure of the material
- Dimensions (shrinkage)
- Tendency to warp
- Filling ratio
- The strength of the weld lines

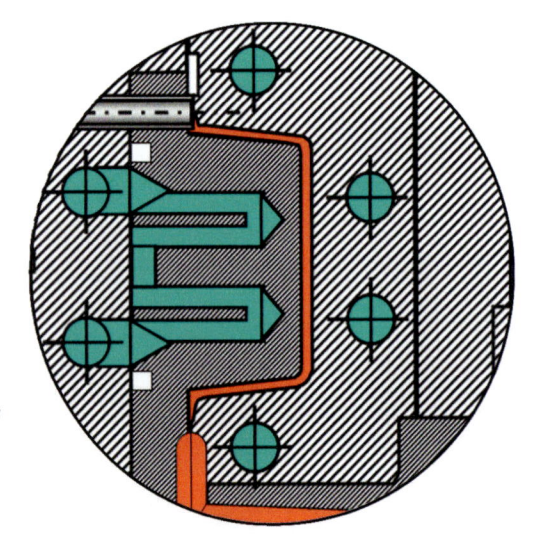

Figure 15.11 The cooling system (blue circles) is located across the view of the image. In order to get efficient cooling of the core in the cavity, so-called facing cooling baffles are used.

To achieve the correct mold temperature, external temperature-control units or a central tempering system for the entire production hall are used. When the mold needs to be adjusted to a temperature above room temperature, one should ideally have one unit for each mold half or tempering circuit. The most common type of tempering liquid is water or oil. With low-pressure systems a temperature of 95 °C and with high-pressure systems a temperature of up to 200 °C can be reached.

By using oil a temperature of 350 °C can be reached, which is necessary for good quality of, for example, PEEK materials. Effective temperature control of the mold cores is especially difficult to achieve, and some solutions for this are shown in Figure 15.13.

15.5 Tempering or Cooling Systems

Figure 15.12 The picture shows two units for water tempering.

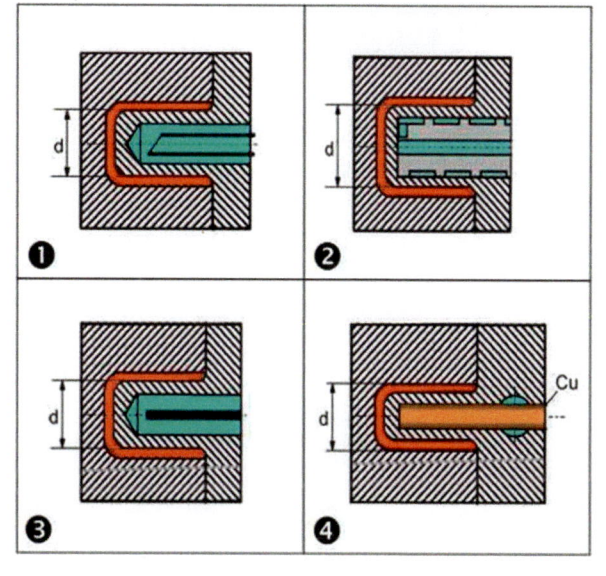

Figure 15.13 The cores in the mold will be very hot and are therefore difficult to cool. Here are some different principles for tempering the core inside the cavity;
❶ By using a cooling pipe
❷ Spiral cooling
❸ By using a cooling baffle
❹ Using cooling pins made of copper

Chapter 15 – Structure of Molds

15.6 Venting Systems

When filling a mold with a plastic melt, the entrapped air in the runners and cavities must get a chance to get out. In order to do this you have to build in air vents at the spots where the air otherwise would have been trapped.

Figure 15.14 shows venting channels presented in yellow color. If the venting is inadequate there is a risk of filling problems of the cavities. Surface burns may also occur on the parts where the compressed air is trapped.

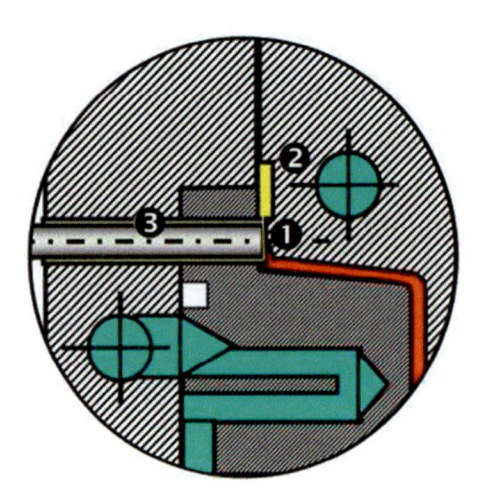

Figure 15.14 The picture shows a narrow slot ❶ leading from the cavity (in red) to a venting channel ❷. The ejector pin ❸ has ground venting grooves that also lead directly out of the plate.

There are certain rules of thumb for the dimensions of the venting channels regarding depth and width.

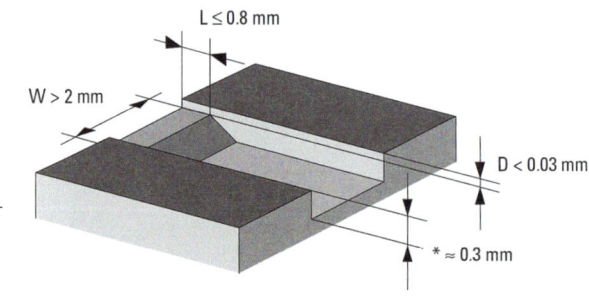

Figure 15.15 The width of the venting channels is 2 mm and the depth is 0.3 mm. The length of the groove is less than 0.8 mm, and the depth must be adapted to the viscosity of the material to be used. D < 0.03 used in this figure is an example suitable for acetal. [Source: DuPont]

15.7 Ejector Systems

The most common type of ejector systems are ejector pins, as shown in Figure 15.16.

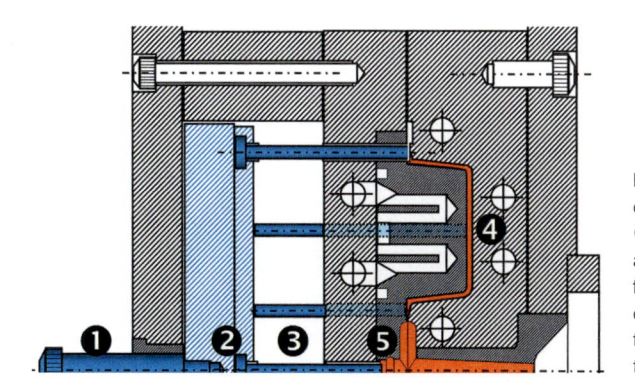

Figure 15.16 When the movable half of the mold is opened, the ejector rod ❶ pushes against the machine frame and presses the ejector plates ❷ forward. These in turn push the four ejector pins ❸ against the part ❹ and the cold slug pocket ❺ which cause the runner and the part to drop.

If a ring-shaped part is made, a number of ejector pins can be replaced by a tubular ejector. If you have a different geometry, the ejector pins can be replaced by ejector rails or by an entire ejector plate. The basket shown in Figure 15.17 is made by using a mold with an ejector plate.

Figure 15.17 The basket shown here is ejected from the mold, as the entire edge acts as an ejector rail while the bottom acts as an ejector plate.

15.8 Draft Angles

In Figure 15.17 it is obvious that the basket has a certain draft angle. This is required for facilitating the ejection. Ideally, the draft angle should be as large as possible. A rule of thumb is that you need to have 1–2° angles for smooth surfaces. If the surface is etched you should add 0.6° for every 0.01 mm of etching depth.

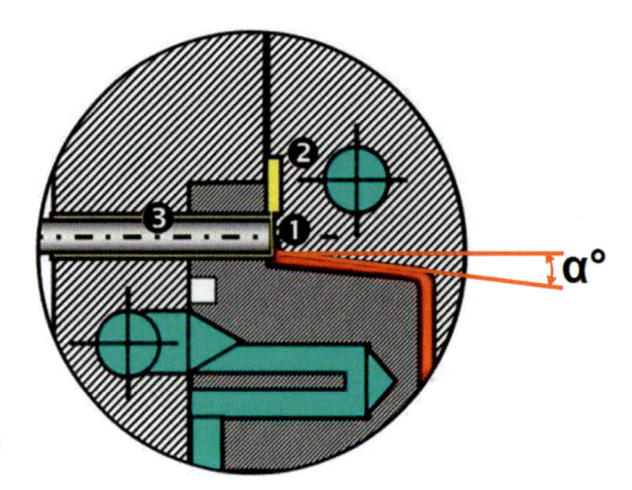

Figure 15.18 The draft angle on the part of the mold shown here (in red) is described by the angle $\alpha°$.

CHAPTER 16
Mold Design and Product Quality

In Chapter 11 problems that may occur on injection-molded parts due to defects in the plastic raw material are described. In this chapter we will look at problems due either to badly designed molds or improper part design. In Chapter 29 we will describe process-related problems.

16.1 Mold-Related Problems

These types of problems are not always as easy to detect by visual inspection as material- or process-related problems are. Many of these problems are only discovered when the parts are mechanically tested or when the parts break under normal stress loads.

Below are some common problems due to:

- Too-weak mold plates
- Incorrect sprue/nozzle design
- Incorrect runner design
- Incorrectly designed, located, or missing cold slug pocket
- Incorrect gate design
- Incorrect venting
- Incorrect mold temperature management

16.1.1 Too-Weak Mold Plates

If you get a flash around the sprue or runners during the injection phase, it may indicate too-high injection speed, too-low lock pressure on the machine, or too-weak mold plates.

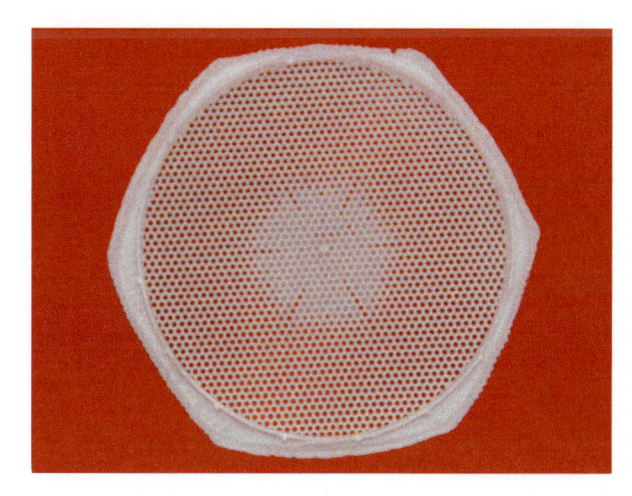

Figure 16.1 This figure shows a fan shroud in acetal.
The gate is in the middle, and it is clear that the mold plate has become deformed so that a flash around the gate has been formed even though the cavity is not completely filled. Moving the mold to a larger machine with higher clamping force did not help in this case.

Chapter 16 – Mold Design and Product Quality

In the case above (Figure 16.1), the first attempt to solve the problem was choosing an acetal grade with a lower viscosity. However, this did not solve the problem entirely (see Figure 16.2). Another option to solve the problem would have been to increase the wall thickness of the shroud or change the grid thickness in the round hole. Neither of these solutions were chosen; instead a solution using flow directors was used. The wall thickness was increased by using a honeycomb pattern (as shown in Figure 16.3).

Figure 16.2 A less viscous grade of acetal with slightly less impact resistance was chosen to fill the fan shroud entirely, but still flash in the middle could not be avoided.

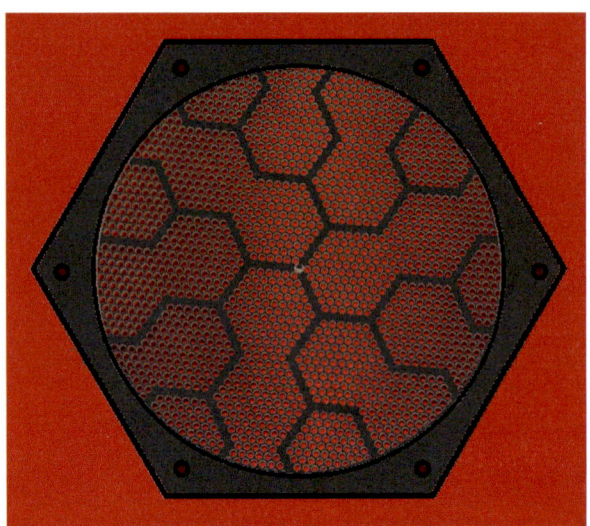

Figure 16.3 By applying a pattern of flow directors one succeeded to fill the shroud without getting the flash in the middle, nor did the cycle time need to be extended.

16.1.2 Incorrect Sprue and Nozzle Design

With incorrect dimensions of the sprue we generally mean that the sprue is either too small relative to the wall thickness of the part or that the diameter of the nozzle is too small. This is especially true for semi-crystalline plastics where correct dimensions are critical. Below is an example of a product that was exposed to a high impact and did not pass the mechanical tests because the setter had forgotten to change to a larger nozzle diameter when the mold was set. Problems may also arise if the nozzle size is too large, causing leakage between the mold and the cylinder.

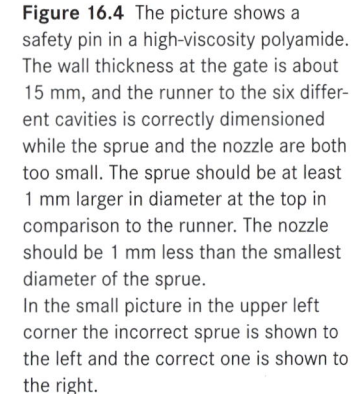

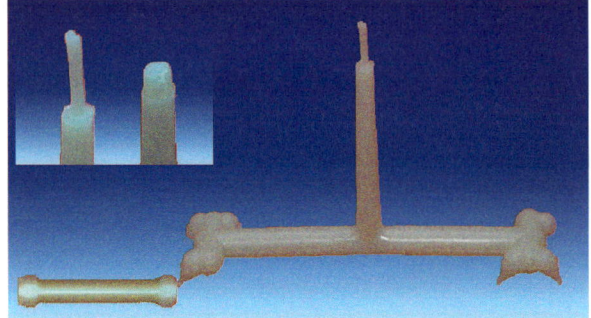

Figure 16.4 The picture shows a safety pin in a high-viscosity polyamide. The wall thickness at the gate is about 15 mm, and the runner to the six different cavities is correctly dimensioned while the sprue and the nozzle are both too small. The sprue should be at least 1 mm larger in diameter at the top in comparison to the runner. The nozzle should be 1 mm less than the smallest diameter of the sprue.
In the small picture in the upper left corner the incorrect sprue is shown to the left and the correct one is shown to the right.

16.1.3 Incorrect Runner Design

The most common problem is that the runners are too small in relation to the wall thickness of the part. This leads to problems as the semi-crystalline plastic freezes in the runner before the parts have been sufficiently packed. Another common problem occurs when there are unbalanced channels causing uneven filling and packing. Figure 16.5 shows an example of a poorly designed runner.

Figure 16.5 In this figure an example of a runner with disadvantageous design is shown. The six cavities that have the same shape and size are filled very unevenly. [Photo: DuPont]

16.1.4 Incorrectly Designed, Located, or Missing Cold Slug Pocket

In the previous chapter, we mentioned that a cold slug pocket has two different functions in the mold. It is supposed to capture material that has frozen in the nozzle, and it should be designed so it facilitates the pulling of the sprue. Cold slug pockets are of greatest importance in molds used for semi-crystalline materials as the risk for cold slugs is very high. If the part design does not allow a cold slug pocket it is possible to use a special procedure instead to eliminate the risk that material is freezing in the nozzle. In such cases the cylinder must be moved back after dosing. Then you must have sufficient decompression (sucking material back into the cylinder) so potential cold slugs may re-melt. Another alternative is to use a hot runner system.

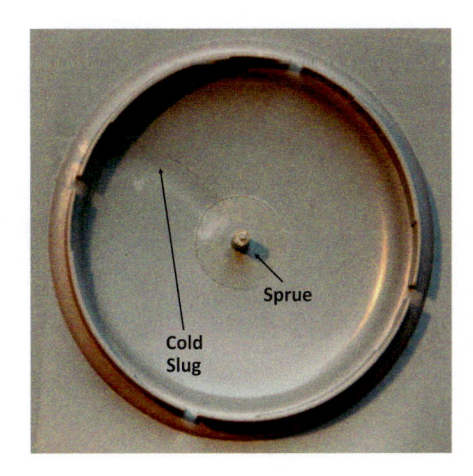

Figure 16.6 The picture shows the center ring with the gate on a 15" wheel cap made of mineral-reinforced PA66.
The sprue is located at the back of the wheel cap because a sticker with the car brand will be placed on the front. This will prevent the need to make a cold slug pocket.

Chapter 16 – Mold Design and Product Quality

16.1.5 Incorrect Gate Design

A too-small gate is the most common gate design problem. In such cases semi-crystalline material freezes in the gate before it has been packed and the shrinkage compensation has been completed. This in turn causes voids, sink marks, or incorrect dimensions to occur.

If you also have a large volume to fill, the amount of shearing can become too high (especially if you have a fast injection speed). This causes the material to degrade in the gate. This can also occur if you have too-small radii in the gate.

Figure 16.7 This picture shows a sprue (cut) sitting on a cover made in impact-modified PA66.
Due to too-small radii in the transition between the sprue and the cover as well as too-high injection speed, the material has been sheared and delamination has formed around the gate.

In some cases the gate can also be badly designed. In Figure 16.8 you can see a conical shaped gate to the right. This design, which is very common, works well for amorphous plastics but is inappropriate for semi-crystalline plastics as they freeze too early. The gate to the left in the picture shows how the gate is supposed to be designed when semi-crystalline materials will be used.

Figure 16.8 To the left is an example of a gate that is designed for semi-crystalline materials.
The size of gate d should be at least half the wall thickness T, and the distance between the runner and part should be less than 0.8 mm. The diameter D should be at least 1.2 × T.
The figure also shows that the gate should be located on the thickest wall of the part. [Source: DuPont]

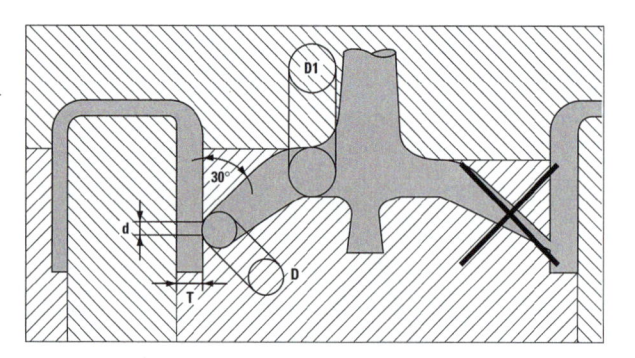

16.1 Mold-Related Problems

A good way to check if a semi-crystalline material has been packed sufficiently is to saw the parts in pieces at several locations to determine if they are free from voids or pores.

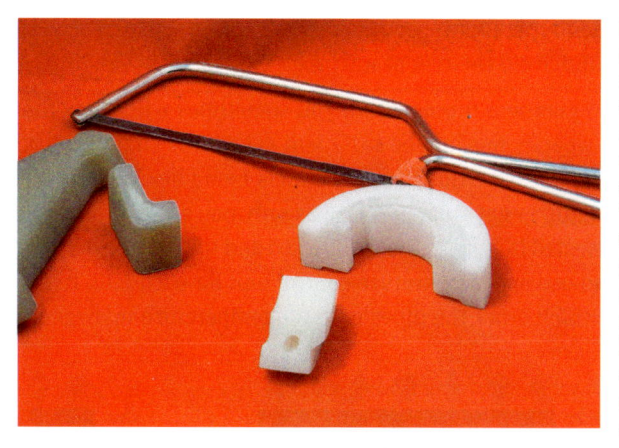

Figure 16.9 This picture shows a railway insulator made of glass fiber reinforced polyamide. When a glass fiber reinforced material is not sufficiently packed, micro-pores will occur, which appear as lighter areas in the sawn surface.
The right part is the starting material for a worm gear made in acetal. In the first cut no pores were detected, as shown on the upper wheel half. The next cut showed a big void in the wall. It is therefore important that you make several cuts when you want to check if a part is free of voids or pores. If voids are present they cause lower mechanical strength.
The weight of the void-free parts should be recorded and used as a quality control tool.

16.1.6 Incorrect Venting

If venting channels are missing, have been blocked by mold deposit (degraded polymer), or have too-small dimensions, the air in the mold can be trapped during the filling phase.

The air will then be compressed and heated to above 1,000 °C, and this causes degradation of the polymer. We call this a "diesel effect". Normally this problem is discovered by a discoloration on the surface in combination with the part not being completely filled.

On black materials this can be hard to see. The burn marks that occur are, however, slightly lighter than the black plastic.

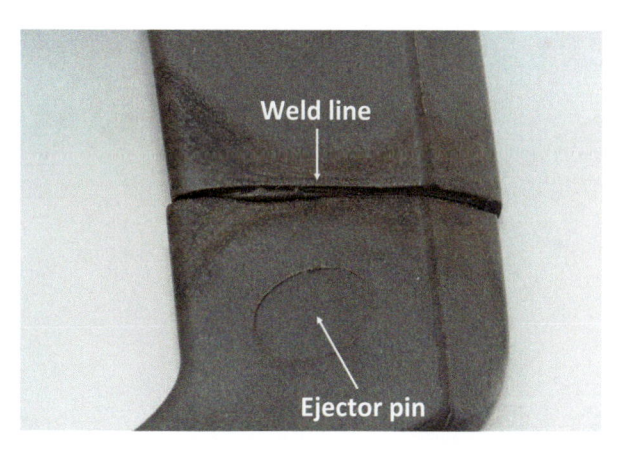

Figure 16.10 The picture shows a part that is filled from two different directions, causing a weld line. It was intended that the weld line should be located on the ejector pin because slots for venting have been ground on it.
One can also see that the plastic around the weld line has been exposed to high temperatures because it has become degraded and discolored.

143

16.1.7 Incorrect Mold Temperature Management

In the previous chapter we looked at the mold temperature-control system and concluded that it is important to have a correct temperature in the cavity to achieve high-quality properties such as surface finish, mechanical strength, dimensions, and avoid the risk of warping.

A common problem that you sometimes will see is that the temperature-control capacity may not be sufficient even if the channels have the correct dimensions. This may be due to the channels being blocked due to corrosion. Sometimes the setup is not done properly as only one temperature-control unit is used and connected in series to both the fixed and the moving halves. It is always recommended to use one temperature-control unit for each half. It may also be of importance in which order the pipes with the temperature-management fluid are connected.

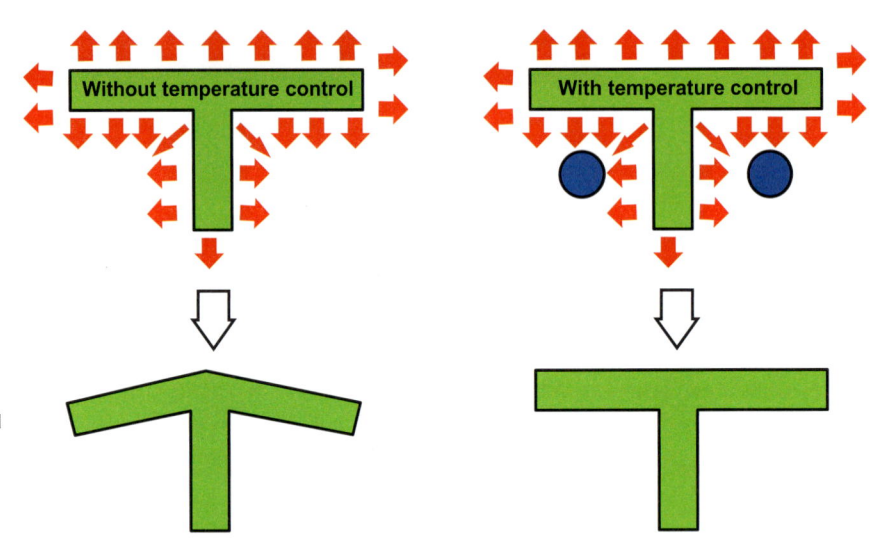

Figure 16.11 This picture shows the cross section of a T-profile in a semi-crystalline plastic injection molded with or without temperature-management control. Where the cooling is lacking, an increase in temperature occurs in the angle, which in turn causes the solidification process to slow down and crystal formation and shrinkage to increase.
A good rule of comparison is: Plastics are very similar to cats: both are drawn toward the heat.

CHAPTER 17
Prototype Molds and Mold Filling Analysis

In the previous chapter we described various errors that depend on either an incorrect part or mold design.

When designing new parts or starting the molding of a new product, you will get a number of new questions and challenges:

- Will the part get the correct dimensions?
- Will it warp?
- Are the runners too long? Will the part be completely filled?
- Where should the gate be placed in order to make the part as strong as possible?
- Are the temperature-control channels correctly dimensioned and located?

17.1 Prototype Molds

In order to avoid unpleasant surprises when starting the production in a new mold, you can use a prototype mold to see what the part would look like once it has been molded. Another option is to complete only one of several cavities in a production mold. These procedures could save both money and time. But they are not always completely reliable as runners, and mold temperature systems seldom correspond to the final production mold. Producers are using prototype molds when the plastic part is very complex or when the production mold is very expensive. Within the automotive industry these kinds of molds sometimes are named "soft molds" as they often are made in aluminum or in soft steel. When it comes to more simple part or mold design most prototypes have been replaced by a mold filling analysis.

Figure 17.1 Prototype mold (highlighted in red) in aluminum for the cams shown in Figure 17.2. Here only one part is made in each shot. Below is the production mold in steel with 16 cavities. This mold is about 30 times more expensive to produce compared to the prototype mold in aluminum.

Chapter 17 – Prototype Molds and Mold Filling Analysis

Figure 17.2 Here we can see a common component used in assembly of furniture. AD-Plast in Sweden developed this component in PPA with glass. They won the prestigious price "Plastovationer 2009" by successfully replacing metal with plastic. The cam is stronger than the former zinc one without needing to change the outer dimensions. [Source: AD-Plast AB]

17.2 Mold Filling Analysis

Mold filling analysis is a computer-based tool that facilitates the ability to get accurate plastic parts in less time when producing a new part or modifying an existing mold.

Figure 17.3 On the image you can see a mold designer in front of his PC working with Moldflow, a mold filling software. Such software is able to run on standard PCs but requires a lot of computing power. In order for the calculations to run as fast as possible it is necessary to have a large internal memory as well as a fast processor.

With mold filling analysis you get:

- The ability to reach the right result in less time
- A powerful tool for successful "lean production" with focus on continuous improvement
- The correct process parameters with a major influence on the property profile of the part
- Knowledge that leads to an increased process window and more robust production
- In general, lower development costs compared to a prototype mold

Figure 17.4 Here we can see a product in the shape according to the design that the designer had in mind. In Figure 17.5 we can see what it actually looked like when it left the mold.

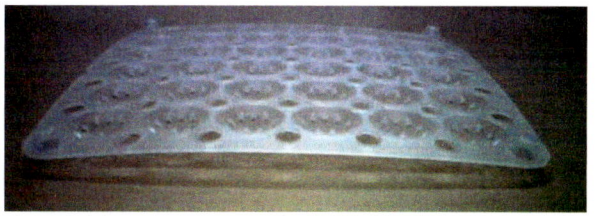

Figure 17.5 Here we can see that the product is heavily bent. The reason is internal stresses probably dependent on an uneven wall thickness or unfavorable temperature control.
By using the shrinkage & warpage modulus in the Moldflow software this could have been predicted before production and could have been corrected so that the defect never occurred.

17.3 Workflow

17.3.1 Mesh Model

Mold filling simulation should be a natural step in the development process of new plastic products. Many designers are using various CAD software packages such as Pro-E, Catia, or Solid Works when creating a 3D model of new products. The STL or Igesfile STL or just IGES file that is generated is then used to create a so-called mesh model.

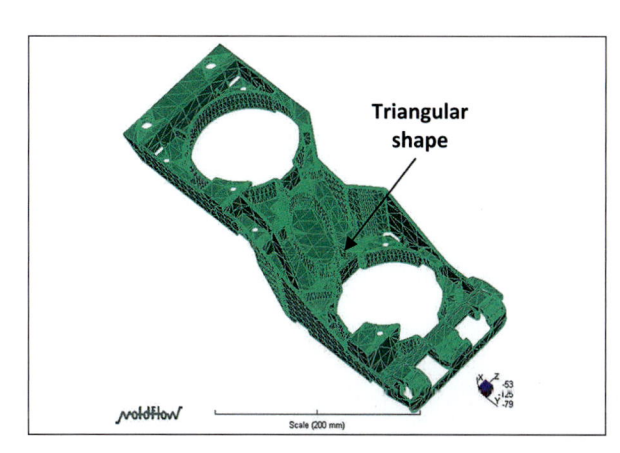

Figure 17.6 This is a 3D model in STL format of a headlight housing used for cars. The model consists of a large number of small triangles.

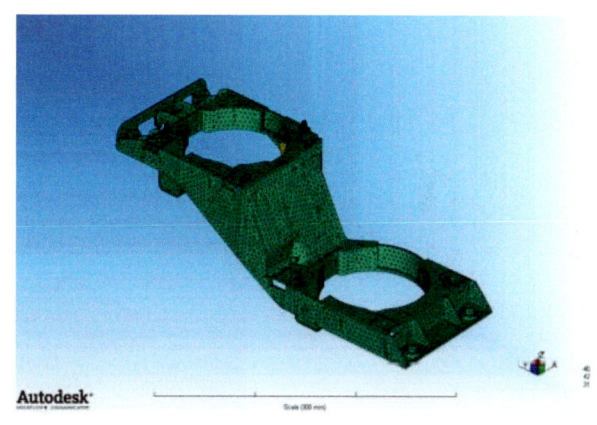

Figure 17.7 Here we see the same headlight housing as shown in Figure 17.6. By using Moldflow analysis you can create a mesh model that is used as "input" into the continuing simulation process.

Chapter 17 – Prototype Molds and Mold Filling Analysis

17.3.2 Material Selection

Once the mesh model is created, the next step is to choose which plastic resin to work with. In Moldflow there is a database with a large number of materials to choose from. If the material of choice is not found among those listed in the database, it is possible to add some needed (rheological) parameters yourself.

17.3.3 Process Parameters

As the melt viscosity of the material is affected by melt and mold temperature and shear rate you must enter these parameters into the database if the material is not found among the preselected materials listed in the database.

Figure 17.8 The picture here shows the window in which the process parameters are to be entered in Moldflow. If the values for injection speed, hold pressure switch, hold pressure, or cooling time are missing, you will be able to select the "automatic" choice for these parameters.

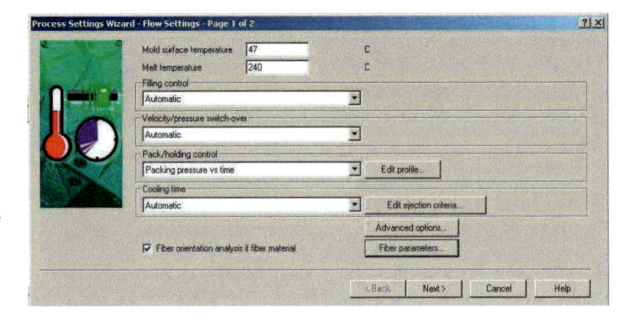

17.3.4 Selection of Gate Location

Before any calculations in the simulation program you must select a possible gate location. If you have no idea where it should be, Moldflow is able to suggest a suitable location.

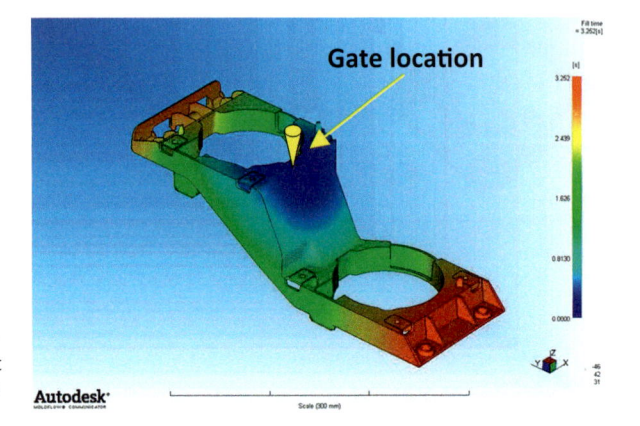

Figure 17.9 The image shows the selected gate location by the yellow cone. It is easy to move the gate location and then recalculate if you are not satisfied with the filling or if weld lines occur in places with high stress level.

148

17.3.5 Simulations

When starting the calculation process, which may take many hours to complete, depending on how complex the part is (how dense the mesh model is) or how powerful your computer is, the calculations can provide answers to the following questions:

- Is this the best possible solution (optimization)?
- Where is the best gate location when considering weld lines and sink marks?
- What will the process window look like?
- How to balance a multi-cavity mold?
- How does my choice of material affect the final product?
- Why are there quality problems (when analysis is performed later in the process)?

17.3.6 Results Generated by Simulations

The results of the various simulation calculations are usually presented as graphs. Below are some that you can analyze:

- Filling sequence
- Pressure distribution
- Cooling time
- Temperature distribution
- Levels of shear stress
- Necessary clamping force
- Location of weld lines
- Air traps
- Process parameters
- Glass fiber orientation

Chapter 17 – Prototype Molds and Mold Filling Analysis

17.3.7 Filling Sequence

In Figure 17.10 you can see the filling time for the headlight housing shown as a function of different colors. Blue is the shortest time and red the longest time. It is also possible to see where the part will be filled last (gray color). Normally the filling process is presented by animations. It is also possible to view the process in Flash format instead by using Moldflow. The Flash format can be used on any computer that has a Flash-enabled browser (available free of charge) installed.

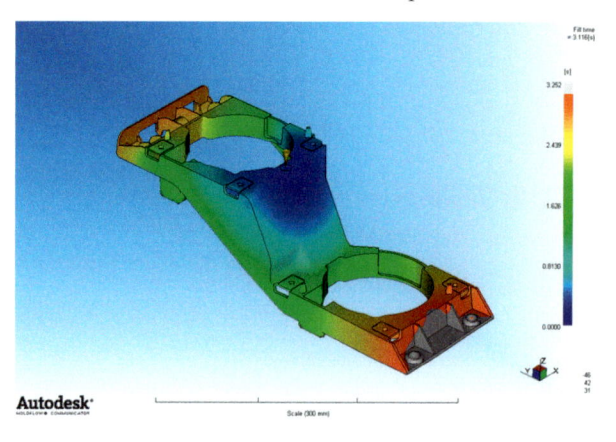

Figure 17.10 Here you can see the mold filling time displayed as a color spectrum. The axis on the right shows what the different colors represent in seconds. Within the gray area, that has not yet been filled, you will most likely get air entrapments and weld lines. Besides being able to determine this by yourself there are specific graphs for this generated with risk analysis.

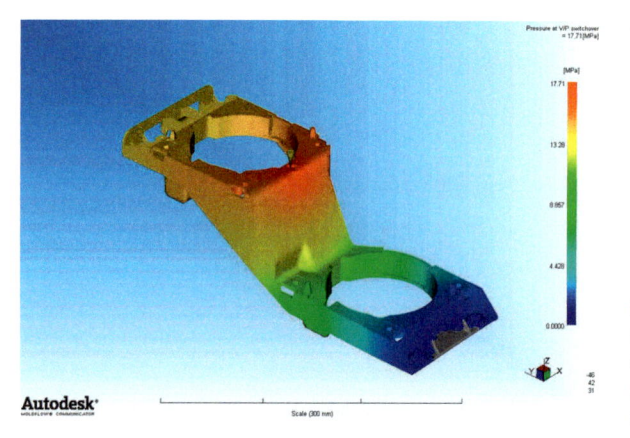

17.3.8 Pressure Distribution

Figure 17.11 This image shows the pressure distribution as a color spectrum. The axis on the right displays the meaning of the different colors in MPa. You can see that a pressure of 17.71 MPa (which is the maximum value on the scale) is not sufficient to fill the headlight completely as you still have a gray area at the bottom.

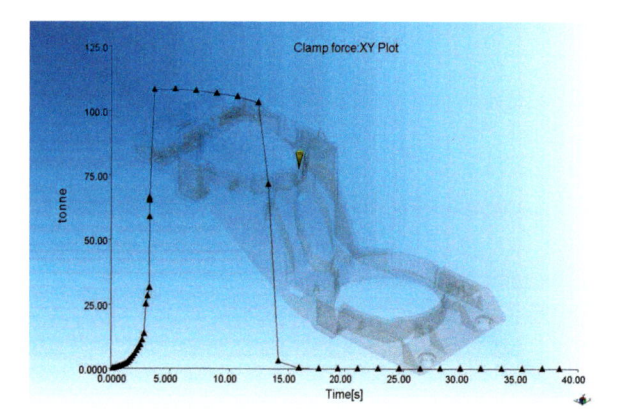

17.3.9 Clamping Force

Figure 17.12 By using Moldflow you are able to analyze how much mold clamping force is needed. The image shows necessary clamping force over the entire injection-molding cycle. In the case of the headlight housing a clamping force of at least 110 tons is needed.

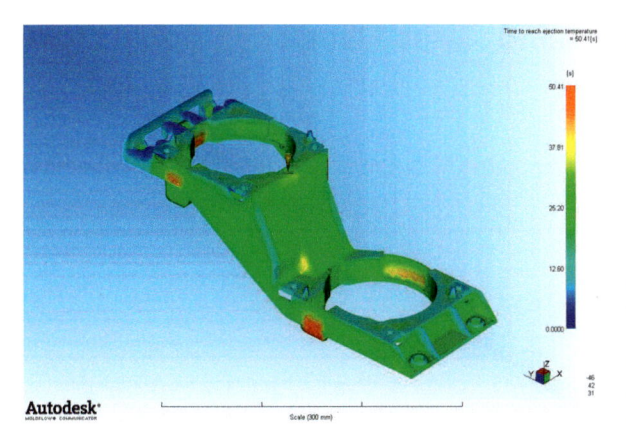

17.3.10 Cooling Time

Figure 17.13 This image shows necessary cooling time before ejection of the product. The green area corresponds to approximately 25 seconds.
However, there are some red areas that require 50 seconds. By optimizing the temperature control in these areas from the beginning you will avoid costly surprises.

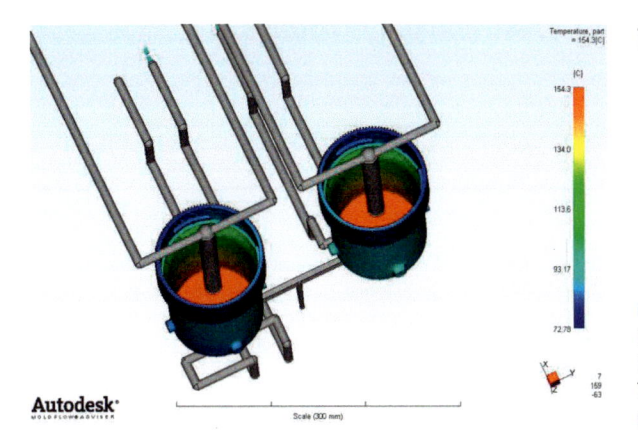

17.3.11 Temperature Control

Figure 17.14 If you are able to conclude that the temperature control is not sufficient, it is possible to modify the cooling channels using Moldflow in order to study the temperature distribution.
The image shows the cooling channels in a two-cavity mold used for production of jars.

17.3.12 Shrinkage and Warpage

Most plastics are anisotropic, meaning that properties such as strength and shrinkage vary with the material's flow direction compared to the cross direction. If you are not aware of this you can run into major surprises concerning shrinkage and warpage when a new mold is used initially. The mold shrinkage that is common in injection molding of thermoplastics depends on several factors, such as:

- The shrinkage properties of the material in various directions (the gate location is a major factor of importance here)
- Molecular chain orientation and fiber orientation in the cavity
- Variations of wall thickness of the part
- Hold pressure and hold pressure time in the molding process (shrinkage compensation)
- The temperature distribution in the cavities during the cooling process

If the shrinkage varies within the part depending on any of the factors above, you will get internal stresses. These stresses cause the part to warp when being released.

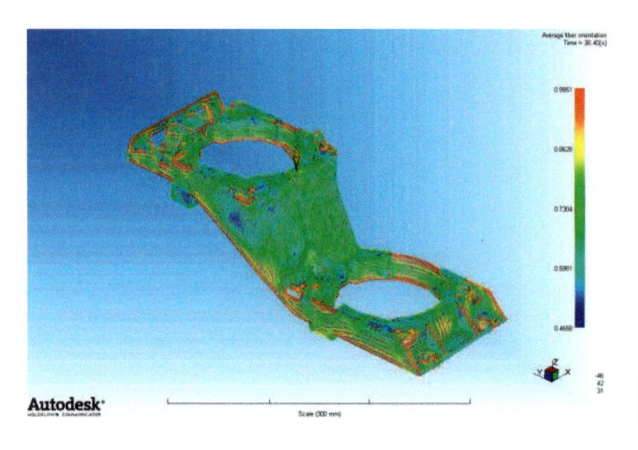

Chapter 17 – Prototype Molds and Mold Filling Analysis

17.3.13 Glass Fiber Orientation

Figure 17.15 By using Moldflow it is possible to see how the glass fibers are oriented within the cavity during the filling process.

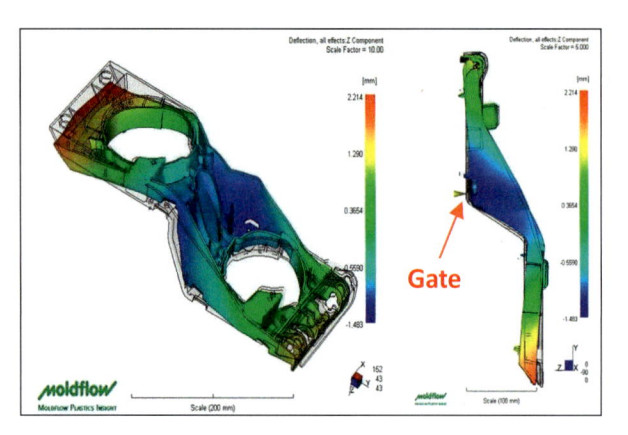

17.3.14 Warpage Analysis

Figure 17.16 One of the most advanced modules of Moldflow is the shrinkage and warpage analysis. Here you can see the results for the headlight housing. If you place the gate in the center of the part as shown in the image, you will get a deviation of the dimensions that is greater than 2.2 mm. This will be on the left short side and a deviation of approximately 1 mm on the right short side.

17.3.15 Gate Location

If the gate is moved from the middle of the headlight housing to the left-hand side shown in Figure 17.16, you get a considerably lower value for warpage, as shown in Figure 17.17.

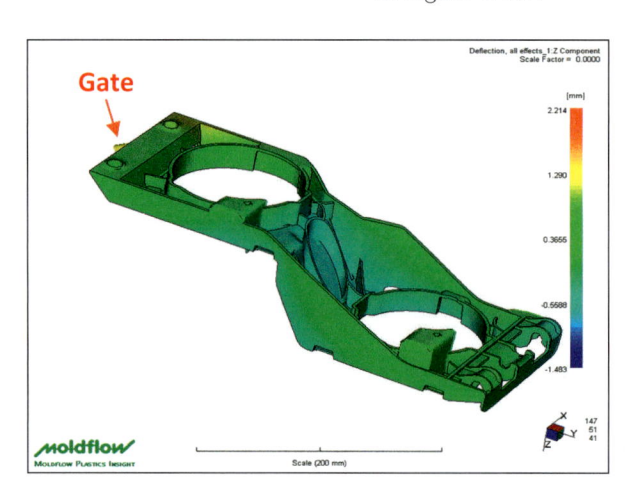

Figure 17.17 By moving the gate to the left short side, a smaller deviation of approximately 0.4 mm on each of the short sides is obtained. This is a major improvement. To make this kind of change in a mold that has already been produced is very costly. This is therefore something that is well justified to do early in the simulation process whenever there is a possible risk of warpage.

17.3.16 Material Replacement

An alternative to moving the gate sometimes can be switching to a different material. In Figure 17.18 the gate has been located at the original location, in the center of the headlight, and by replacing the semi-crystalline polypropylene with glass fiber with the unreinforced amorphous material PC/ABS, you can see a significant reduction in warpage.

Figure 17.18 By keeping the original location of the gate but making a material replacement you get a considerably smaller deviation than that in Figure 17.16.

17.3.17 Simulation Software

Below are a few links to the simulation software producers:

- Moldflow (Autodesk, USA): www.moldflow.com
- CadMould (SimCon, Germany): www.simcon-worldwide.com
- Moldex3D (CoreTech, Taiwan): www.moldex3d.com

CHAPTER 18
Rapid Prototyping and Additive Manufacturing

In the previous chapter we described various prototyping tools. In this chapter we will look at methods to produce prototypes or small production series without use of molds made of metal.

18.1 Prototypes

The reason for using a prototype or model during the development process of a new product is that it:

- Shortens the development time so that the marketing process can start earlier
- Often facilitates communication between parties during the development process
- Enables opportunity to test various functions and/or interactions with other components
- Emotionally and physically cannot be fully replaced by virtual models

Figure 18.1 Before computers were used in the development process of new products, prototypes and models were handmade. This picture was taken at the Naval Museum in Karlskrona, Sweden. Here two admirals in the navy in 1779 make the decision to build a new battleship using a very detailed wooden model.

Producing models is something that humans have done throughout history. Most kids today get their first contact with models by playing with Lego or modeling clay. The advanced computerized technology of 3D manufacturing used today was developed in the late 1980s and has taken steps through CAD/CAE/CAM/CNC.

Which technique you choose depends entirely on the complexity of the part. If it is a part with simple geometry it is usually cheaper to produce it using cutting methods such as milling, laser, or water cutting. If the part is more complex, rapid prototyping (additive technology) may be the only possible solution or a much cheaper solution compared to cutting techniques even if the material is significantly more expensive (about 50 SEK/kg (5 US$/kg) of plates in polyamide and 3,000 SEK/kg (300 US$/kg) for photopolymer in the SLA method). However, you must take into account that the production of milled models requires removal of up to 90% of the material, while the amount of material waste when using rapid prototyping is negligible.

18.2 Rapid Prototyping (RP)

This kind of additive technology is fairly new and goes under a variety of names. When performing a search the following terms may be helpful: rapid prototyping (RP), rapid tooling (RT), rapid application development (RAD), additive manufacturing (AM), or 3D printing.

We will take a look at the following methods:

1. SLA – Stereolithography
2. SLS – Selective Laser Sintering
3. FDM – Fused Deposition Modeling
4. 3DP – Three-Dimensional Printing
5. Pjet – PolyJet

All methods are based on a computer 3D model (CAD) that is then converted to an STL file (stereolithography). The computer program then "cuts" the model in layers, followed by the rapid prototyping equipment adding on each layer until the prototype is completed (see Figure 18.2). The evaluation process from CAD model to STL file usually take a few minutes to complete.

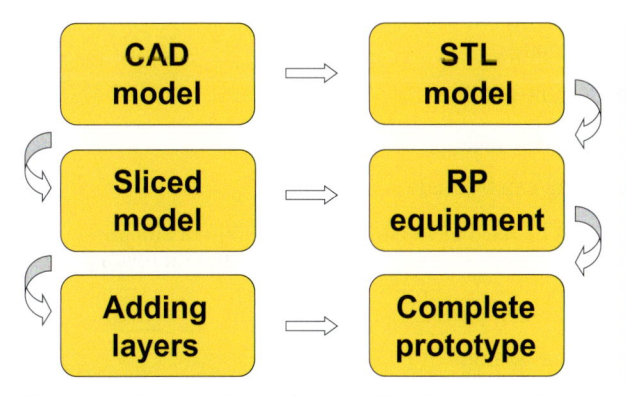

Figure 18.2 The procedure for the production of prototypes with additive manufacturing.

Figure 18.3 Here is an example of an SLS model of an elastic bellows composed of a large set of layers.

18.2.1 SLA – Stereolithography

This method was the first to be launched on the market in the late 1980s. The principle of this method is to have a light hardening photo-polymer in a tank. A computer-controlled mirror allows a UV laser beam to sweep over the surface of the polymer. When the beam hits the surface, the polymer hardens and forms a layer of approximately 0.1 mm that is bonded to the underlying layers.

The part stands on a platform, and when a new layer is added the platform is lowered 0.1 mm. It takes about one hour per cm to build parts with SLA.

The parts will then be cured in an oven.

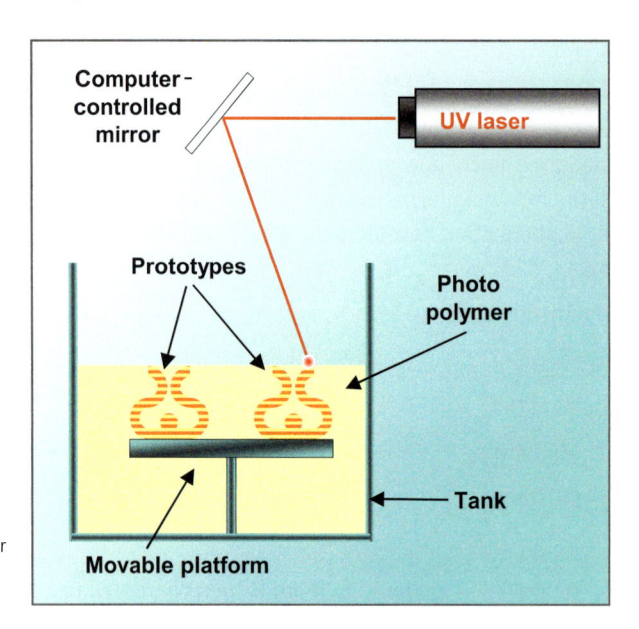

Figure 18.4 This picture shows the principle for the SLA method. The laser beam hits the surface and causes the liquid photopolymer to harden.

Advantages of Using the SLA Method

- It is quick. The entire process can be accomplished in a few hours
- The parts get a better surface finish compared to other RP methods
- Transparent parts can be made
- Tolerances of ±0.1 mm can be obtained
- Normally no problems with warpage
- Thickness down to 0.35 mm can be made
- Parts can be made that can be used as masters in production of silicone molds
- Negligible waste in production of parts
- Ceramic-like materials are available

Figure 18.5 Accura Bluestone is a stiff, hard, heat-resistant material used for the SLA.

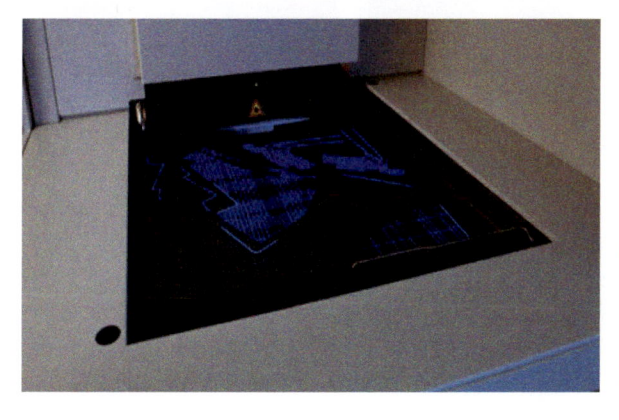

Figure 18.6 The blue pattern on the perforated platform, in the tank of an SLA machine, is the pulsating laser sweeping across the surface of the photopolymer and drawing a layer of the part.

Limitations of Using the SLA Method

- There are only about 20 different UV light hardening epoxy grades available
- In most cases the parts cannot be used in functional tests (they may be too brittle)
- The surface normally needs to be polished
- The part may require supporting pillars
- Some SLA grades are moisture sensitive
- Most SLA materials cannot be used for temperatures above 50 °C; 170 °C is the maximum temperature, and higher temperatures cause more brittle parts

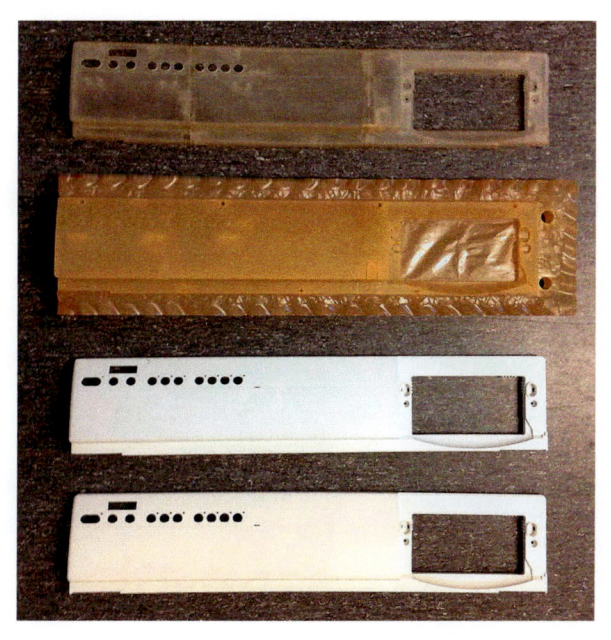

Figure 18.7 At the top of this picture we see a panel made of epoxy using SLA technology. Under the SLA panel there is the lower half of a silicone mold made by using the SLA panel as a master. Under the silicone mold there is a prototype made in polyurethane (thermoset) produced in the silicone mold. At the bottom there is the finished molded panel in ABS.

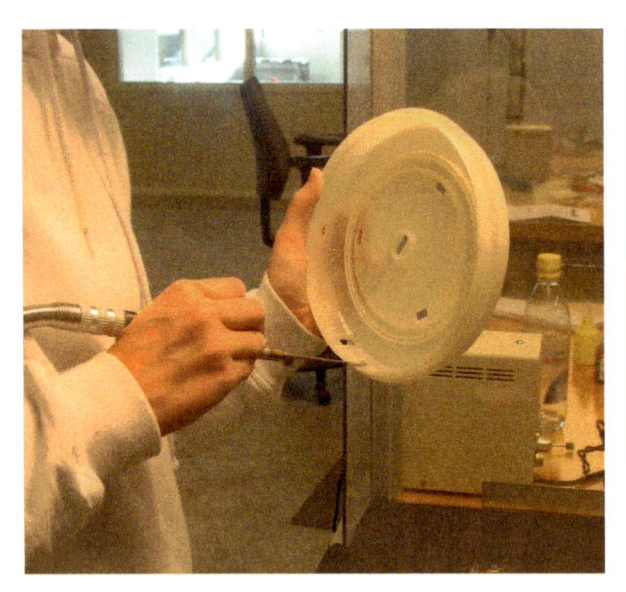

Chapter 18 – Rapid Prototyping and Additive Manufacturing

Figure 18.8 After the SLA parts have been cured they are subjected to manual polishing. This finishing step greatly affects the price of any SLA part. [Photo: Acron Formservice AB]

Figure 18.9 In this picture to the left we can see a transparent part with micro-sized support pillars immediately after curing. To the right there is the same part after trimming. At the bottom the part has been lacquered with a clear coat. The parts are initially crystal clear (remaining so for up to two years), but will become more yellow as the UV resistance of the material is not good.
[Source: Acron Formservice AB]

Figure 18.10 In the top picture we can see a part with a raw surface produced by the SLA method. In the middle the part has been trimmed and polished. At the bottom we can see the finished part, painted and fitted with stickers.
With SLA you get a nearly perfect preview of what a full series produced part will look like. It is also possible to make functional tests. [Source: Acron Formservice AB]

18.2.2 SLS – Selective Laser Sintering

This method came one year later than SLA and differs from the former by using a CO_2 laser that melts and sinters together a semi-crystalline polymer in powder form. The first step involves the use of a counter-rotating roller spreading a thin layer of powder onto a moving platform. The powder is then heated to a temperature just below its melting point.

By using a computer-controlled mirror the laser beam is sweeping across the powder surface. The laser beam is pulsating, and when it hits the surface the polymer melts and forms a 0.1 mm layer that melts together with underlying layers. For each finished layer the platform is lowered 0.1 mm, allowing the roller to spread a new layer of powder onto the surface.

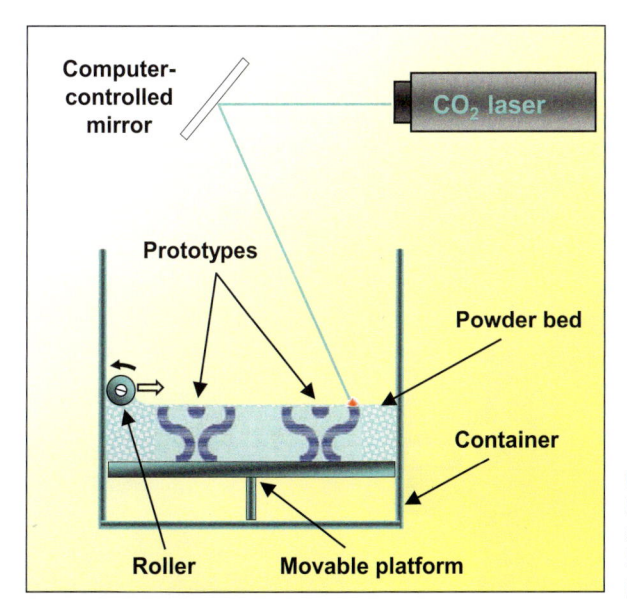

Figure 18.11 The dark walls of the bright powder container are made of sintered fused material. When the last layer is sintered any remains of loose powder are then removed.

Advantages of Using the SLS Method

- More choice of polymers such as polyamide with or without glass fibers, PP, PEEK, and thermoplastic elastomers are available
- Parts with complex geometry can be made without any other support than the powder
- Tolerances of ±0.2 mm can be obtained
- Wall thickness down to 0.5 mm can be made
- High resistance to elevated temperatures
- Suitable for functional tests
- Suitable for small scale series production

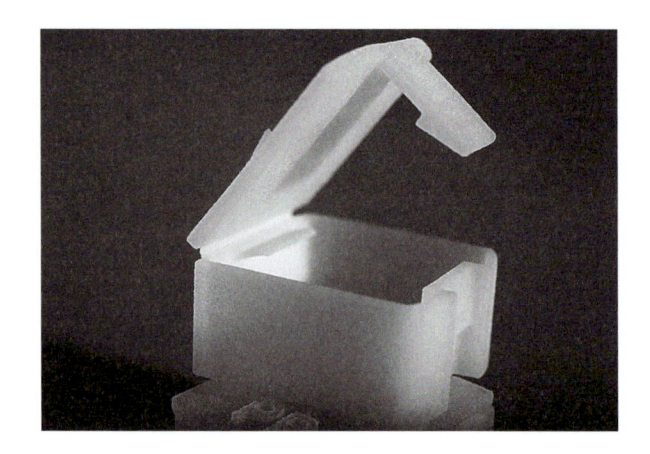

Chapter 18 – Rapid Prototyping and Additive Manufacturing

Figure 18.12 SLS model in PP where functional tests of both hinges and snap-fits are possible. [Photo: Acron Formservice AB]

Limitations of Using SLS

- Less surface finish compared to SLA (the parts get a rougher surface)
- Internal stresses (the parts may warp)
- The prototypes need to be polished

Figure 18.13 The picture shows an SLS machine capable of producing 500 × 500 × 750 mm sized parts. The machine costs about 720,000 Euro and is used both for prototypes and small-scale production (<2000 parts). [Source: Acron Formservice AB]

Figure 18.14 The picture shows a floor drain made in polyamide 11. All SLS materials become porous and hygroscopic. If you want to do a functional test of the floor drain, you must seal the porous structure by coating it with a waterproofing solution.

18.2.3 The FDM Method

FDM stands for "fused deposition modeling", both which are registered trademarks of Stratasys Ltd. It uses Ø1 mm thermoplastic filaments that are delivered in coils with a price per kg above 200 Euro. The filaments are heated in a nozzle, causing the filaments to melt. The nozzle is controlled by a computer and builds layers in horizontal directions. A vertical movement equal to the thickness of the layers is made before the next horizontal layering starts. Each layer melts together with the underlying layers. To reduce the risk of the part collapsing under its own weight, an additional nozzle can be used to make a support layer that does not melt together with the thermoplastic. The support layer will be produced simultaneously with the thermoplastic and will be removed when the part is finished.

Typical nozzle diameters are 0.127, 0.178, 0.254, and 0.330 mm, which also are the thickness of each layer. The following amorphous thermoplastics are commonly used for FDM: ABS, PC, PC/ABS, PEI, PLA, and PPSU.

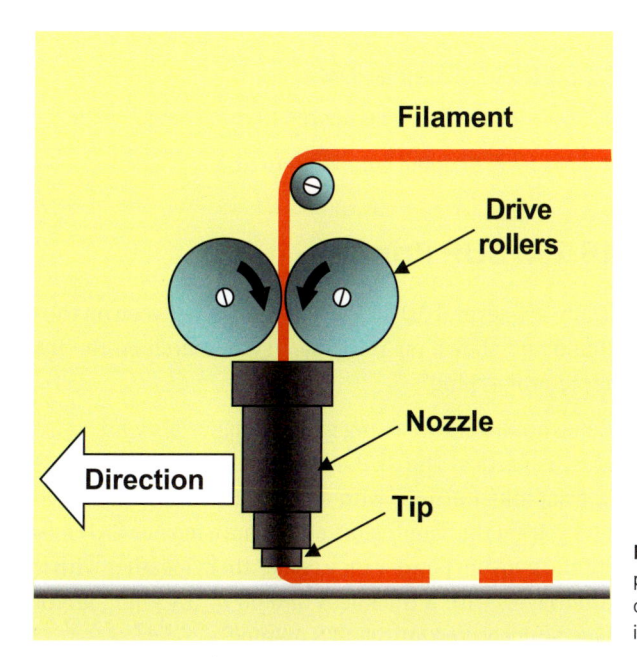

Figure 18.15 Here you can see the principle for the FDM method. A string of melted highly viscous thermoplastic is building layer after layer.

Advantages of Using the FDM Method

- A large choice of amorphous thermoplastics with high strength and good long-term properties are available
- Materials with high temperature resistance, FDA, ISO 10993-1, and V-0 approval are available
- Parts with complex geometry and thin walls can be made without internal stresses at high accuracy
- Tolerances down to 0.127 mm are possible
- Wall thickness down to 0.25 mm is possible
- Good reproducibility makes FDM suitable for small scale series production

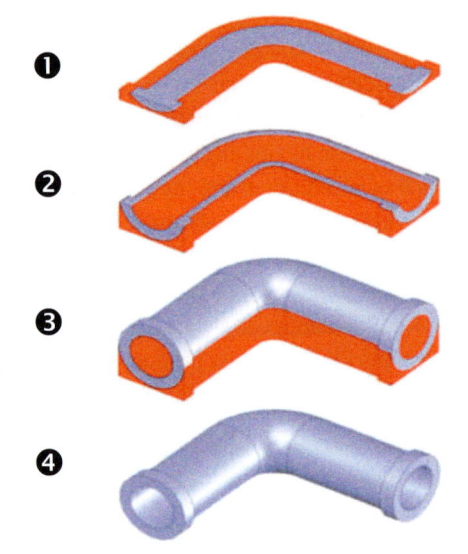

Figure 18.16 In pictures ❶ and ❷ the part in blue and its supporting layers in red are built up layer by layer. Picture ❹ shows what the part looks like when it is removed from the machine. Picture ❸ shows the part after it has been washed and cleared and ready for delivery.
[Source: Digital Mechanics AB]

Limitations of Using FDM

- Less surface finish compared to SLA (rough surface that needs to be polished)
- Less level of detail compared to SLA

18.2.4 3DP Printing

There are several 3D methods to choose between on the market. Many of them are found and shown on YouTube. Try to search on the Internet by using "YouTube rapid prototyping".

Some producers are using gypsum materials while others are using either stiff or elastic plastics. The 3DP method by Z Corporation uses a gypsum powder and a special inkjet printer with a four-color head. The ink contains a glue-like binder, and you get the same colors as you have in your CAD model. When the ink droplets hit the gypsum powder they "glue" them together with the underlying layer. Layer after layer, with a thickness of 0.08 or 0.15 mm, are built up in much the same way as described for the SLS method.

Characteristics of Using 3DP

+ Low production costs
+ Fast method with a high level of detail
+ Excellent for visual purposes

− Bad surface finish (needs to be polished)
− Not suitable for mechanical tests

18.2 Rapid Prototyping (RP)

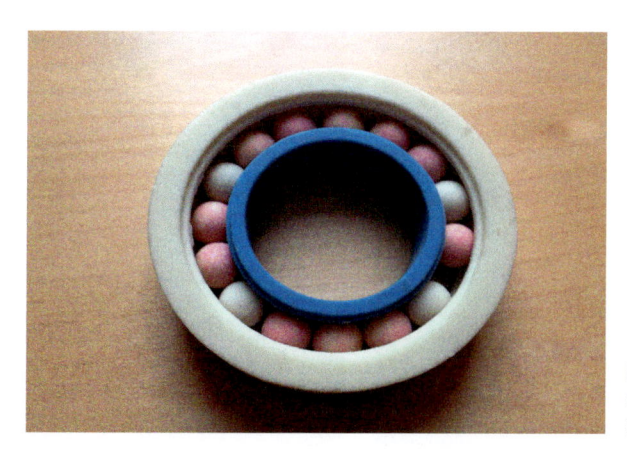

Figure 18.17 Here is a working ball bearing made in gypsum by using the 3DP method.

18.2.5 3D Printing

During recent years several small 3D printers have emerged on the market. They are available in price ranges from under 1,000 Euros to above 20,000 Euros. However, they can only use ABS. The simple low-cost models lack temperature-control chambers and a nozzle for support materials.

Characteristics of Using 3D Printing

+ Low production costs
+ Space saving (desktop models are available)
− ABS is the only material
− Slower printing speed versus regular printers
− Less precision and level of detail
− Risk of internal stresses if lacking heating chamber

Figure 18.18 Desktop model of a 3D printer from Stratasys in the 10,000 Euro class. In the front you can see the ABS filament bobbins as well as some of the panels in various colors that have been produced in the machine.

163

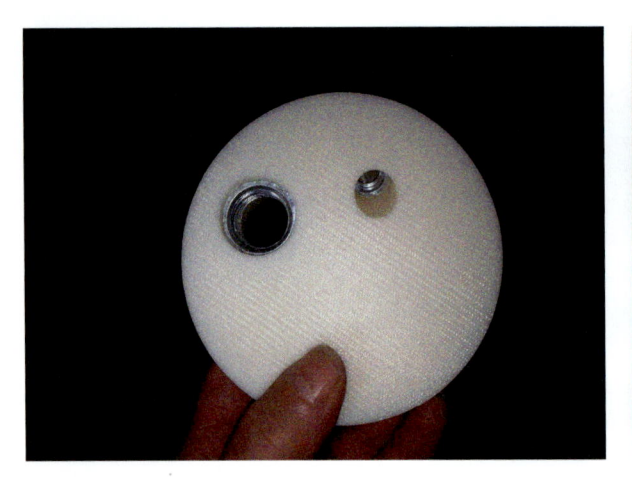

Figure 18.19 By using the FDM method it is possible to add metal inserts that will be embedded by the molten plastic filaments.
[Source: Protech AB]

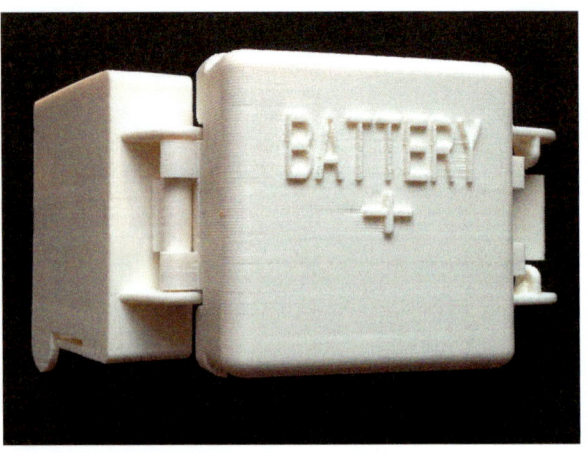

Figure 18.20 Functional model of a battery box with hinges and snap-fits made in the 3D printer shown in Figure 18.19.
[Source: Protech AB]

18.2.6 PolyJet

The PolyJet method was developed in the early 2000s by the Israeli company Objet Inc. The method uses a similar process to inkjet printers, but the ink has been replaced by a liquid acrylic-based photopolymer. On the side of the printing block UV lights are placed that cause the photopolymer drops to harden immediately as they hit the underlying layer.

This is one of the most rapid methods available. It enables precision down to 0.016 mm for each layer. It produces excellent "styling prototypes" with a high level of detail, and just as with the FDM method a gel-like, water-soluble support material will be used. The support material will then be washed away once the processing of the part is completed.

The support material is "printed" simultaneously with the photopolymer. Any subsequent curing, trimming, or polishing will not be needed.

Characteristics of Using the PolyJet Method

+ High speed with a high level of detail
+ Wall thickness down to 0.3 mm is possible
+ Clean and suitable in office environment
+ Two different materials in different colors or hardness can be used simultaneously in up to 30 different combinations

− Only acrylic plastics are available
− Less suitable for functional tests
− The material will creep under load
− Lack of long-term material data

18.2 Rapid Prototyping (RP)

Figure 18.21 Here is one of the PolyJet machines at the company Digital Mechanics AB. Behind the PolyJet you can see FDM machines that take up substantially more space.

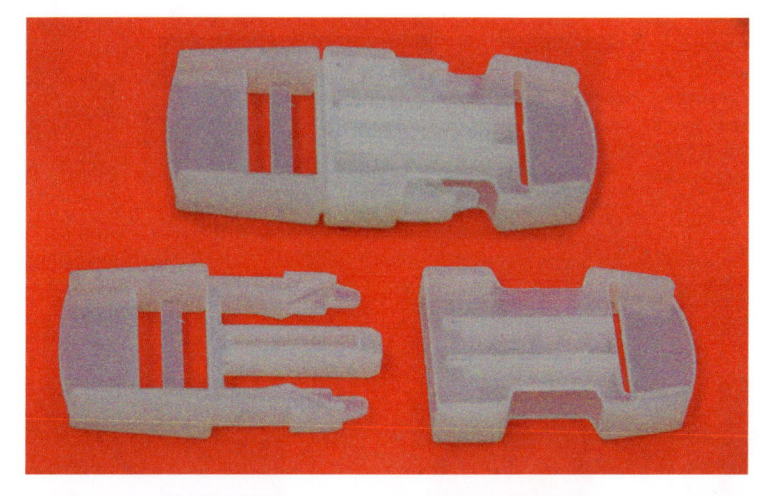

Figure 18.22 Here you can see the prototype of snap locks made by PolyJet. The production time is about half an hour. Whether you fill the whole plate or make just a single snap lock the time will be the same. The material can withstand testing of the snap-fit function a few times but does not withstand pulling forces.
[Source: Digital Mechanics AB]

Figure 18.23 The picture shows the printer block of the PolyJet machine in Figure 18.21. The block has 8 printer heads with 96 channels. Four of them are used for the supporting material and the other four for the photopolymer. When making two-component parts every other head is used for the two material combinations.
[Source: Digital Mechanics AB]

165

Chapter 18 – Rapid Prototyping and Additive Manufacturing

18.3 Additive Manufacturing

The different rapid prototyping methods are not only used to produce prototypes and models. They can be used for one-piece or small scale series production. There are several names and abbreviations for this type of production. Additive manufacturing (AM), direct digital manufacturing (DDM), and rapid manufacturing (RM) are some of the most common names. Regardless of which method you will use, the starting point is always 3D data, and you can divide the production into four different groups:

- Principal models
- Functional models
- Fixtures and grip tools
- End-use products

Figure 18.24 Here is a principal model of a wheel suspension. The model has been made in one piece using the two-component technique of a PolyJet machine. By mixing two different acrylics simultaneously, different colors and degrees of hardness have been obtained. The wheels are made of a soft material (Shore A 70) while the beams are made of a stiff one. [Source: Digital Mechanics AB]

Below you can see a number of different products, everything from ornaments to advanced aviation instruments. They have all been made by additive manufacturing.

Figure 18.25 A bride and groom used for decoration on a wedding cake. This is a one-piece product that has been produced in full color using the 3DP in gypsum. A computer program creates a 3D model from four photos of each bride and groom, and about two hours later the process has been completed. The finished product is sold at a price of about 250 Euro.
[Source: Digital Mechanics AB]

Figure 18.26 This is a picture of a rear lamp of a car. It has been made in transparent ABS by the use of the FDM technique.
This is an example of "manufacturing on demand", which is steadily increasing when it comes to spare parts for older cars or MCs as well as for small series of parts used for luxury cars. [Photo: Protech AB]

Figure 18.27 The instrument handle shown here is made in polycarbonate by the FDM method. The handle was made standing to achieve the best impact strength. It took about 15 hours to produce it. This is an example of a small scale series production with 50 handles per year.
[Source: Digital Mechanics AB]

18.3 Additive Manufacturing

Figure 18.28 In the picture to the left is a gripper to a robot and to the right a part used for mounting equipment. Both are produced by SLS in polyamide. The brass bushings are screwed in afterward and secured with glue. [Photo: Acron Formservice AB]

Figure 18.29 This picture shows machine parts made in small scale series by FDM. [Source: Protech AB]

Figure 18.30 On the left is shown a very advanced gyro-stabilized electro-optical system. This type of equipment is used in unmanned aerial vehicles and helicopters. The camera system has components made in ABS with very high precision using the FDM technique. [Source: DST Control AB]

Figure 18.31 The picture shows an electronic component made in the thousands by FDM. The part has a 0.2 mm wide groove. When injection molded the part failed as the core that forms the groove could not be cooled well enough. [Photo: Digital Mechanics AB]

167

CHAPTER 19
Cost Calculations for Moldings

Most molders are using advanced computer-based software to calculate costs or post-costs of injection-molded parts. Unfortunately, it is very seldom that injection machine setters have insight into or get the opportunity to use such software, even though they have great potential to affect the costs by adjusting the injection-molding parameters.

How often does it happen that setters add a few seconds of extra cooling time when they have a temporary disturbance of the injection-molding cycle? And then forget to change back to the original settings before the parameters are saved for the next time the mold will be set up? Those extra seconds can mean thousands of Euros or Dollars in unnecessary production costs per year and may also reduce the company's competitiveness.

The purpose of this chapter is to show how a fairly detailed cost calculation for injection-molded parts can be made. The setter also gets a tool that enables him/her to see how changes that are made in the process can influence the cost of the molded part. This tool is based on Microsoft Excel and is available for downloading at www.brucon.se. The user does not need any extensive knowledge of Excel in order to fill in the input values required to immediately obtain the final cost picture at the bottom of the page.

The rest of this chapter will explain how to use the Excel file and what the different input values mean.

When you open the file called Costcalculator.xls you must first make a copy of this file to your computer's hard drive, otherwise the macro functions will not work. Depending on how the default values are set for your own Excel program, it may be necessary to make modifications of the security settings. Detailed information of how this is to be done can also be found on the author's homepage. The Excel file is also in "read-only" mode, so it should be saved under a different name once you have completed it.

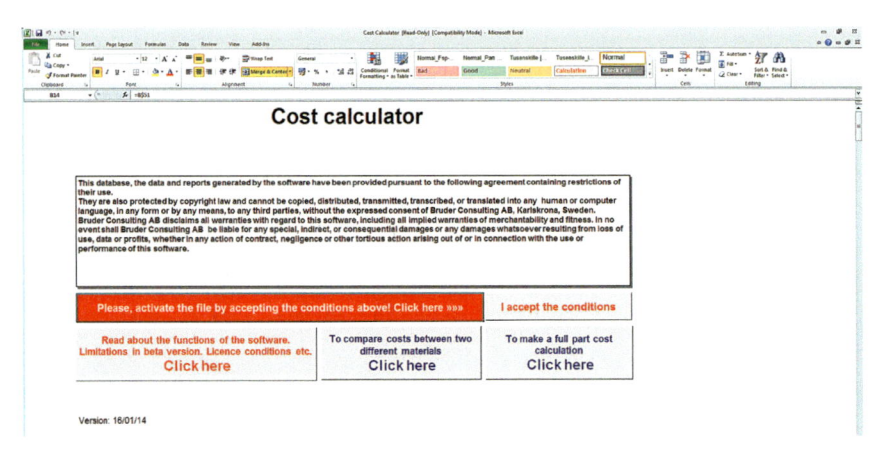

Figure 19.1 The start menu once the Excel file has been opened.

There are three different functions to choose between:

1. Read about the functions of this software
2. Compare the costs between two different materials
3. Make a full part cost calculation

Before you click on the key "I accept the conditions" you are only able to "Read about the functions of the software". The two other keys will only display blank pages.

Figure 19.2 Once you have clicked on "I accept the conditions" you will see "The file is active", and all the different functions can now be used.

19.1 Part Cost Calculator

We will start with the "Part cost calculator". This is the most advanced function, and we will go through all input values before we end the chapter with the "Material comparison calculator".

In "Part cost calculator" you can make a relatively complete cost calculation for a single part, total delivery volume, or annual volume. When filling in the white input fields with blue text a quick way to get to the next field is to use the "Tab" key on your computer keyboard.

The final result is obtained at a given sales price but it is also possible to get the sales price using a predetermined profit that you wish to achieve.

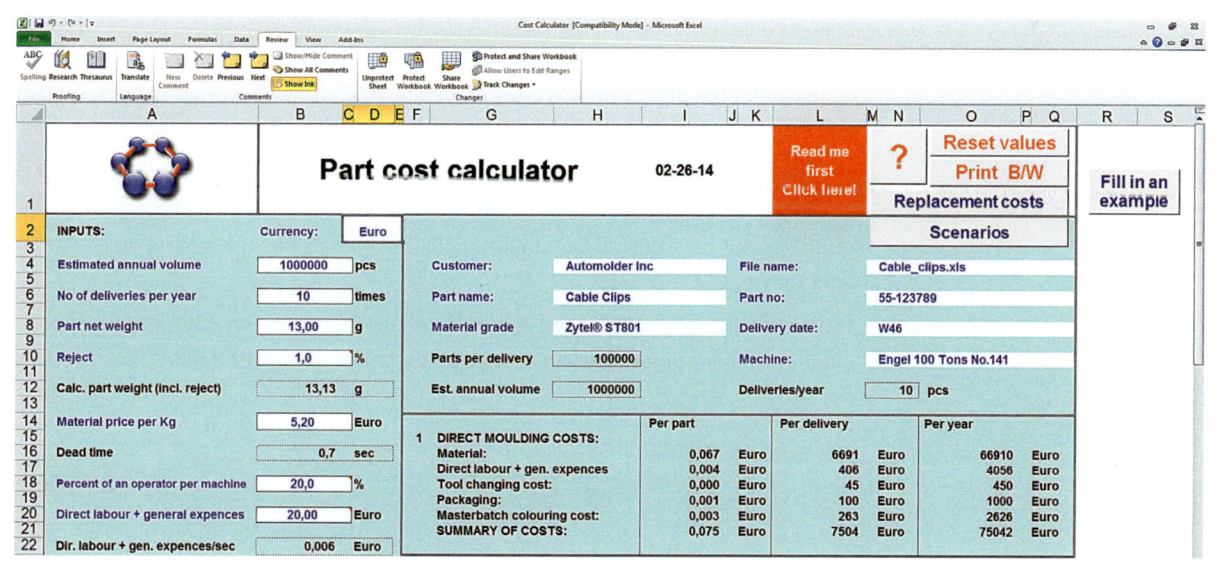

Figure 19.3 If you wish to practice with the same values that are shown above, just click on the key "Fill in an example" and the spreadsheet will automatically be filled with the values.

When you start with a blank spreadsheet you will be in the input field *Currency* (top left). Once you have filled in the currency, i.e. USD, Euro, or whatever you wish to use, the selected currency will appear everywhere in the spreadsheet where it is used, once the number for the annual volume has been filled in.

To get to the next field use the "Tab" key and you will come to *Estimated annual volume*.

Then go to and fill in the fields *Customer* and *Filename*.

The next input field is *Number of deliveries per year*. Once a value has been added to this field the program is able to calculate *Parts per delivery* that is shown as a dashed box in the display window.

The next input fields to fill in are *Part name* and *Part no*.

The next numeric field is *Part net weight*. Here you fill in the actual weight of the part if it has been produced or an estimated value of the weight.

Other data fields to fill in are *Material grade* and *Delivery date*.

The next numeric field is *Reject*. In the upper-right corner of the field there is a small red triangle. This red triangle will be seen in several of the input fields. It signals that if the requested value is unknown you are able to get a suggested value by holding the mouse pointer over that particular field. In this example the value for Reject is estimated to lie between 0.5-2%. Once the value for Reject has been added, the program calculates the *Calculated part weight (incl. reject)*. In this weight we have taken into account that some additional material is needed due to scrap. This gives an intermediate result that is displayed in black within a dashed frame marked in blue.

The last field in the information window is *Machine*, meaning which machine will be used.

The next numerical input field is *Material price per kg*, and once this value has been entered you can see the value for *Direct moulding costs* in the result window. Here the results for the material costs per part, per delivery, and per year have been calculated. Every time you enter a value in a numerical input field the result will be calculated and show up in the result window in the spreadsheet.

The next input field is *Percent of an operator per machine*. Hold the mouse pointer above the field and you can see "If one operator looks after one machine the percent of an operator = 100%. If one operator looks after 5 machines the percent of an operator = 20%" etc.

In the fields *Direct labour and general expenses* the total payroll including social expenses for the operator are calculated. The program then calculates the expenses per second.

19.1 Part Cost Calculator

Administration expences	5,0	%	2 DIRECT POST-OPERATION COSTS:							
			Material:	0,020	Euro	2000	Euro	20000	Euro	
No of parts per packaging box	1000	pcs	Direct labour + gen. expences	0,056	Euro	5556	Euro	55556	Euro	
			Non-direct post-operation cost	0,001	Euro	76	Euro	756	Euro	
Price of packaging box	1,00	Euro	SUMMARY OF COSTS:	0,076	Euro	7631	Euro	76311	Euro	
Percent of masterbatch	2,0	%								
			3 COMPUTED DIRECT COSTS:	0,151	Euro	15135	Euro	151353	Euro	
Masterbatch price per Kg	10,00	Euro								
Net cycle time (measured)	14,6	sec	4 GROSS EARNING (9-3):							
			Totaltally:	0,091	Euro	9115	Euro	91147	Euro	
No of cavities	4	pcs	Per hour:	84,55	Euro	84,55	Euro	84,55	Euro	
			In % of net price:	60,2	%	60,2	%	60,2	%	
Parts per hour (theroretical)	986,3	pcs								
Parts per hour (real)	927,6	pcs	5 MACHINE & TOOL COSTS:							
			No of machine hours:	-		107,8	hours	1078,0	hours	
Usability	95,0	%	Costs per machine hour:	25,00	Euro	25,00	Euro	25,00	Euro	
			Machine cost:	0,027	Euro	2695	Euro	26951	Euro	
Time per part (theoretical)	3,65	sec	Amortised tool cost	0,001	Euro	50	Euro	500	Euro	
Time per part (real)	3,88	sec								
			6 MANUFACTURING COST:	0,179	Euro	17880	Euro	178804	Euro	
Machine cost per hour	25,00	Euro								
Tool cost to be amortised	500	Euro	7 ADMINISTRATIVE COSTS:	0,009	Euro	894	Euro	8940	Euro	
			Administrative costs in %	5,0	%	5,0	%	5,0	%	
Tooling change time per delivery	1,0	hours								
Cost for tooling change	45,00	Euro	8 TOTAL COSTS (6+7):	0,188	Euro	18774	Euro	187744	Euro	

Figure 19.4 The input fields are in white with a blue font and the calculations fields in blue with a black font.

In the field *Administration expenses* the percentage of the company's estimated operating expenses are entered. These include administration, energy costs, marketing costs, communication costs, etc. Depending on the size of the company, this is typically in the range of 5–10%.

In order to make a complete calculation there are fields for *Number of parts per packaging box* and *Price of packaging box* (see Figure 19.4). The results will appear in the result window *Direct moulding costs* under *Packaging* (see Figure 19.3).

If you color your resin with a masterbatch, the next two input fields *Percent of masterbatch* and *Masterbatch price per kg* are to be filled in. If you have a resin with natural or a fully compounded color just skip these fields.

The next input field is *Net cycle time measured*. If the part has never been molded before, an estimated cycle time must be filled in.

In the field *Number of cavities* you should enter either the actual number or the planned number of cavities. The program will now calculate the theoretical value of *Parts per hour*.

Once the percentage for *Usability* (the hours when the machine is in full production divided by 24 hours) the actual value of *Parts per hour (theoretical)* is calculated. Normally there are several stops per day depending on maintenance, failure corrections, and changes. For three shifts of production 90–98% is normally accepted.

For *Parts per hour (real)* the calculation takes into account that you need some extra shots to compensate for the rejects. When parts per hour have been calculated you will also get the values for *Time per part (theoretical)* and *Time per part (real)*.

Chapter 19 – Cost Calculations for Moldings

In the field *Machine cost per hour* you use the sum for machine amortization, maintenance, and energy consumption, plus the costs for robots etc. is to be added to this cost.

When comparing calculations from different molders there tend to be large deviations depending on how they calculate these costs.

If you own the mold *Tool cost to be amortized* including maintenance should be filled in. If you are a subcontractor not owning the mold yourself, you normally will be responsible for the maintenance of the mold and must take these costs into account as well.

In *Tooling change time per delivery*, you should fill in either the actual or the estimated time for mold change. The result *Cost for tooling change* will now be calculated by multiplying the value for *Tooling change per delivery* with the real costs for labor. When this has been added the result pane for *Direct moulding costs* in Figure 19.3 will be completed.

Post-operation cost per part	0,020	Euro		9	SALES PRICE (to customer):	0,250	Euro	25000	Euro	250000	Euro
Post-operation handling time	10,0	sec			J. Freights	0,003	Euro	250	Euro	2500	Euro
					J. Rebates	0,005	Euro	500	Euro	5000	Euro
Non-direct post-operation cost	1,0	%			NET PRICE :	0,243	Euro	24250	Euro	242500	Euro
Sales price (gross)	0,250	Euro		10	PROFIT:	0,055	Euro	5476	Euro	54756	Euro
Freight per delivery	250,00	Euro			PROFIT PER HOUR:	50,79	Euro	50,79	Euro	50,79	Euro
					PROFIT IN %:	29,2	%	29,2	%	29,2	%
Rebate	2,0	%									

Figure 19.5 The bottom part of the calculation sheet is shown with a positive profit highlighted in green.

If you have any costs of post-operation involved with adding something to the part such as a label, grease, paint, or chrome, you can add the costs for this in *Post-operation costs per part*.

If you have manual handling for the above costs you can enter the time into the field named *Post-operation handling time*. The result will be found in *Direct labour and general expenses* within the *Direct moulding costs pane* in Figure 19.4.

Cost for tooling change	45,00	Euro		8	TOTAL COSTS (6+7):	0,188	Euro	18774	Euro	187744	Euro

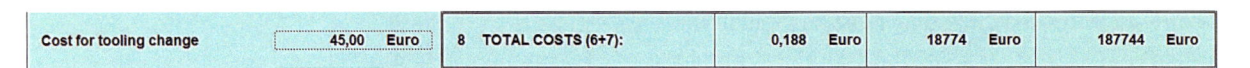

Figure 19.6 When the "Non-direct post-operation costs" have been calculated you will also get the "Total costs".

The last part to fill in is *Sales price gross* whereby the *Profit* is calculated once the costs for *Freight per delivery* and/or *Rebate* have been taken into account.

If the field Sales price is left empty, a new field named *Net profit in %* pops up at the bottom of the spreadsheet (see Figure 19.7).

Post-operation cost per part	0,020	Euro									
			9	SALES PRICE (to customer):	0,235	Euro	23468	Euro	234680	Euro
Post-operation handling time	10,0	sec		./. Freights	0,003	Euro	250	Euro	2500	Euro
				./. Rebates	0,005	Euro	469	Euro	4694	Euro
Non-direct post-operation cost	1,0	%		NET PRICE :	0,227	Euro	22749	Euro	227487	Euro
Sales price (gross)	0,000	Euro								
			10	PROFIT:	0,040	Euro	3974	Euro	39742	Euro
Freight per delivery	250,00	Euro		PROFIT PER HOUR:	36,87	Euro	36,87	Euro	36,87	Euro
				PROFIT IN %:	21,2	%	21,2	%	21,2	%
Rebate	2,0	%								
Net profit in % (-> sales price)	25	%								

Figure 19.7 Please note that "Sales price (gross)" and "Sales price (to customer)" are the only fields with three decimals.

If the *Net profit in % (→ sales price)* is filled in as a percentage, it is possible to see what the price will be to the customer. See *Sales price (to customer)*.

An example: If the profit has been set to 25% the total profit will be 21.2% as the shipping costs (Freight per delivery) and the rebate reduce the total profit with 3.8% in this case.

At the beginning of this chapter we mentioned that setters sometimes increase the cooling time when they have process problems and forget to readjust the time when saving and closing down the job. Normally an increase of the cooling time will increase the total cycle time with the same amount. Let's see an example of how two extra seconds of cooling time will affect the profit in the spreadsheet above. To do this you can replace the actual value (14.6 sec) in the field *Net cycle time measured* (see Figure 19.4) with 16.6 sec and see how this affects the *Profit*.

A better alternative is to click on the "Scenarios" key in the upper-right corner of the worksheet. Before you do this, choose to go back to the original sales price set at 0.250 Euros, which then will give you a profit of 29.2%.

19.2 Part Cost Scenarios

In Figure 19.8 we can see that the original *Annual Loss/Profit* is 54,756 Euros and the *Loss/Profit in %* margin is 29.2%. If we increase the *Net cycle time (measured)* by two seconds from 14.6 to 16.6 seconds in the *Alt. 1* column, the profit will be reduced by 4.460 Euros down to 50,296 Euros and the profit margin drops to 26.2%. All fields in the column that are affected by this cycle time change are highlighted in yellow.

In the *Alt. 2* column we let the new cycle time remain and will see how much the material price must be reduced in order to obtain the original profit. The result is that the price must be decreased by about 0.33 Euros to 4.87 Euros.

With these two examples you should understand that the spreadsheet "Part cost scenarios" can be a useful tool both for calculating the price of a new part or for a machine setter to get a better understanding of how the changes that he/she can control may affect the economics.

Chapter 19 – Cost Calculations for Moldings

	Part cost scenarios						01-16-14	Read me first Click here!	Currency: Reduce no of decimals	?	Reset values
											Print B/W
											Part cost calculator
		Part cost calc.	Unit	Alt. 1	Alt. 2	Alt. 3	Alt. 4	Alt. 5	Alt. 6	Alt. 7	Alt. 8	Alt. 9	Alt. 10
Annual Loss / Profit	54756	Euro	50296	54755	54756	54756	54756	54756	54756	54756	54756	54756
Loss / Profit in %	29,2	%	26,2	29,2	29,2	29,2	29,2	29,2	29,2	29,2	29,2	29,2
Difference Loss / Profit			-4460	-1								
Estimated annual volume	1000000	pcs	1000000	1000000	1000000	1000000	1000000	1000000	1000000	1000000	1000000	1000000
No of deliveries per year	10	times	10	10	10	10	10	10	10	10	10	10
Part net weight	13,00	g	13,00	13,00	13,00	13,00	13,00	13,00	13,00	13,00	13,00	13,00
Reject	1,0	%	1,0	1,0	1,0	1,0	1,0	1,0	1,0	1,0	1,0	1,0
Calc. part weight (incl. reject)	13,13	g	13,13	13,13	13,13	13,13	13,13	13,13	13,13	13,13	13,13	13,13
Material price per Kg	5,20	Euro	5,20	4,87	5,20	5,20	5,20	5,20	5,20	5,20	5,20	5,20
Dead time	0,7	sec	0,8	0,8	0,7	0,7	0,7	0,7	0,7	0,7	0,7	0,7
No of operator per machine	20,0	%	20,0	20,0	20,0	20,0	20,0	20,0	20,0	20,0	20,0	20,0
Direct labour + general expences	20,00	Euro	20,00	20,00	20,00	20,00	20,00	20,00	20,00	20,00	20,00	20,00
Dir. labour + gen. expences/sec	0,006	Euro	0,006	0,006	0,006	0,006	0,006	0,006	0,006	0,006	0,006	0,006
Administration expences	5,0	%	5,0	5,0	5,0	5,0	5,0	5,0	5,0	5,0	5,0	5,0
No of parts per packaging box	1000	pcs	1000	1000	1000	1000	1000	1000	1000	1000	1000	1000
Price of packaging box	1,00	Euro	1,00	1,00	1,00	1,00	1,00	1,00	1,00	1,00	1,00	1,00
Percent of masterbatch	2,0	%	2,0	2,0	2,0	2,0	2,0	2,0	2,0	2,0	2,0	2,0
Masterbatch price per Kg	10,00	Euro	10,00	10,00	10,00	10,00	10,00	10,00	10,00	10,00	10,00	10,00
Net cycle time (measured)	14,6	pcs	16,6	16,6	14,6	14,6	14,6	14,6	14,6	14,6	14,6	14,6

Figure 19.8 If you click on the key "Part cost scenarios" you will get the spreadsheet above. Under the green cell "Part cost calc." you will get the same values as in the previous spreadsheet. In the following columns these values are repeated, and it is here you can enter your alternative values.

Below is a simplified calculation that may be used as an example of the economic consequences effected by a change of parts.

19.2 Replacement Cost

You will get the last function in the Cost calculator by clicking on the key "Replacement costs" at the start page. The spreadsheet is shown in Figure 19.9. Here you can also click on the key "Fill in an example" to get values to exercise with.

In the "Replacement costs sheet" you can compare some material and machine costs of two different materials. By holding the mouse pointer over the red cell "Read me first Click here" you will get a quick manual of what you can do in this spreadsheet.

NOTE: In order for the spreadsheet to work, all of the white fields containing a small red triangle in the upper-right corner need to be filled in.

19.2 Replacement Cost

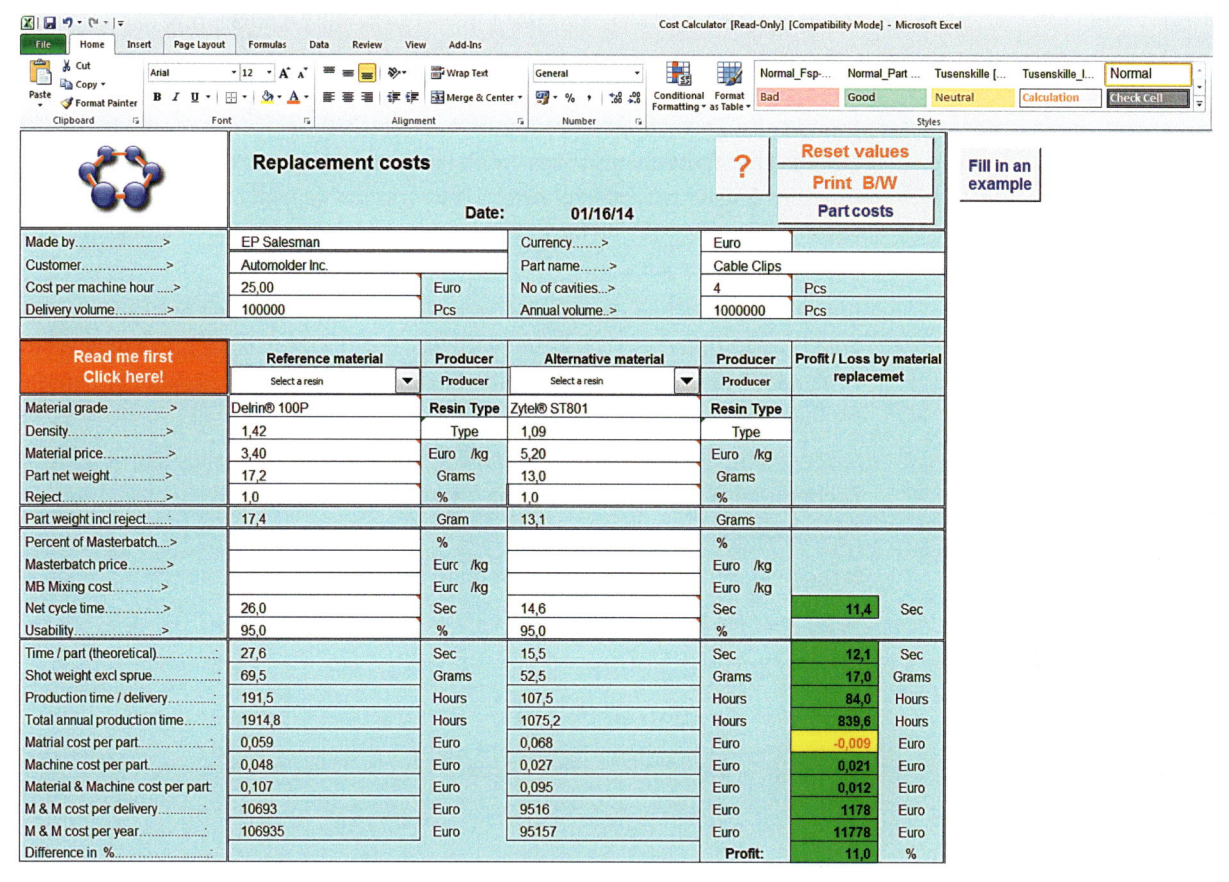

Figure 19.9 The "Replacement costs" spreadsheet enables you to easily compare material and machine costs between a large number of specified material grades.

When you are selecting the materials there is a choice of 3,130 pre-programmed material grades to choose from. See the drop-down menu or use the key *Select a resin*. Once a *Material grade* has been selected its *Density* will be shown automatically. If your desired material is not found in the list you can fill in the name and the density of the grade manually.

Once the *Part net weight* has been entered for the *Reference material* the weight of the *Alternative material* will automatically be calculated based on its density. If this weight deviates from the actual measured value you can overwrite manually.

Typical values of *Reject* are 0.5 to 2% and for *Usability* between 90 to 98% at 3 shifts.

In the lower-right corner of the spreadsheet you can see the results in *Profit* or *Loss* when replacing the reference material with an alternative one.

In the example in the spreadsheet above a cable clip has been produced by using a tough acetal grade from DuPont called Delrin 100. This material is used as the reference material in our comparison with other materials. The end user was not completely satisfied with this choice of material and would like to test other materials that are even tougher. The choice fell on a "super tough" PA66 grade, also made by DuPont and known as Zytel ST801.

When Zytel ST 801 was suggested, there were objections as this material was considered to be too expensive with a cost of 1.80 Euros more per kg. But after testing, the results showed that despite the higher material price for Zytel ST801 you were still able to get a profit increase of 11,778 Euros. The reason for this is that the alternative material has both a lower density and a shorter molding cycle. Only the density reduction of Zytel ST801 was not enough to compensate for the difference in price of 1.80 Euros when compared to the acetal. It resulted in a loss of 0.009 Euro per part.

But as Zytel ST801 could run 11.4 seconds faster, as it does not need such a long hold pressure time as acetal, the profit was 0.021 Euro per part. The total profit of about 0.012 Euros per part resulted in an annual profit of 11,778 Euro on the production of one million parts.

NOTE: If a field in the column *Profit/Loss by material replacement* shows a green background with a black font that means that the alternative material has an advantage.

If the field is yellow with a red font there is an advantage for the reference material.

CHAPTER 20
Extrusion

This chapter was developed together with Talent Plastics AB. Their competent staff at the factories in Gothenburg and Alstermo in Sweden provided the information and images that were crucial to creation of the chapter.

Extrusion is the second largest processing method after injection molding.

Figure 20.1 The material feed in an extruder is in principle the same as in a meat grinder. The difference is, when the plastic material is transported through the cylinder, it is also heated and melted.

Figure 20.2 In the picture you can see a modern extruder. This is a so-called single-screw extruder. The material filling hopper is replaced by a material handling system that is connected via tubes in the ceiling. The tool (die) is to the left.

20.1 The Extrusion Process

It is a continuous process where you can make "endless" tubes, pipes, profiles, sheets, film, cable, and monofilament. You can also coat metal (e. g. cables), paper, or fabrics. Another method is film blowing.

In the extrusion line you can also add different post-processing stations such as printing, punching, milling, cutting, coiling, flocking, and muffing in order to get fully completed products.

In compounding of plastic granules or producing monofilament, extruders are also used in the production line.

20.1.1 Advantages (+) and Limitations (−)

+ A wide range of thermoplastics can be used
+ Tools are significantly cheaper than in injection molding
+ So-called multilayer tubes and profiles can be made
+ Wide sheets can be made
+ Corrugated pipes and tubes can be obtained
+ Foamed products can be made
+ Tight tolerances can be obtained
+ Good surface appearance can be obtained

Chapter 20 – Extrusion

+ Thin-walled products (e.g. films) can be made
+ An inner core of foam can be obtained
+ Metal cores can be coated (e.g. electrical cables & wires)

− Space requirements in length for extrusion lines
− Recycling of multi-layer pipes or profiles

Figure 20.3 That good surface finish can be obtained, such as you can see on this pipe in ABS.

Figure 20.4 The picture shows an extrusion hall with 19 extruder lines in a row at Talent Plastics in Gothenburg, Sweden. The width and length of a typical extrusion line is about 4 × 25 meters. In the foreground you can see a winder with a yellow hose. To the right at the back of the hall profiles are rolled up on large wooden drums. [Photo: Talent Plastics]

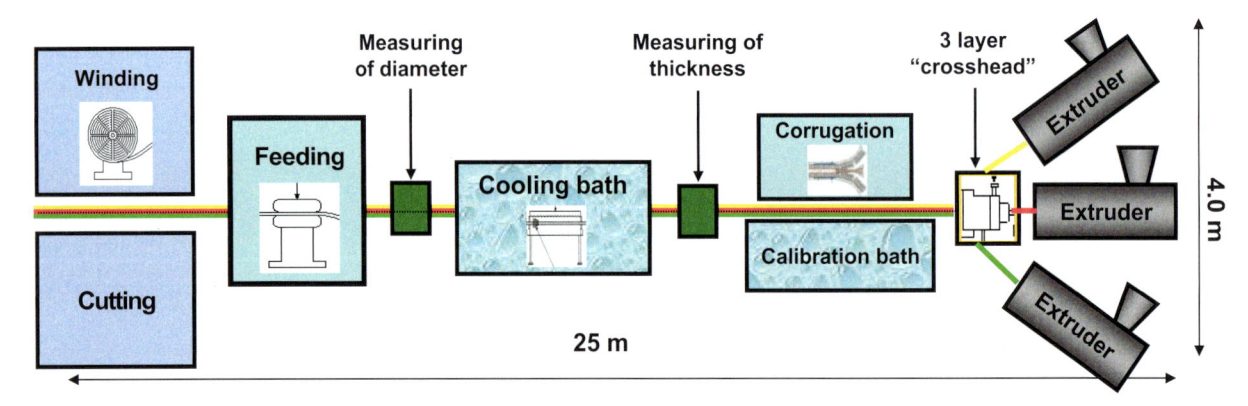

Figure 20.5 This diagram represents an extrusion line for producing a profile with three different color layers. You need an extruder for each color. A typical extrusion line starts with the extruder/extruders with the tool that shapes the profile. After the tool, there is usually a calibration bath where the profile is fine-tuned under vacuum. If you want a corrugated profile a corrugator replaces the calibration bath. The next station is the cooling bath. One or more feeding units are used before any post-processing stations as printing, cutting, etc. The last stations are either cutting and packing or winding up the profile on a drum. In most cases, the line is horizontal, but for some products such as large cables or blown film an angle tool is used and the output of the products will be vertically in buildings with very high ceilings.

20.2 Materials for Extrusion

Many different thermoplastics can be used for extrusion. Extrusion grades generally have higher viscosity and no surface lubrication compared to injection molding grades. The reason for high melt viscosity is to avoid that the extruded profile collapse before it reaches the calibration bath. The most common materials in extrusion are:

Polyethylene (LD, HD, MD, PEX)	TPE-O	ABS
Polypropylene (PP)	TPE-S	SAN
Polyvinylchloride (PVC)	TPE-V	PMMA
Polystyrene (PS)	TPE-E	PEEK
Polycarbonate (PC)	TPE-U	PTFE
Polyamide (PA)	TPE-A	POM

Figure 20.6 Maintenance-free window profiles in rigid PVC are very common in many countries. They are increasingly replacing wooden profiles during renovation and new construction.

Figure 20.7 As for rigid PVC, plasticized PVC is also a common material for extrusion. Garden hoses with or without textile reinforcement are common products in flexible PVC.

Chapter 20 — Extrusion

20.3 The Extruder Design

Figure 20.3 shows the principle of how an extruder is constructed. The usual way to show extruders is with the flow from right to left, although it is opposite to the direction you are reading the text.

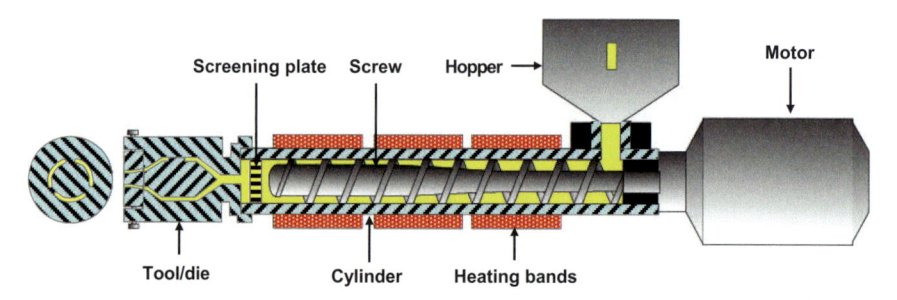

Figure 20.8 Inside the cylinder there is a screw (or twin screws) driven by a motor which can either be located behind the screw/screws or underneath. The plastic material is fed at the rear of the cylinder through a hopper or an automatic material feeder via tubes. In the picture, the plastic material is yellow. The screw rotates all the time and feeds the plastic through the cylinder while it is heated and melted. The heat is obtained by using heating bands and by the frictional heat generated between the cylinder and the screw flights. At the front of the cylinder there is a screening plate which makes the melt more homogeneous. The tool is located after the screening plate and shapes the profile.

20.3.1 The Cylinder

The cylinder is basically a steel pipe with high-gloss inside. The simplest and least expensive cylinders are single-screw extruders. Twin-screw extruders are more complicated and therefore have a significantly higher price.

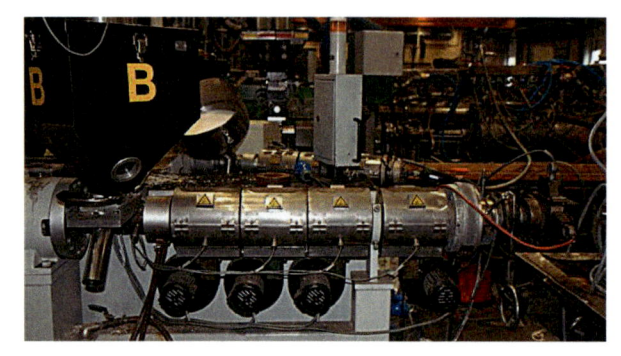

Figure 20.9 On older machines you can often see cylinder with heating bands and cooling fans, whereas they generally are encapsulated on newer machines. Each heating band has a temperature sensor and is controlled from the control panel of the extruder.

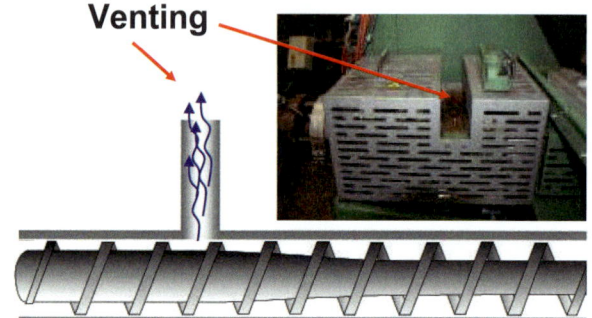

Figure 20.10 It is quite common to have a venting zone on twin-screw extruders, but they do also exist on single-screw extruders. Water vapor can be removed, which is an advantage if the plastic material has been stored in cold storage during the winter months and may have condensation on the granules.

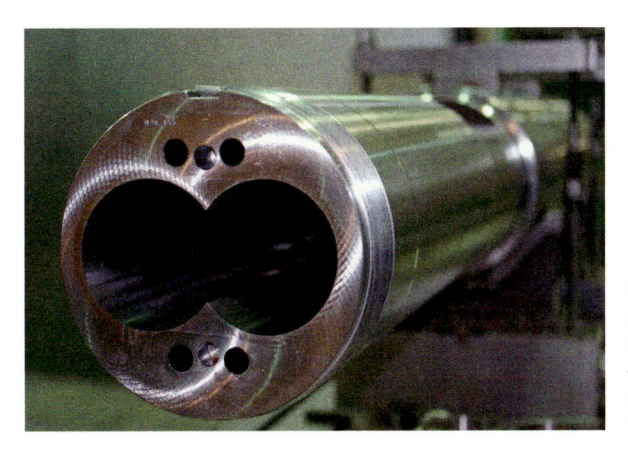

Figure 20.11 The picture shows the barrel of a twin-screw extruder. If you use glass fiber-reinforced plastics that cause high wear it is appropriate to provide the cylinder with a bi-metal lining.

20.3.2 Single-Screws

Single-screw extruders are mainly used for polyolefins, ABS, and polycarbonate, while other materials usually work better with twin-screw extruders. Figure 20.12 shows the geometry for the most common type of single-screws. When you specify the size of a single-screw you use the L:D ratio, that is, the length to diameter ratio, which can be in the range from 16:1 to 40:1. A standard screw that can handle many different materials generally has L:D ratio of 25:1.

Unlike twin-screw extruders, you can use very high speed (up to 1,000 rpm) and thereby obtain high capacity.

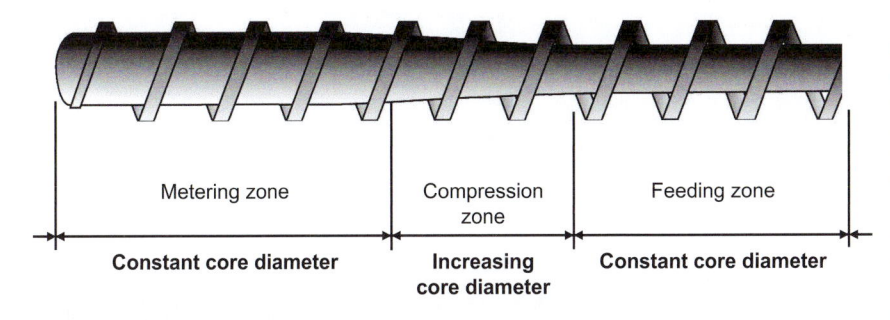

Figure 20.12 An extrusion screw does not have constant core diameter and flight height. This varies along the screw so you divide it into different zones. The first zone is called the feeding zone, and here the plastic starts to melt. In the next zone the core diameter is increasing and it is called the compression zone, where the melt gets compressed and homogenized. The last zone is called the metering or melt pumping zone.

20.3.3 Barrier Screws

Another way to increase the capacity of single-screws is to give them a special geometry with double threads. This type of screw is called a barrier screw, and it is used mainly for PE and PP.

Figure 20.13 The picture shows a barrier screw having a more efficient processing capability than a standard screw and thereby with significantly higher capacity. The zone at the front of the screw is a special mixing zone which facilitates dispersion and homogenization of the melt. [Photo: Weber]

20.3.4 Straight Twin-Screws

Straight twin-screws are mainly used for PVC in powder form. The disadvantage is that the screws and thus the extruder are relatively long. The reason is that they generate low frictional heat and therefore require a long melting zone to make the melt completely homogeneous.

Figure 20.14 The picture shows about 2/3 of the length of a pair of straight twin-screws designed to rotate towards each other (see Figure 20.16).

20.3.5 Conical Twin-Screws

Conical twin-screws are very gentle to the material. They are mainly used for PVC, but also PE and PP can be processed. They produce a homogeneous melt at half the distance compared to straight twin-screws with the same capacity.

Figure 20.15 You get very efficient processing with conical twin-screws, with both powder and granulates, as the inlet of the cylinder has a much larger area than the outlet. This provides a higher compression but usually requires a closed-loop cooling system in the screws to disperse the frictional heat. If the screws for example have a diameter of 40 mm at the outlet of the cylinder, a typical inlet diameter will be about 75 mm. [Photo: Weber]

20.3.6 Rotational Direction

Normally the screws rotate in the direction towards each other, but there are also screws with the same rotation, as illustrated in Figure 20.16.

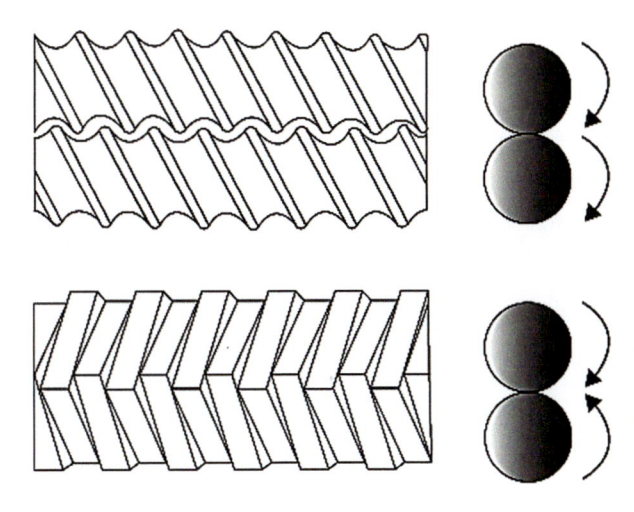

Figure 20.16 In the upper picture, the screws are turned in the same direction. They fit to each other and have the same direction of rotation. These types of screws are gentle and are used for wood-filled plastics. The lower picture shows screws that are, respectively, left and right twisted. They are interlocked and rotate towards each other. This is the most common type of screw.

20.3.7 Comparison of Single-Screws and Twin-Screws

When choosing the type of extruder to use for a particular job you make an assumption how much material usage there is per hour (output) and which resin you will use.

Table 20.1 Comparison of Different Screw Types

	Single-Screw	Straight Twin-Screws	Conical Twin-Screws
Melt process	Friction between the screw flight and the cylinder. A barrier screw with grooved feeding zone improves the processing.	Knead the melt with friction between the screws themselves and between the cylinder and the screw flights.	The same as for straight twin-screws but the reduced area along the cylinder also means higher compression and better processing.
Processing cost (kW per kg material)	Higher than for twin-screws. With the barrier screw, the cost is comparable to twin-screws.	Lower than standard single-screws.	Similar to straight twin-screws.
Extruder length	Similar length as straight twin-screws.	Similar length as single-screws.	Much shorter length.

Chapter 20 – Extrusion

20.3.8 Tool/Die

After the screening plate in the cylinder there is the die that shapes the product.

The die is normally heated by external heating bands. When processing wood-filled bioplastics, some cooling might be required.

Figure 20.17 The picture shows a die for production of 110 mm plastic tubes

Figure 20.18 The different plastic tubes in this picture can be made in the kind of die that is shown in Figure 20.17.

20.3.9 Calibration

After the die the melt goes into a calibration zone where it cools down and gets its shape.

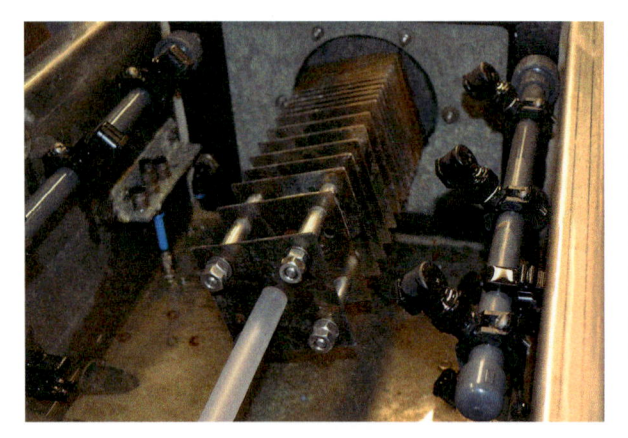

Figure 20.19 The picture shows the calibration zone in production of a tube in PP. The outer diameter is calibrated here.

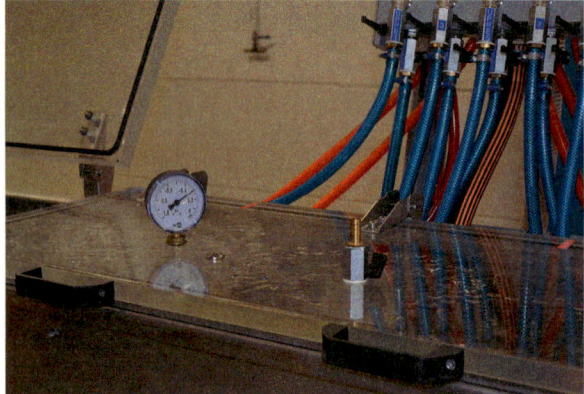

Figure 20.20 The calibration zone is usually located in a water bath in a sealed tank under pressure.

20.3.10 Corrugation

When producing corrugated hoses, e.g. vacuum cleaner hoses, the calibration zone is replaced with a corrugator, which is a chain of mold blocks, as shown in Figure 20.21.

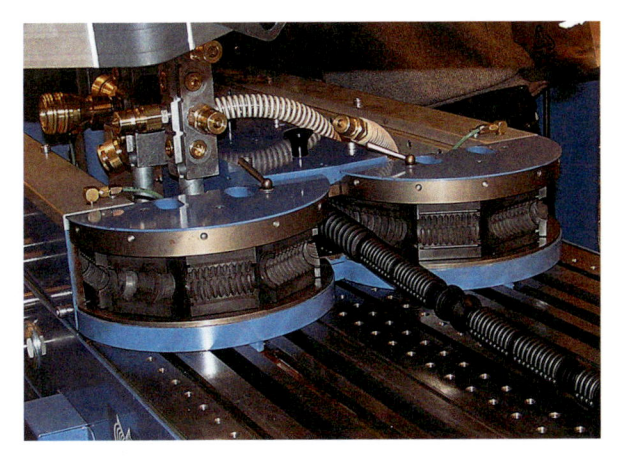

Figure 20.21 By shifting the interchangeable mold blocks you can get endless variations of the tube. The mold blocks have a vacuum that gives the shaping of the products.

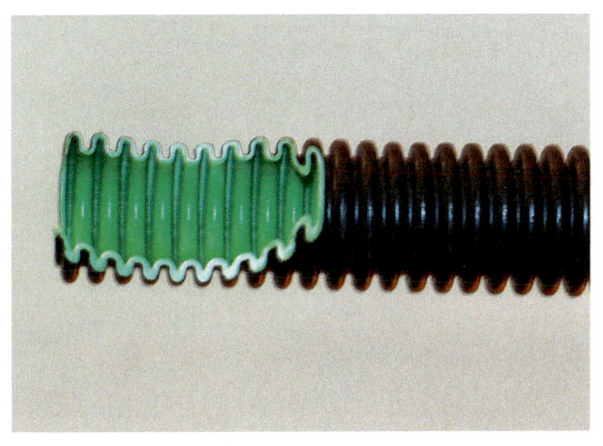

Figure 20.22 This pipe has been produced in a co-extrusion die and then been profiled in the type of corrugator shown in Figure 20.21.

20.3.11 Cooling

The step after the calibration zone is the cooling. In general a water bath, as shown in Figure 20.23, is used. When a corrugator replaces the cooling bath air cooling will be sufficient as shown in Figure 20.24.

Figure 20.23 The image shows two cooling tanks after another. The cooling distance is 10 meters. To reduce leakage where the product comes out there are profiled brass plates.

Figure 20.24 Between the corrugator and cutting station or the winding drum the profile will be air-cooled at a distance of several meters.

20.3.12 Feeding

To obtain a constant speed in the extrusion line one or more feeding stations are used. The speed is adjusted to the plastic, thickness, and tolerance of the profile plus the extruder capacity.

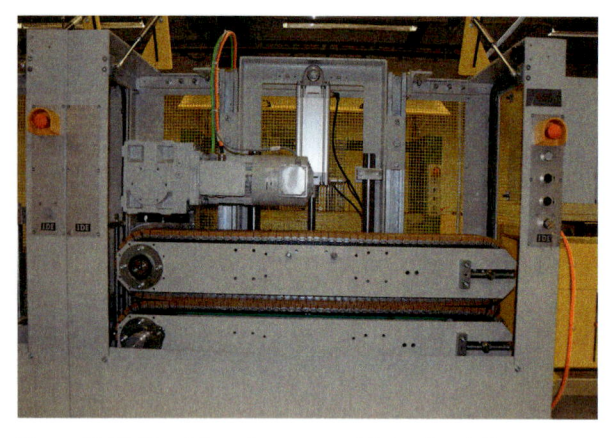

Figure 20.25 The picture shows a complete feeding station.

Figure 20.26 Here you can see a profile in the feeding station.

20.3.13 Marking

Many products are ID-marked "in-line". The marking is usually done by inkjet or laser technique, but hot stamping by foil also is possible.

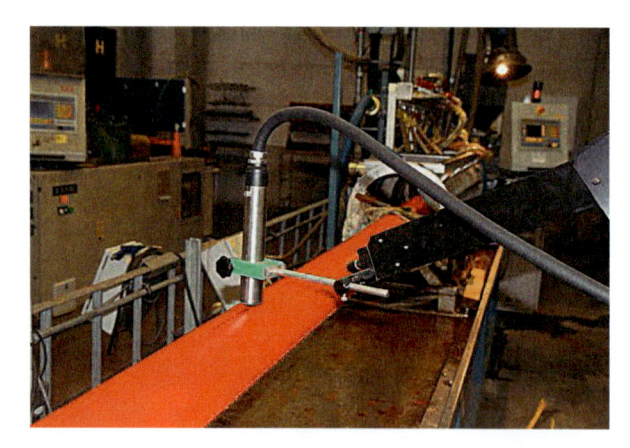

Figure 20.27 The picture shows an inkjet head in the printing process of cable ducts.

Figure 20.28 Laser marking provides a more permanent marking but requires more expensive equipment.

20.3.14 Further Processing

Some products require a special type of processing in the extrusion line. Examples of this are milling, punching, and drilling. These processes are normally automated.

20.3 The Extruder Design

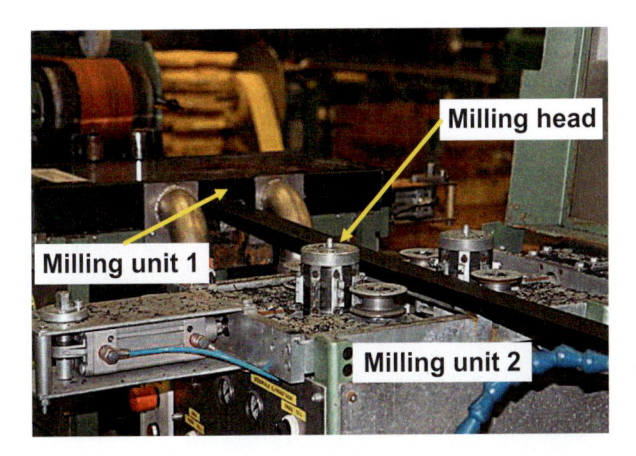

Figure 20.29 The picture shows precision milling of a profile that requires extreme tolerances, which in this case means ±0.05 mm. The black profile goes through the milling unit 1, which is in use with the protective cover closed. The two metal bends are exploiters. In the foreground you can see the milling unit 2, which is used for a different profile. Milling unit 2 has the protective cover open so you can see the two milling heads with cutting blades.

20.3.15 Cutting

Cutting is the last processing step. This is done either with circular saw blade or other cutting equipment. The shape of the product determines the choice of cutting equipment. A typical cut is 90° but other angles also exist.

Figure 20.30 Here you can see 90° cutting by a circular saw blade.

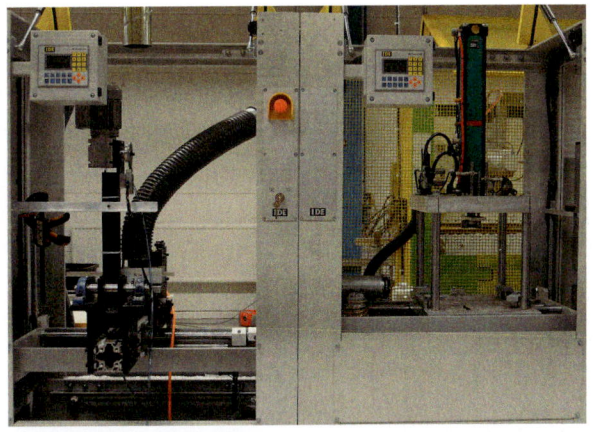

Figure 20.31 The picture shows a processing unit with milling to the right and cutting to the left.

20.3.16 Winding

Many products that require long lengths can be rolled on drums by logistical reasons. These can be up to 3 meters in diameter in order to be transported in the most cost-effective way.

Figure 20.32 Here you can see winding to a drum. The equipment in the foreground is a punching unit that punches holes at regular intervals.

Figure 20.33 Here you can see a winder with a drum having a diameter of 3 meters, which is an appropriate size for road transport.

Figure 20.34 Here you can see a profile that is delivered on a drum. It is used as a filling profile in the large sea cables and welded together by mirror welding so that it can be several miles long (see Figure 20.45).

20.4 Extrusion Processes

Depending on the type of tool that is used after the extruder you can divide the extrusion process in different categories:

- Straight extrusion
- Extrusion with angled tool (i.e. coating)
- Extrusion of plates and sheets
- Co-extrusion
- Film blowing
- Cable production
- Monofilament production
- Compounding

20.4.1 Straight Extrusion

Most extruded products are made in straight tools regardless if the outer dimension is from a few tenths of a millimeter or up to 3 meters.

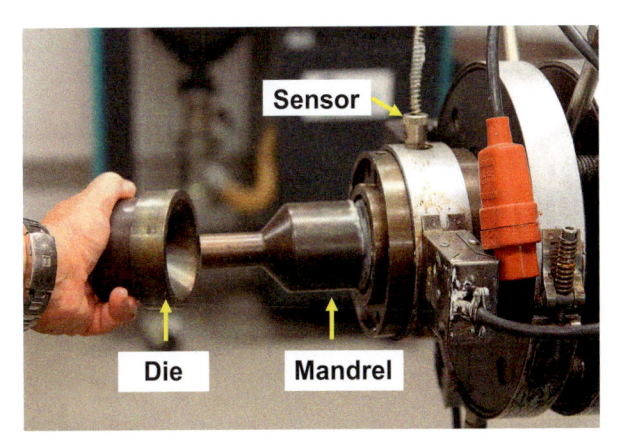

Figure 20.35 The picture shows a typical pipe tool. A mandrel that forms the inner diameter can be seen to the right and the die which shapes the outer diameter can be seen to the left. You can also see the heating bands and the sensor on the top that measures the temperature.

Figure 20.36 The pipes in this image have been made in the die you can see in Figure 20.35.

20.4.2 Extrusion with Angle Tool/Coating

When coating fabric, paper or metal sheets you normally use an angle tool as shown in the coating line depicted in Figure 20.37.

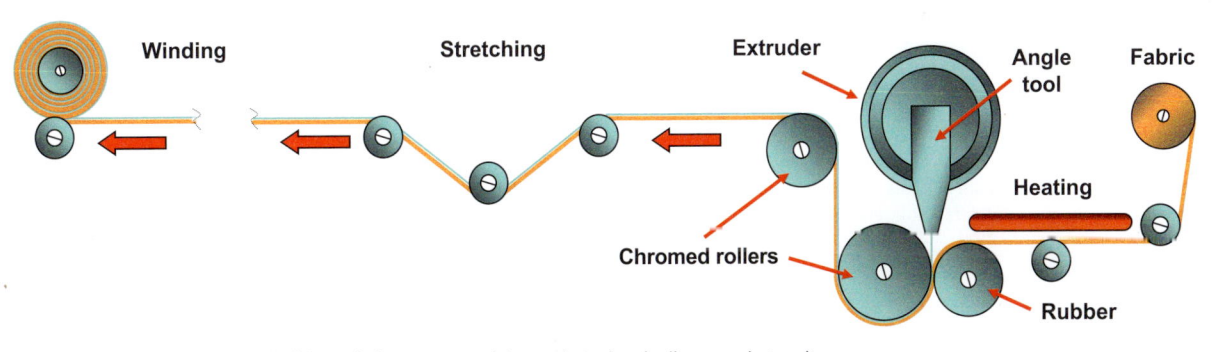

Figure 20.37 The molten plastic (blue color) comes out of the angle tool and adheres as hot melt glue to the preheated substrate that may be metal, paper, or fabric (yellow) as in this figure. The rubber roller presses the plastic melt towards the fabric and is then first cooled towards the chromed rollers and then air cooled before the coated fabric is rolled up. The thick red arrows indicate the flow direction.

20.4.3 Extrusion of Plates and Sheets

In the production of flat film or plates, a slit tool is used which can have a width of over 2 meters. Normally amorphous thermoplastics such as ABS, PC, PC/ABS, PETG, PMMA, PVC, and SAN are employed, but the semi-crystalline thermoplastics PE, PP, and TPE-O can also be used.

The thermoplastic resin is extruded between two rollers and formed into a plate or thick foil between these and a third roller. The rollers have cooling and the thickness depends on the slit thickness and the distance between the rollers. After the plate has solidified a little, a protective foil is added and the plate will then be cut after 7 to 8 meters in the correct length by the use of a circular saw blade.

If you want a patterned or textile surface finish instead of a smooth one (as in car interiors), you have to replace one of the smooth rollers to one that has an etched surface.

Plates of amorphous materials can then be further processed by thermoforming. Thick, soft PVC film can be cut into strips and used for shielding against wind, cold, or heat at industrial doors that require passage of trucks, etc.

Figure 20.38 Here you can see a 2,050 mm wide slit tool with a number of thermal sensors. It is important that the temperature is uniform over the entire width and this is especially important when co-extruding the plates of different layers. [Source: Arla Plast AB]

Figure 20.39 The image shows a number of 10 mm thick PC sheets in the standard dimension 1,250 × 2,050 mm available for delivery. Arla Plast in Sweden produce plates in different materials and lengths in thicknesses from 0.75 to 20 mm. [Source: Arla Plast AB]

20.4.4 Co-extrusion

Co-extrusion is used when you have multiple layers of different colors or materials in your product. In order to achieve this, you need more than one extruder. In order to obtain good adhesion between the different materials they must be compatible with each other.

In some case you can improve both the environment impact and the economy by using recycled materials in any of the layers.

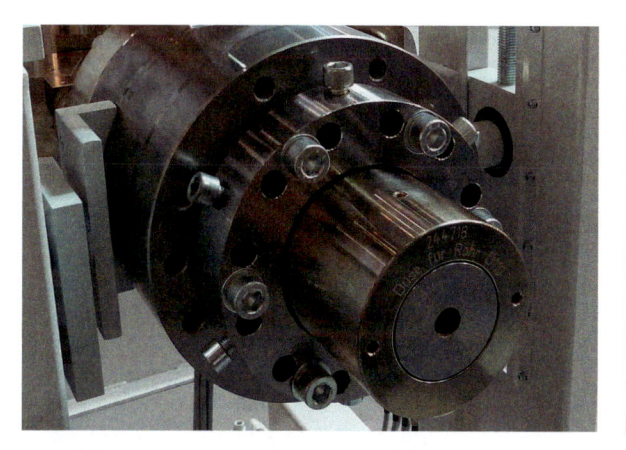

Figure 20.40 In the depicted pipe die, three layers are made by two extruders. The second extruder is located above the main extruder with a vertical inlet seen in the background.

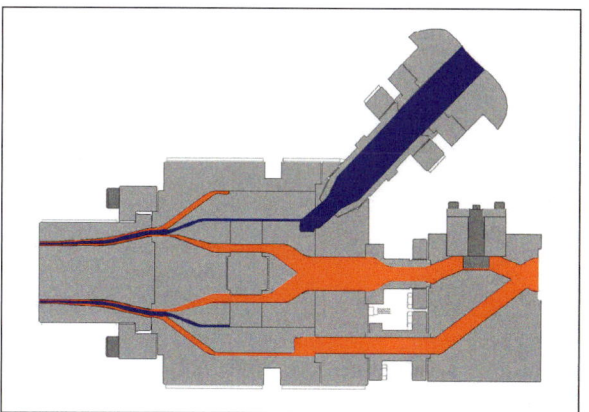

Figure 20.41 The principle of the three-layer die shown in Figure 20.40 is shown here. In the red outer and inner layers there is a virgin resin and in the blue middle layer recycled regrind can be used. [Image: Weber]

20.4.5 Film Blowing

Figure 20.42 depicts a film blowing line. After the extruder a tubular head which extrudes a thin film tube is placed. The film is then blown, cooled, and squeezed with the help of rollers. At the final station the film can then be chosen to be printed before it is rolled on bobbins, cut into sheets or welded and punched out into plastic bags. In order to prevent the melted foil hose to stick on the blow head the film is kept at a distance with the help of an air gap. You can change the thickness of the film by varying the gap of the blowing head. The width of the film can be varied by the air bubble size (air pressure) and the orientation (biaxial) by the speed of the rollers. This offers a very short development time at low cost.

The plastics commonly used in film blowing are: PE, PP, PET, PA, EVA, EBA, and EMA.

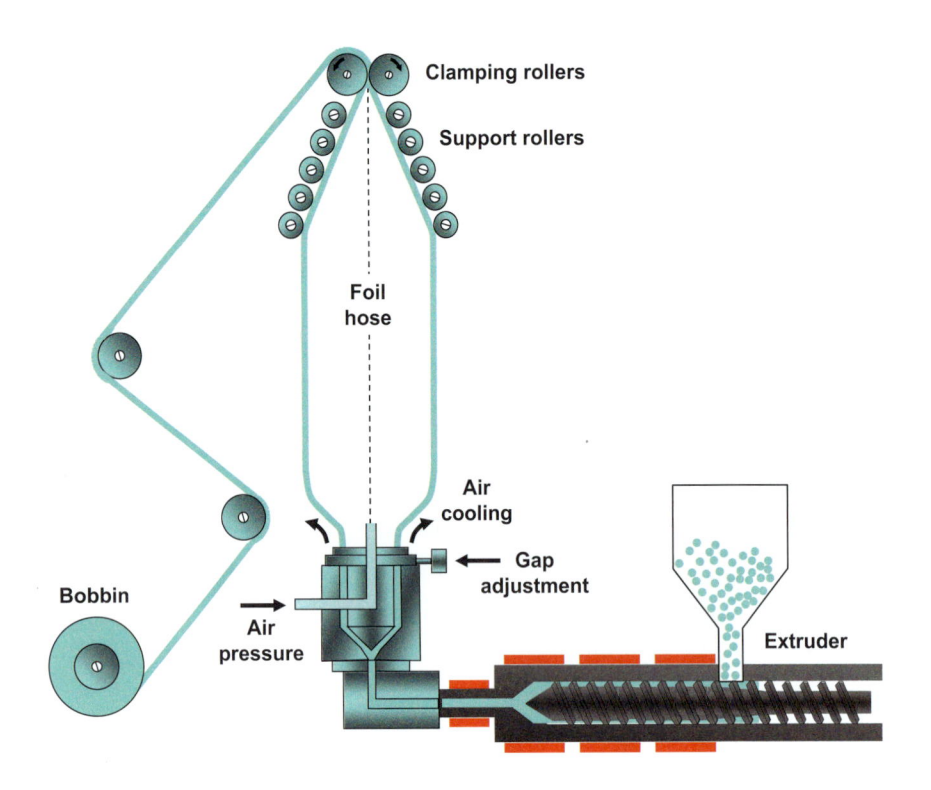

Figure 20.42 The figure shows the principle of film blowing.

20.4.5.1 Advantages (+) and Limitations (−)

+ Fully automatic mass production of low-cost products (plastic bags and household film)
+ Very thin-walled products can be made
+ Many materials are approved for food contact

− Restricted choice of materials (only high viscosity)
− High investment costs and space requirements in height for blown film lines

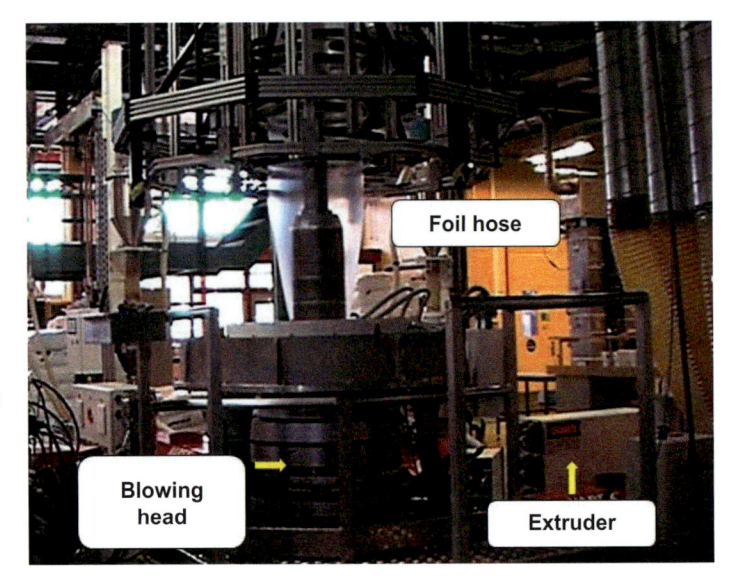

Figure 20.43 The picture shows a seven-layer film blowing head available on DuPont's technical center in Geneva. The extruders stand in a ring around the blowing head which is about 75 centimeters in diameter. The film, which is blown vertically, can be seen in the middle of the tool.

20.4.6 Cable Production

Cable production is an advanced manufacturing process in several steps. The wire may be either metal or optical fiber. Copper is the dominant electrical wire, but aluminum is also used. The electrical wire is supplied as a single thread in a large diameter on a drum. Then it is drawn to the right dimension (diameter) and annealed to improve strength and conductivity before it is twisted. In the next step an insulating housing, which may be of rubber or of various thermoplastic resins, is extruded around the wire. The material selection is dependent on the requirements of electrical insulation, heat resistance, and flame protection.

The thermoplastic resins commonly used are:

- Plasticized PVC for low voltage and installation cables
- Polyethylene (LDPE, LLDPE, or HDPE) for insulation and jacketing of high voltage cables
- PEX for high voltage cables with extra high voltage transmission
- Different TPEs, e.g. TPE-E for rodent resistance and TPE-V for low voltage outdoor use
- Polyamide 11 and 12 for chemically resistant and termite resistant cables
- FEP and PVDF in cables with very high temperature and flammability classification

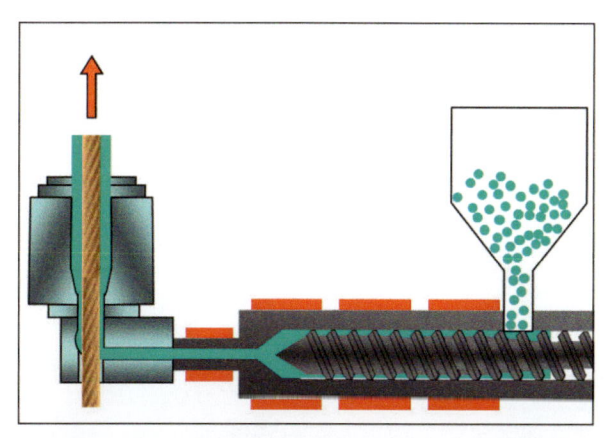

Figure 20.44 Large cables are made in angle tools.

Figure 20.45 The picture shows a sea cable with three twisted copper wires with a diameter about 30 mm, which is isolated in several layers. In order to stabilize the cable, extruded profiles have been added. The cavities (see red arrow) can be used as protection for the optical fibers or tubes for transport of liquids.

Figure 20.46 The picture shows ABB's cable factory in Karlskrona, Sweden. Below the high cooling tower the extruders with vertical angle tools are located. After the cables have been air cooled in the tower they will be wound on drums or directly delivered on the kind of specialized vessel that is seen in front of the factory. Some of these ships have the capacity to transport up to thousands of kilometers of cable without the cable being spliced.

20.4.7 Monofilament

The extrusion of fibers and thin threads is called monofilament production. Normally this is made in a tool with a large number of holes. After the tool, the monofilaments will be stretched during cooling before they are wound onto bobbins and then cut for use in different products.

Figure 20.47 The bristles in a toothbrush are normally made of polyamide 612 monofilament fixed into a shaft of SAN.

Figure 20.48 Fishing line is another example of monofilament made of polyamide.

20.4.8 Compounding

In the production of plastic granules, various additives such as release agents, heat stabilizers, UV additives, and pigments are added to the plastic material in powder form before it runs through an extruder. This process is called compounding (see more in Chapter 8).

In one of the two compounding methods monofilaments are made in nozzles with a large number of holes (see Figure 20.49, left). A rotating knife cuts the monofilament direct after the nozzle and granules with a lens-shaped geometry will then be created and air-cooled. This method is called "melt-cut".

In the second method a number of round strands are extruded and cooled in a water bath (see Figure 20.49, right). After the strands have solidified they will be cut in a cutter. The shape of the granules will then be cylindrical. This method is called "strand-cut".

Figure 20.49 To the left we can see the nozzle in production of lens-shaped granules, flying around in the air after the rotating knife has cut them when they come out. To the right there are five strands coming out of the second type of nozzle. The granules are cooled in a water bath before being cut.

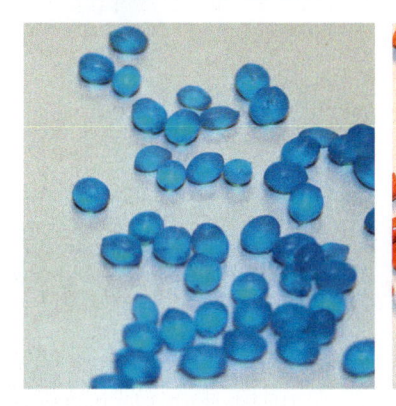

Figure 20.50 To the left we can see lens-shaped granules. This type is usually used for material with low processing temperature such as PE and PP. To the right there are granules with a cylindrical shape. This is the predominant form for glass fiber reinforced plastics.

20.5 Design for Extrusion

In the design of a profile you will have endless possibilities to vary the geometry, surface appearance, color, material, and function. The more functions you can integrate in the same profile, the better the profitability will be. To reduce the complexity of the profile if it is made in one piece, you can also use different joining methods.

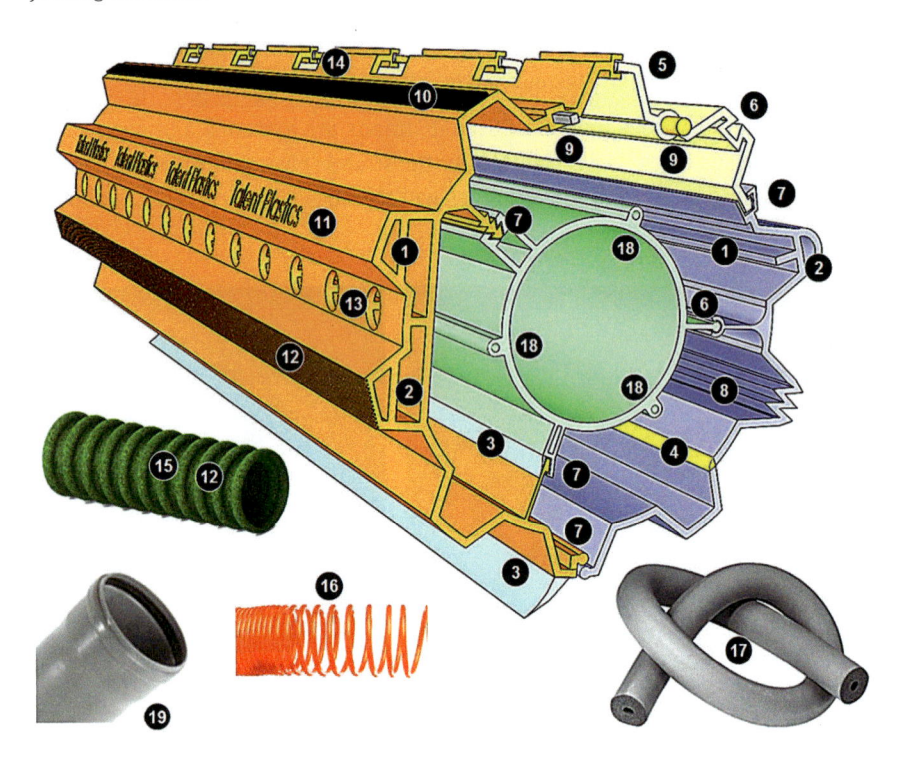

Category 1 – Economic optimization
(1) Ribbing – stiffening
(2) Cavity

Category 2 – Fixation
(3) Sealing lip
(4) Hinge
(5) Guide
(6) Sliding joint
(7) Snap-fit joint

Category 3 – Flexibility/reinforcement
(8) Bellow
(9) Insert/reinforcement

Category 4 – Surface functions
(10) Friction tape
(11) Printing/stamping
(12) Decoration surface (brushed/flocked)

Category 5 – Others
(13) Round side hole (milling/drilling)
(14) Irregular hole (punching)
(15) Corrugation
(16) Spiral forming
(17) Foaming
(18) Extruded hole for self-tapping screw
(19) Muffing
(20) Hot plate welding

Figure 20.51 The profile shown here does not exist in reality. It is the result of a workshop by students at KTH, Stockholm in Sweden performed to illustrate some of the possibilities that designers should keep in mind when they develop extruded profiles.

20.5 Design for Extrusion

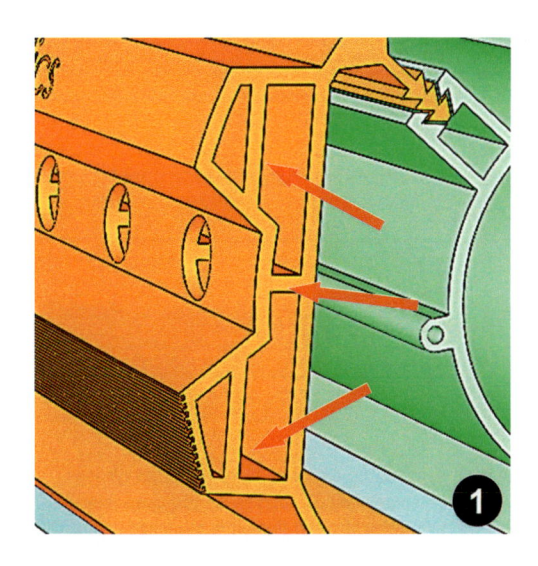

20.5.1 Ribbing – Stiffening

Figure 20.52 Ribs and the type of framework shown in the picture improve both torsional and bending stiffness. Just as in the injection molding process you must dimension the thickness of the ribs in proportion to the wall thickness to eliminate the risk of sink marks. In the design of the framework you must take into account the difficulty of cooling the internal ribs.

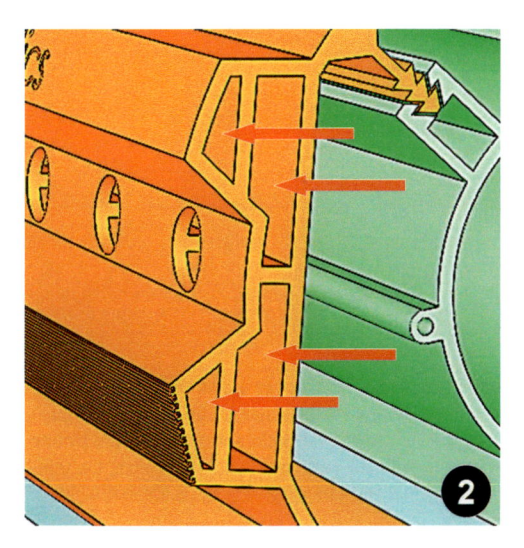

20.5.2 Cavity

Figure 20.53 By coring out the thick sections of a profile and make it hollow, you can maintain the stiffness while making weight savings. Usually it will also lower the total cost and reduce the risk that the own weight of the profile will contribute to too high deflection. Sometimes you want the cavities for transport of liquids or air or using them as protection of cables.

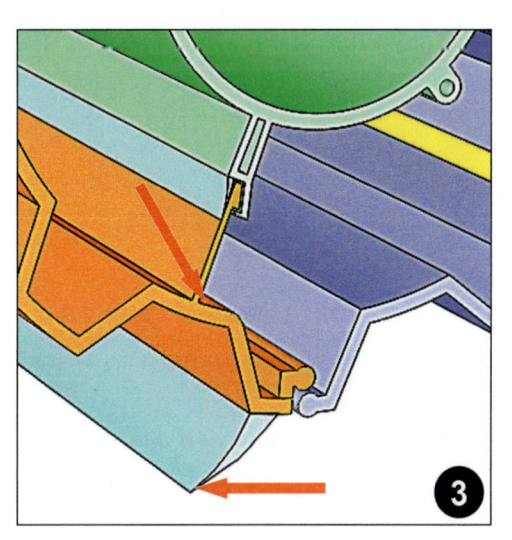

20.5.3 Sealing Lip

Figure 20.54 By connecting several extruders to the same tool you can simultaneously co-extrude different materials or the same material in different colors. The picture shows a soft (light blue) sealing lip that has been co-extruded on a stiff (yellow) profile. It is important to have mechanical adhesion if the chemical adhesion is not sufficient.

Chapter 20 – Extrusion

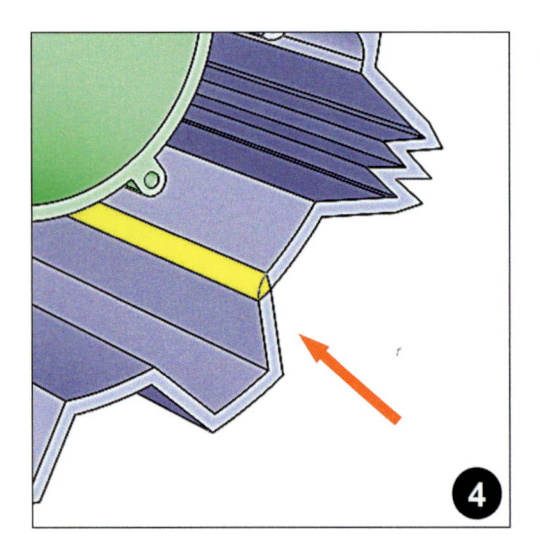

20.5.4 Hinge

Figure 20.55 Hinges can be provided by co-extruding flexible plastics (yellow) between two stiff walls (blue). With some plastics, such as PP, you can achieve hinges without co-extrusion just by making the wall thinner in the hinge itself.

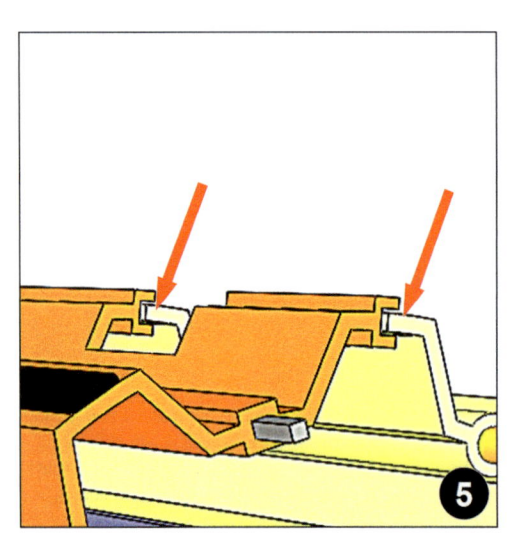

20.5.5 Guide

Figure 20.56 When it comes to complex or very large profiles it can sometimes be appropriate to design them into several sections that will be fit together. You will then also have the option to combine different colors or materials. The picture shows a type of guide that is used to fix the sections to each other along the side. For permanently joining them together you can use glue in the guide slot, which can be applied or sprayed in-line.

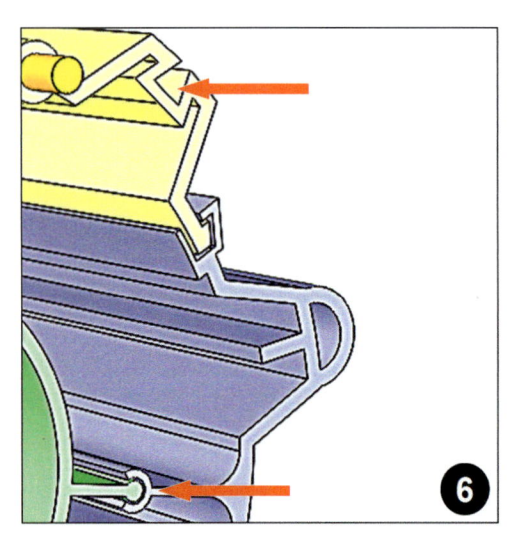

20.5.6 Sliding Joint

Figure 20.57 Here you can see a profile that consists of several sections. If these have to be pushed into each other or must be able to move relative to each other in the longitudinal direction you can use different slots with or without snap-fit function. If you would like to have a low friction between the sections you need to select a material with a low coefficient of friction in one of the profiles.
NOTE! You should avoid the same material in both sections even if the material has a low coefficient of friction because the wear generally increases.

20.5 Design for Extrusion

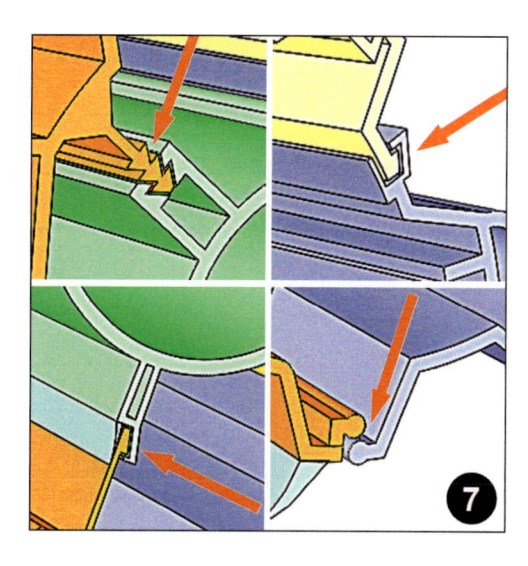

20.5.7 Snap-Fit Joint

Figure 20.58 The picture shows four different types of snap-fit joints, where the top two ones are for permanent assembly while the bottom two ones can be disassembled. In the bottom left corner you can see a snap-fit with a sealing lip (light blue).

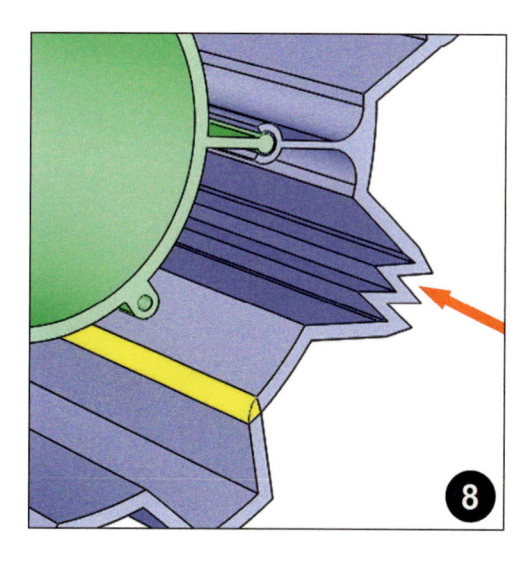

20.5.8 Bellow

Figure 20.59 If you need a profile with high stiffness but at the same time need good flexibility, the latter requirement can be improved by designing the profile with a bellow. Bending stress and spring back is otherwise a limitation if the yield strength of the material risks being exceeded.

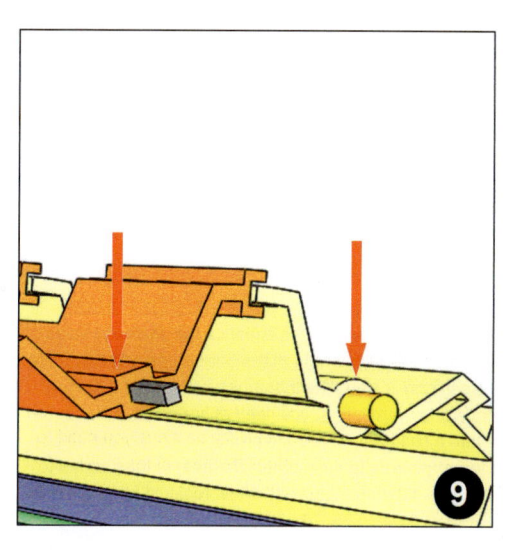

20.5.9 Insert/Reinforcement

Figure 20.60 Sometimes you need to reinforce your profile, if it is made of a soft plastic, to reduce the suspension in the longitudinal direction. You can encapsulate a steel or aluminum wire, carbon, glass, or aramid fiber or polyester yarn. If you want to integrate an optical or electrical cable, this can also be done by encapsulation.

Chapter 20 — Extrusion

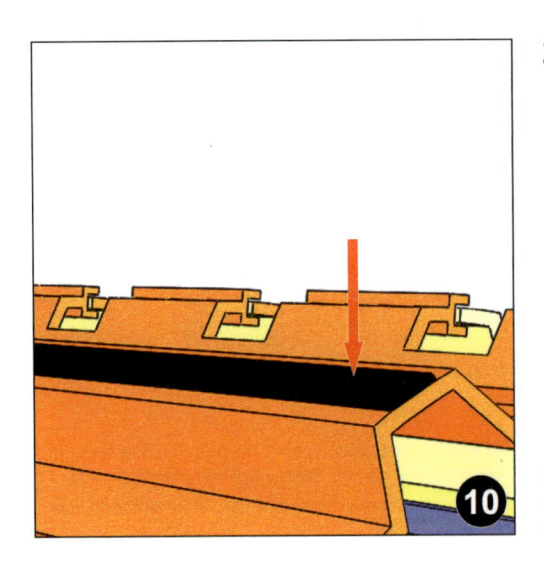

20.5.10 Friction Surface

Figure 20.61 There are several methods to get surfaces with differing friction compared to the base material. Sometimes you can add a friction tape "in-line" or co-extrude a plastic that has lower or higher friction.

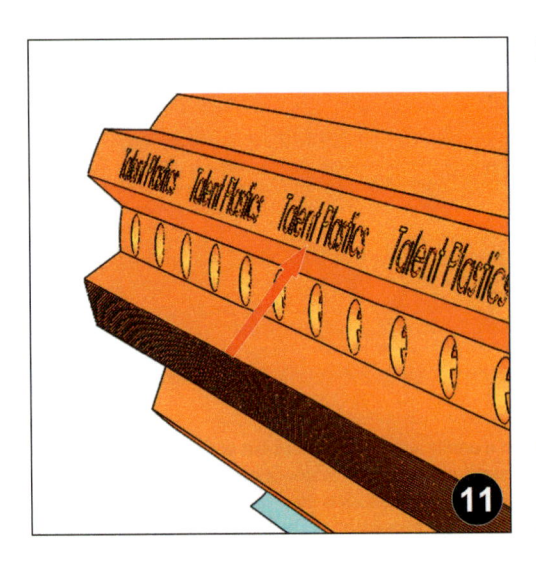

20.5.11 Printing/Stamping

Figure 20.62 If you want text, logos, or symbols on your profile, this can be obtained "in-line" by hot stamping with a special tape or inkjet/laser marking. You can also print afterwards with the above methods or use tampon or screen printing (see Chapter 13).

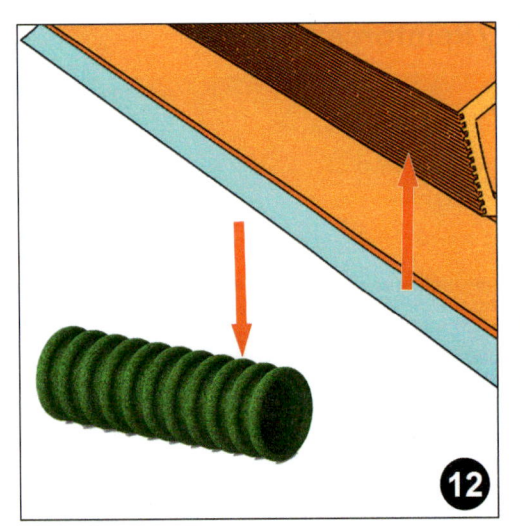

20.5.12 Decoration Surface

Figure 20.63 Designed decoration surfaces often enhance the aesthetic value of the product. But it is important to select the right material as some soft materials can cause problems. The green corrugated pipe in the picture is flocked "in-line" but flocking can also be done afterwards. First you need to spray the surface with glue. Then, put the profile into a chamber, where the profile and fibers are electrostatically charged with different polarity to each other. The fiber is then added to the surface and fixed by the glue. When the glue has dried, residual fibers will be blown off and reused.

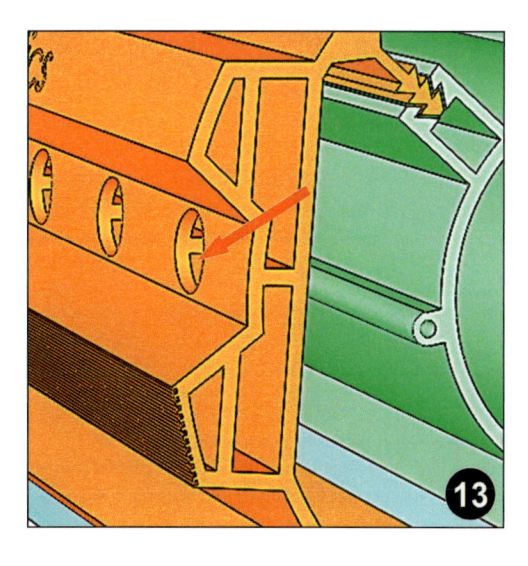

20.5.13 Drilled Side Holes

Figure 20.64 If you want holes that do not go completely through the wall, drilling or milling is normally done automatically "in-line".

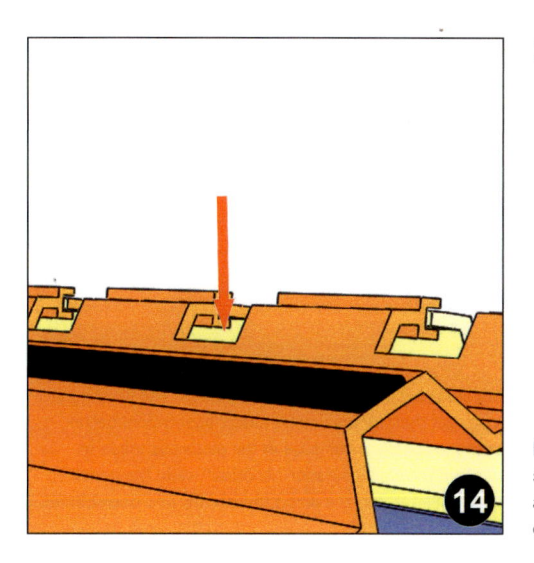

20.5.14 Irregular Holes

Figure 20.65 Irregular holes can be made both by stamping and milling. You can stamp even without an anvil. If you only want to stamp through one wall you must accept getting scrap inside the profile. To avoid this, drilling or milling is the option.

20.5.15 Corrugation

Figure 20.66 Both stiff and soft plastics can be corrugated. Most often, weight savings and improved flexibility will be the reasons to corrugate. Corrugated hoses are widely used in the electrical and automotive industries to facilitate assembly. They can replace textile reinforced rubber hoses and provide improved durability.

Chapter 20 – Extrusion

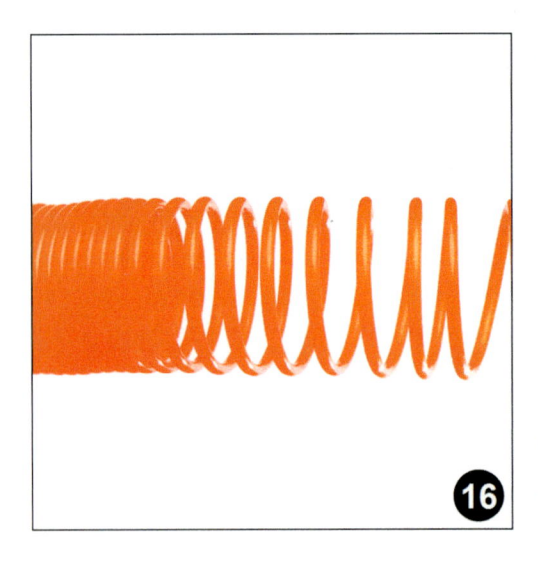

20.5.16 Spiral Forming

Figure 20.67 Spiral forming of tubes or cables is as a rule not performed "in-line". It will be achieved in a later step. You preheat the tube or cable in a water bath after which it is wound on a conical rotating mandrel, which is open in one end.

20.5.17 Foaming

Figure 20.68 Most thermoplastics can be more or less foamed. The foaming will either be to save weight, or to get thermal insulation, shock absorption, better grip, or a combination of these properties. The tube in the picture is made of foamed polyethylene and is used for pipe insulation. Another use in the construction industry is foamed PVC profiles to replace wood for window and door frames and skirting boards.

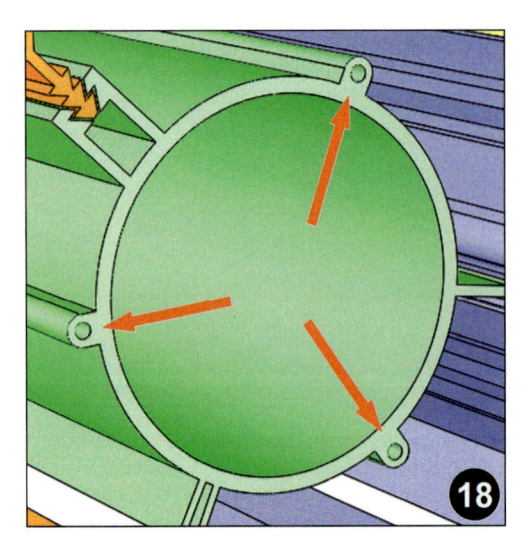

20.5.18 Extruded Screw Holes

Figure 20.69 If you want to put a lid or any other type of end cap after the profile has been cut you can incorporate mounting holes directly into the profile wall. The assembly can then be made by self-tapping screws, pins, press-fits, or by threaded metal inserts for use with machine screws.

20.5.19 Muffing and Hot Plate Welding

If you need to lengthen your pipes, you can use muffing for the assembly (see image 19 in Figure 20.51).

For other types of profiles, you can lengthen them with help of hot plate welding (see Chapter 25).

CHAPTER 21
Alternative Processing Methods for Thermoplastics

21.1 Blow Molding

Blow molding is a fully automated process that produces hollow products from thermoplastic. There are two main variants. The first involves extrusion of a hollow tube, known as a "parison", into a cavity between two mold halves (see Figure 21.1). In the second, an injection-molded "preform" is heated and blown into the cavity (see Figure 21.2). Most PET soft drink bottles are produced in this way.

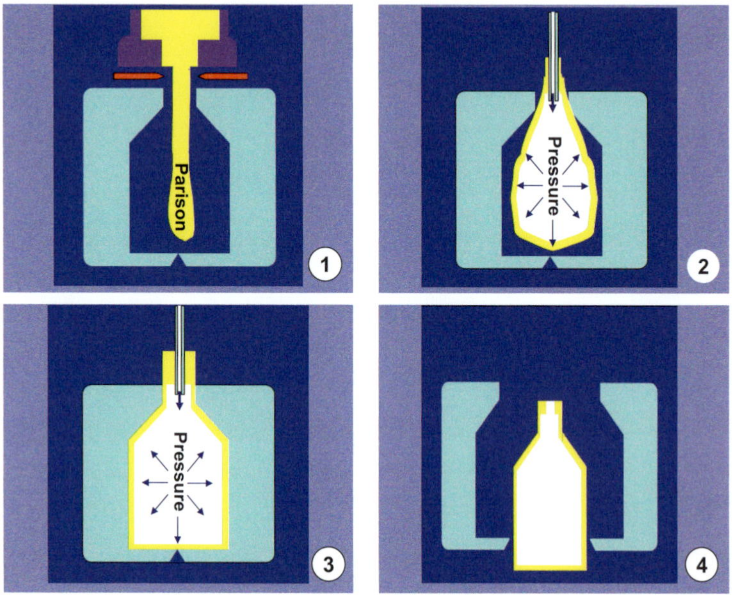

Figure 21.1 Blow molding with parison
Picture ❶ shows the extruded tube coming through the extruder head into the cavity.
In picture ❷ the mold has been moved to the next station, where the tube is blown out toward the walls of the cavity using compressed air.
In picture ❸ the tube has been completely pressed out against the cavity walls and allowed to cool.
Picture ❹ shows the finished part being ejected from the mold.

As a rule, only special grades with relatively high viscosity are used for blow molding, e.g. PE, PP, PVC, PET, PA, and some thermoplastic elastomers.

Multiple extruders can be used for different layers of the parison, for example to improve the barrier properties of a product. The hose can also be extruded in sequence, e.g. soft segments alternating with rigid ones to produce rigid tubes with integrated soft bellows.

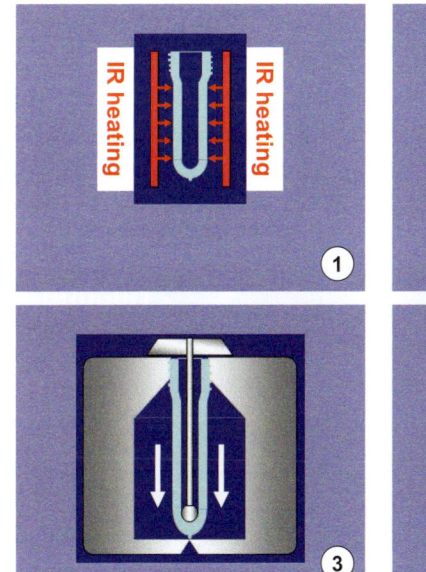

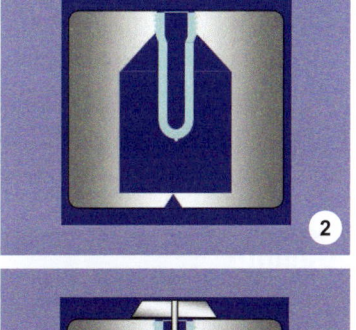

Figure 21.2 Blow molding with preform
Picture ❶ shows the preform being heated by infrared elements.
In picture ❷ the preform has been placed into the mold, which is closed.
In picture ❸ the preform is stretched toward the bottom of the mold cavity using a rod with a ball.
In picture ❹ the preform is blown out against the cavity walls using compressed air, after which the walls are cooled so that the product can be ejected.

Advantages of Blow Molding

+ Large products can be produced
+ Products can have complex shapes
+ Thin-walled goods can be produced
+ Material combinations can be made

Limitations of Blow Molding

− Not suitable for all plastics
− High machine and equipment costs require large batch production
− Surface finish is relatively poor
− Difficult to keep tight tolerances

Figure 21.3 The tank shown here holds about 1 m^3 and is an example of blowing molding with parison.

Figure 21.4 This PET bottle is made from the preform to the left in the picture.

21.2 Rotational Molding

Rotational molding is the least known of all plastic processing methods. It is estimated that there are only about 1,500 rotational molding companies in the world.

The main product is toys (about 40%), followed by tanks and containers (about 30%). The method is partly manual.

Figure 21.5 Picture ❶ shows the open mold, which is fixed on a vertical shaft, which in turn is secured on a horizontal shaft. The plastic material in powder form (shown in orange) is filled into the mold.
In picture ❷ the mold has been closed and starts to be rotated in three directions so that the powder is distributed over the inner surface of the mold, which is heated in an oven.
In picture ❸ the material is molten and evenly spread across the mold, which is then cooled down by fans or compressed air while continuing to rotate.
In picture ❹ the material has solidified, the mold is opened, and the finished product can be lifted out.

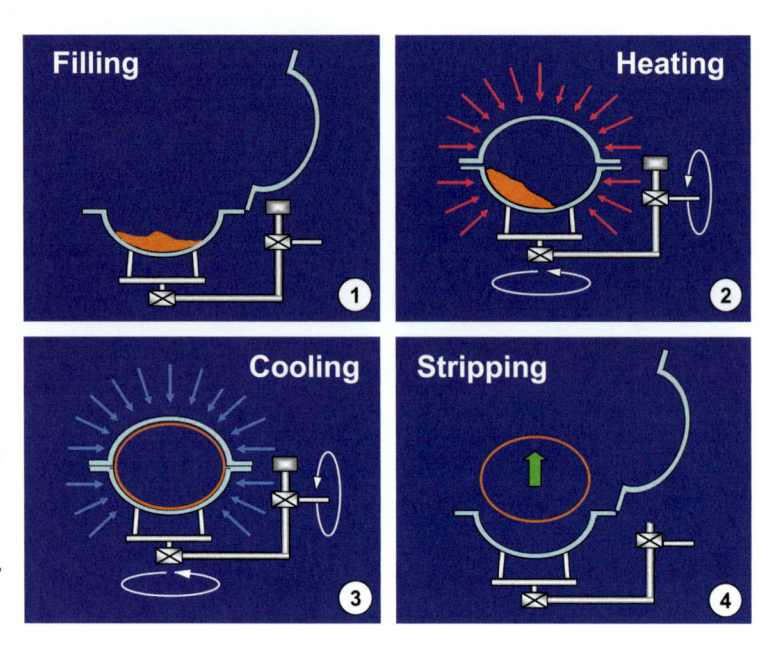

The most common materials used in this process are different types of polyethylene where LLDPE (about 60%) is the most used followed by HDPE (about 10%). The second most common material is PVC (about 15%). Other materials that can be used are PP, EVA, and PA12.

Advantages of Rotational Molding

+ Short development times
+ Low equipment costs
+ Very large products can be made (up to 20 m^3)
+ Profitable even for small runs (> 100 items)
+ Several components (with the same wall thickness) can be produced simultaneously in the same mold rig
+ Components become stress free

Limitations of Rotational Molding

− Long cycle time (30–60 min)
− Limited choice of materials

Figure 21.6 Beacons and buoys are made in PE by rotational molding. An alternative method to the one described above is known as "rock and roll."
This allows the whole oven to rock up and down while a rig with different molds rotates inside it. When the material has melted, the rig is taken out, still being constantly rotated horizontally, and the molds are placed in a separate cold room to cool, all the while maintaining the same pattern of movement.

21.3 Vacuum Forming

One variation of this method is called thermoforming. This method is used almost exclusively for amorphous plastics and is very common in the packaging industry, where it is used to make blister packaging for everything from pills to sophisticated electronic items. Figure 21.7 illustrates the method.

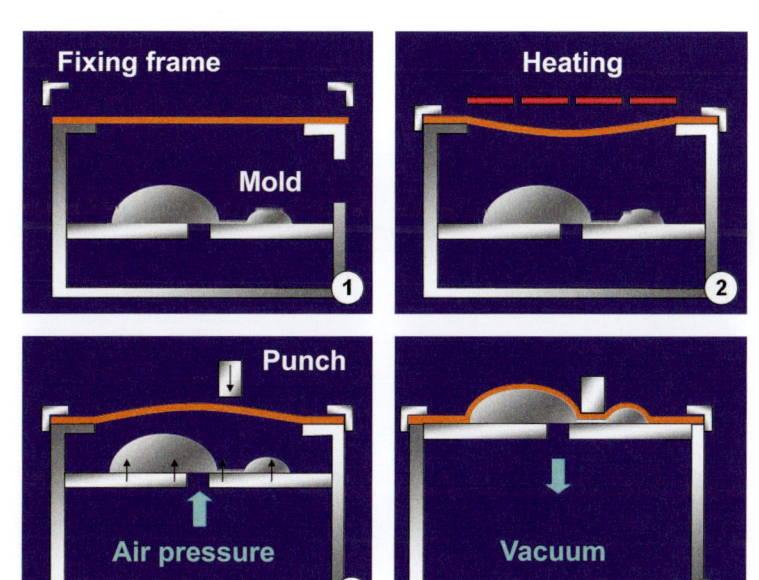

Figure 21.7 Picture ❶ shows the plastic sheet (in orange), which is placed on top of the machine above a hemispherical mold.
In picture ❷ the plastic sheet has been fixed to the machine by a frame so that it fits tightly. The sheet is now heated up by an IR element.
In picture ❸ the mold starts to move upward while creating an overpressure from below with compressed air. A punch is simultaneously applied from above.
In picture ❹ the mold is pressed up against the plastic sheet, which is simultaneously sucked down by the vacuum. The punch helps to shape the disc from above. The plastic is then cooled down with the help of fans.
The component is then taken to the next station where the hemispheric product is stamped or milled out to separate it from the remainder of the plastic sheet. This can result in a lot of scrap, but this material can be recycled and used to produce new sheets.

Chapter 21 — Alternative Processing Methods for Thermoplastics

The most common materials used for vacuum molding are PC, ABS, SAN, PS, PET, PVC, and PMMA.

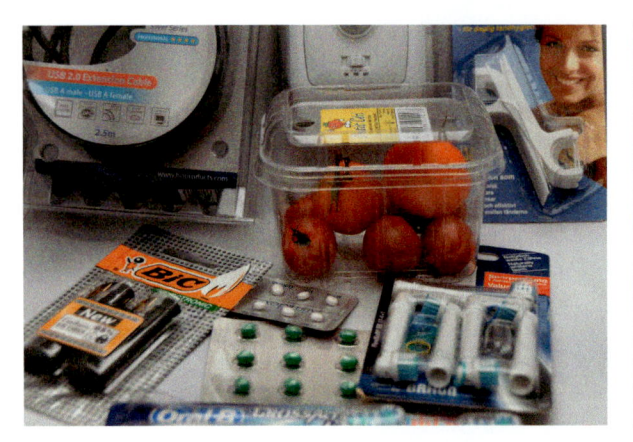

Figure 21.8 Blister packaging for a variety of consumer products. It is very common for foods to be sold in vacuum-formed packaging in transparent PET, such as these cherry tomatoes.

Figure 21.9 Antenna dome for luxury yachts manufactured in small batches. [Photo: Sematron AB]

Advantages of Vacuum Forming

+ Short development times
+ Low equipment costs (wooden molds can be used)
+ Very thin-walled products can be produced
+ Profitable even for small runs (> 10 items)

Limitations of Vacuum Forming

− Relatively long cycle time (several minutes)
− Limited choice of materials

CHAPTER 22
Material Selection Process

A major task of designers and project engineers is selecting the right material for their applications. When the materials under consideration include plastics, this task is particularly difficult, since there are hundreds of different polymers and thousands of different plastic qualities to choose from. Finding the right material requires knowledge, experience (your own or access to that of others), and sometimes a little luck.

If you choose a material that is considered "somewhat too good", this will usually be reflected in the cost being seen as "somewhat too high", which can have an effect on your competitiveness. However, on the other hand, if you choose a material of a "borderline" quality, you run the risk of complaints and a bad reputation in the marketplace, which also affects your competitiveness.

22.1 How Do You Select the Right Material in Your Development Project?

Let's take a new iron as an example. Before the designer starts thinking about what materials to use in the iron, he must be clear about the following:

- What should the new iron look like?
- What different functions should it have?
- What will it cost?

Figure 22.1 So... what should this new iron look like?

When answering the above questions, the designer should make a list of the requirement specifications, as detailed as possible.

22.2 Development Cooperation

A good way to reduce development time when it comes to plastic components is to utilize the collective expertise and experience of a project team consisting of its own development department working with a potential material supplier and potential manufacturers (e.g. molder and mold maker).

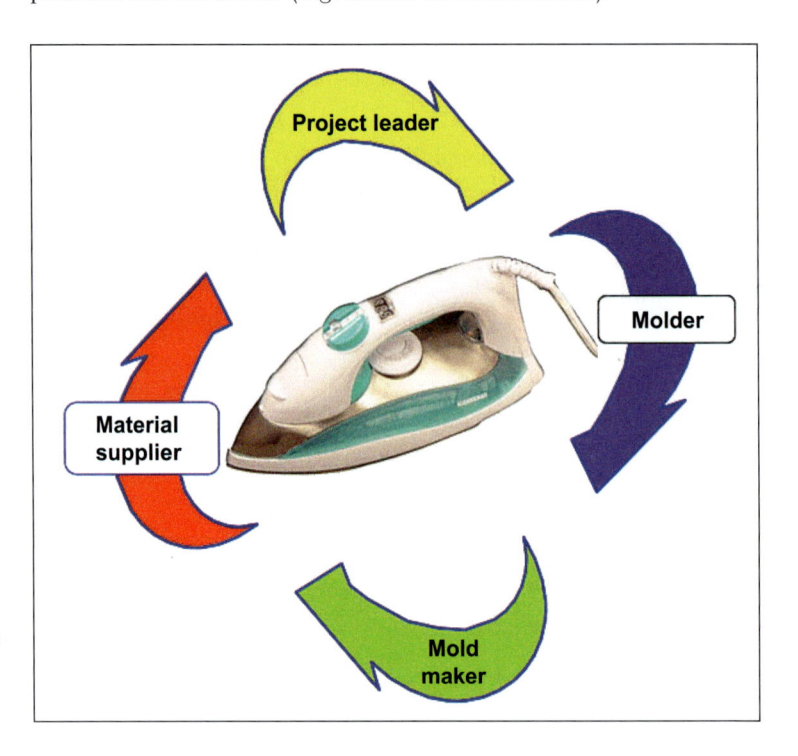

Figure 22.2 Successful collaboration with project subcontractors generally shortens development time significantly.

22.3 Establishing the Requirement Specifications

To establish a complete requirement specification from scratch is very difficult. As a rule, one always encounters new challenges in development work. The requirements can be divided into categories:

1. Market requirements:
   - New functionalities
   - Regulatory requirements
   - Competitive situation
   - Cost objectives

2. Functionality requirements:
   - Integration of multiple functions in the same component
   - Different assembly methods
   - Surface treatments

3. Environmental requirements:
   - Chemical restrictions
   - Recycling (easy to disassemble and to sort)
4. Manufacturing requirements:
   - Processing methods (e.g. injection molding)
   - Molding equipment

22.4 MUST Requirements

When the preliminary list of product requirements has been developed, they should be divided into key requirements ("MUST have") and other requirements ("WANT to have").

Key requirements are usually defined by:

1. Regulatory restrictions (e.g. electrical insulation, flame retardancy, food grade approved); or
2. Industry standards or norms.

With regard to the example of the iron, the key requirements would obviously be that the plastic material must have both good electrical and thermal insulation. In addition, it must not contain toxic pigments or stabilizers. Often the key requirements are determined by previous experience in developing similar products. If you are new and inexperienced to product development work, it is important to quickly build up a network of contacts within and outside your workplace in order to answer these types of questions.

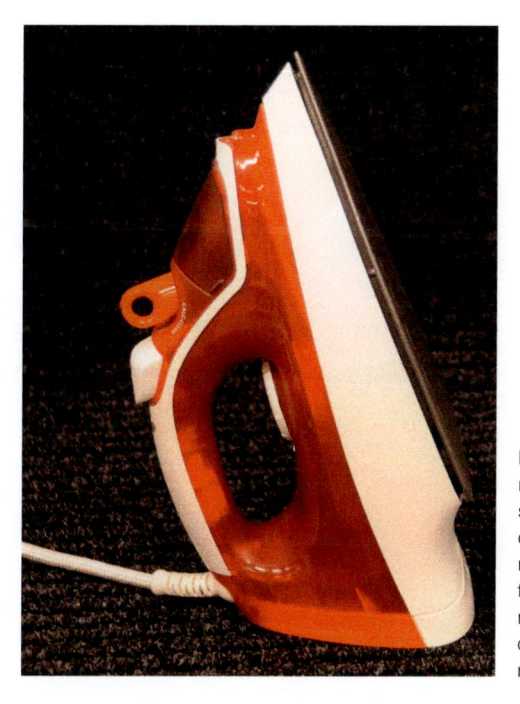

Figure 22.3 The best red color pigments, for use outdoors or at high service temperatures, are based on the carcinogenic chemical element cadmium and are banned in most countries. Therefore, cadmium-free pigments fall into the "must have" category when it comes to regulatory requirements.

22.5 WANT Requirements

Those other requirements that are not absolutely necessary to the functionality of the product are usually referred to as a "wish-list" or "nice-to-have" requirements. For the iron, the wish-list of demands for the plastic material could include the following:

- Bright colors that will not yellow at high temperatures
- Transparent (e.g. visible water level indicator)
- Scratch resistant
- Able to withstand most common household cleaners
- Plastic raw material price under 2.50 Euros/kg

Figure 22.4 The iron in the picture is marketed as having a "self-cleaning function". While this feature is not absolutely necessary to iron fabrics, it is a wish-list specification that fits the "market requirements" category, since it could help facilitate sales in competition with other irons.

Many of the wish-list requirements are also determined by previous experience of developing similar products. But as a rule, there is much more room to let your imagination run free. Here are some ideas that students at one of our technical universities suggested when discussing the wish-list requirements specification for the iron:

- Able to identify the type of fabric being ironed and automatically adjust the temperature
- Be able to check by SMS whether the iron is switched off or not
- Be cordless e.g. powered by a fuel cell or sit in a holder to recharge

Figure 22.5 "There's rarely such a thing as a new idea!"
In the 19th century there were "fuel cell" cordless irons. The fuel consisted of glowing embers of coal that were placed inside the iron. You could also get a "rechargeable" iron, which consisted of a lump of iron with a wooden handle, recharged by placing it on the wood stove to heat up.

22.6 Specify and Sort the Material Candidates

It is important to have quantifiable requirements (i.e. requirements that can be described by numerical values) when setting up the list of key requirements and wish-list requirements. Plastic suppliers' product literature and data sheets can be used to obtain the material data needed when comparing the characteristics considered important in the requirement specification. These can often be downloaded directly from the Internet or obtained in materials databases such as CAMPUS, Material Data Center, or the UL IDES Prospector; see Chapter 10. Many plastic suppliers describe the various applications their materials can be used for and often have a regularly updated newsletter on their homepages. These can be the source of valuable advice and good ideas.

CASE STUDIES

Capless Fuel Filler Locks Out Wrong Nozzles ●

Lighter Engine Cover Components ●

Hytrel® Ski Boot Collar ●

Figure 22.6 DuPont is a leading supplier of engineering plastics. They describe many practical cases on their homepage, http://www.dupont.com/products-and-services/plastics-polymers-resins.html, three examples of which are shown here. [Source: DuPont]

Yet another way to find a suitable material is to examine competing products with similar functions. Some raw material suppliers help by making material identifications and then proposing suitable candidates from their own catalog.

When you have identified the different potential material candidates, it is advisable to try to get two or three different alternatives that you can continue to work on. For example, see Table 22.1.

Table 22.1 This table shows three different material candidates you have identified as suitable to test a prototype tool for manufacturing cable clips for cars. The values that are best for each characteristic identified in the requirements specification are shown as white text on a red background.

Properties	Unit	Impact mod. PA	Impact mod. POM	PBT high viscosity
Material price	Euro/kg	4.5	**3.5**	3.75
Elongation at yield	%	**37**	22	3.8
Tensile strength	MPa	43	**71**	58
Stiffness (E-modulus)	MPa	900	**3,000**	2,700
Charpy notched impact strength 23/−30°C	kJ/m^2	**115/17**	15/10	5.7/3.4
Rel. weld line strength		1	2	**3**
Rel. chemical resistance		2	2	2
Heat deflection temperature (0.45/1.8 MPa)	°C	132/64	**165/95**	160/60
Creep modulus (1/1,000 h)	MPa	1,200/750	2,700/1,500	**2,600/1,800**
Req. hold pressure time	sec	**6**	16	6
Mold shrinkage	%	1.3	**2**	1.6
Suitable for hot runners		1	2	**3**
Req. moisture conditioning		Normally not	No	No

22.7 Make a Detailed Cost Analysis

You can only generate an accurate cost estimate once you have identified suitable material candidates.

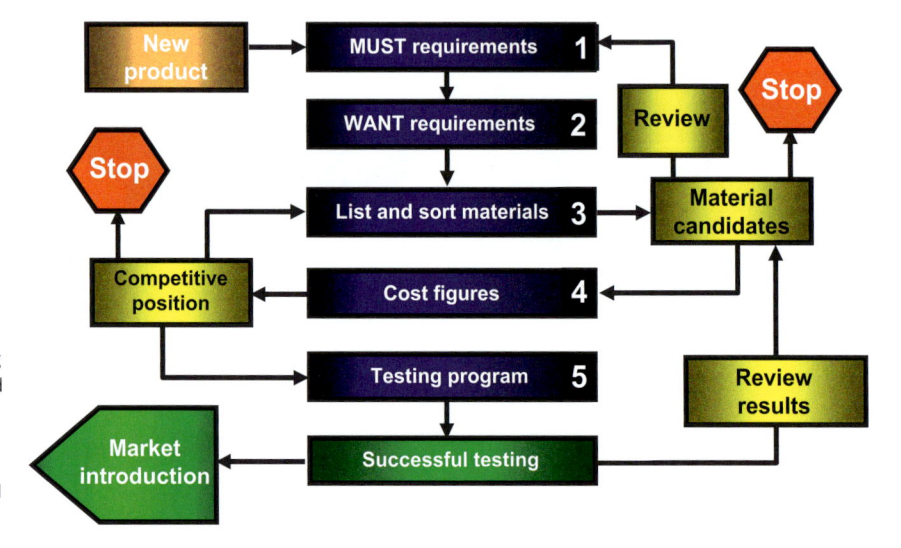

Figure 22.7 This flow chart shows how the material selection process works in many companies. It will usually take several "loops" before coming up with a product that can be launched on the market. For simple products, it may take a few months from idea to finished product, while a complicated product like a new car can take several years.

To keep the costs as low as possible, you should:

- Optimize the design for the plastic material(s) you have selected
- Integrate components and functions where possible
- Minimize the need for assembly or post-production processing
- Optimize the number of cavities and machine size
- Optimize production costs with regard to the material's per-kilo price, density, cycle time, and reject level

22.8 Establish a Meaningful Test Program

Try to simulate real conditions. For example, if the product will be exposed to stress for a very long time, try to interpolate the curves. Often tests are based on OEM requirements (original equipment manufacturer) or industry specifications or standards. Others are developed within the company developing the product. All tests must be reproducible, and it is very important that test results are properly documented.

Many companies are beginning to test products created by a prototype tool. Since both size and strength are influenced by gate location and cooling processes, it is important to take this into account when the prototype is produced. This is in order to avoid unpleasant surprises during the tests of your product when it is subsequently manufactured in the production tool.

Depending on the product and the type of test, it may be necessary to test a large number of parts to set the standard for approved components, since the spread between the various cavities or shot-to-shot can vary significantly. With impact tests, at least 50 parts usually need to be tested to give sufficient accuracy on the spread.

CHAPTER 23
Requirements and Specification for Plastic Products

The requirement specification will differ from product to product depending on what it will be used for. For a spatula, heat resistance and food approval are key requirements; for the blade to an indoor hockey stick, toughness and the ability to shape the blade afterward are the most important properties. This section will address most of the properties that should be included in the requirement specifications of thermoplastic products. It is important to bear in mind that the tougher the requirements for a product, the more expensive it becomes to manufacture.

Below there is a list of the different things that need to be considered when drawing up the requirement specification for a new plastic product:

1. Background information
2. Batch size
3. Part size
4. Tolerance requirements
5. Part design
6. Assembly requirements
7. Mechanical load
8. Chemical resistance
9. Electrical properties
10. Environmental impact
11. Color
12. Surface properties
13. Other properties
14. Regulatory requirements
15. Recycling
16. Costs

23.1 Background Information

This generally refers to a description of the product and its intended usage and is often defined by the following questions:

- Have we developed a similar product?
- What new features will the product have?
- Is this just a new size (upscaling/downscaling) of an existing product?
- Can we modify the geometry of an existing product to create this new one?
- Does the new product require a radical change of materials?
- How do the competitors' products work?
- What tests, studies, or reports already exist regarding this type of product?

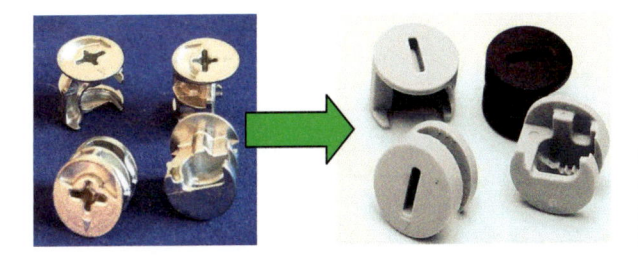

Figure 23.1 The picture on the left shows a zinc cam (furniture assembly screw). The cam in the picture on the right has the same dimensions but there has been a radical change to the material used: i. e. polyamide instead of zinc.

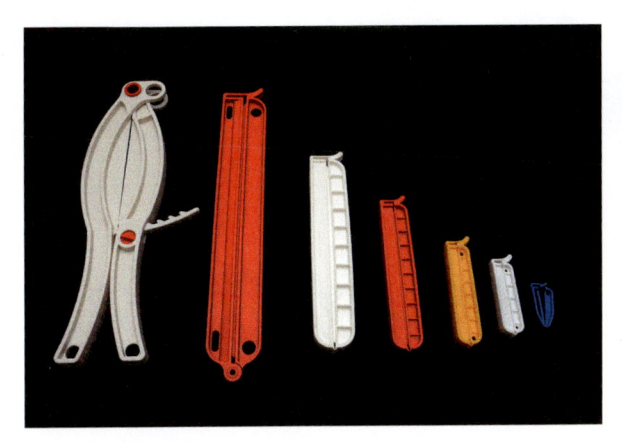

Figure 23.2 This picture shows various sealing clips. Working from the left, clips 3 to 6 are purely upscaled/downscaled versions of the same product, whereas the other three clips have different geometries.

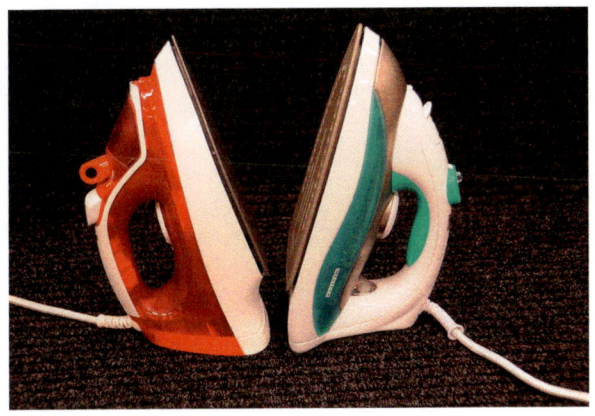

Figure 23.3 When collecting background information for the development of a new product, it is common to compare existing products on the market to try to find possible improvements, new functions, or lower production costs than your competitors.

23.2 Batch Size

It is important to be clear about the potential batch sizes that can be considered for your new product. You need to know this in order to make a good estimate of production costs, where both the tool cost and the size of the machinery are very important. For example, if the batch size is less than 1,000 items per year, maybe injection molding would be too costly, and other manufacturing methods, such as machining of rods, vacuum forming of sheets, or rotational molding, should be considered.

Hundreds of millions of the cams shown in Figure 23.1 are manufactured every year, and the company who developed them makes use of a large number of molds with 32 cavities in each to keep up with demand. Figure 23.4 shows a product made by vacuum forming. With this manufacturing method, a profit can be made even with very small production runs.

Chapter 23 – Requirements and Specification for Plastic Products

Figure 23.4 The picture shows a product made by using vacuum forming. Since the tool that shapes the component can be made of wood, costs can be kept down and a profit made even at batch sizes of 10 parts per year. For batch sizes over 1,000 parts per year, injection molding would be a good alternative manufacturing method.

23.3 Part Size

If the part size is between 5 cm^2 and 50 cm^2 and will be injection molded, there are many molders to choose from, resulting in a competitive price. For products over 50 cm^2, the number of molders with sufficiently large machinery is much reduced.

Figure 23.5 Furniture in talc-filled PP. The number of molders on the market is limited for this size product.

23.4 Tolerance Requirements

Injection-molded parts cannot be manufactured with the same tolerances as machined parts.

Although most part designers know this, it should always be considered in the requirement specification so that time and money are not wasted. Generally, injection-molded parts can be divided into three categories with respect to quality:

- "Normal" moldings
- Technical moldings
- High-precision moldings

The DIN 16901 standard specifies these categories in terms of general tolerances and dimensions and indicates acceptable deviations.

"Normal" moldings have the lowest requirements for quality control and are characterized by short injection-molding cycles and low waste costs.

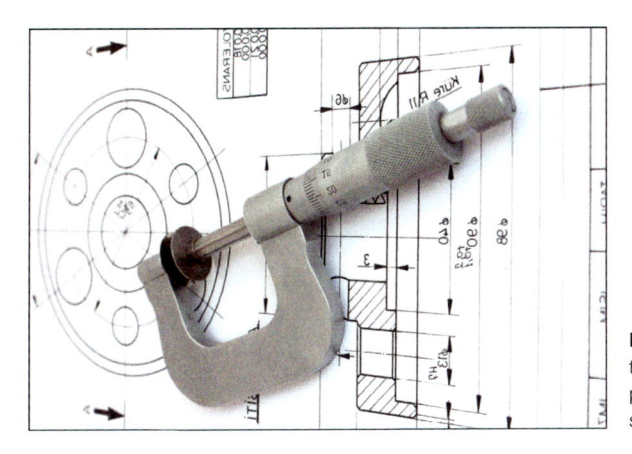

Figure 23.6 A good way to keep down the manufacturing costs for a new product is not choosing unnecessarily strict tolerances.

Technical moldings are significantly more costly to produce because the requirements for tools, production equipment, and correct molding parameters are higher. In addition, tighter quality control is needed, and you may therefore also face higher levels of waste costs.

The last category, high-precision moldings, requires high-precision molds, optimal process parameters, 100% monitoring, and quality control. This affects cycle time and increases production and quality control costs and thus increases the unit price of the final component.

Dimension variations in plastic components can be caused by:

1. Tolerances in the manufacturing of molds
2. Tolerances in the injection-molding process
3. Tolerances in the raw material (e.g. glass fiber content)
4. Warpage, as a result of:
   – Mold shrinkage
   – Post-shrinkage
   – Part design
   – Flow rate and orientation
   – Internal stresses
   – Variable temperatures in the mold
5. Variations in part dimensions due to:
   – Moisture absorption
   – Thermal expansion

23.5 Part Design

A plastic product can be either two-dimensional or three-dimensional. A two-dimensional product can be manufactured by extrusion or injection molding.

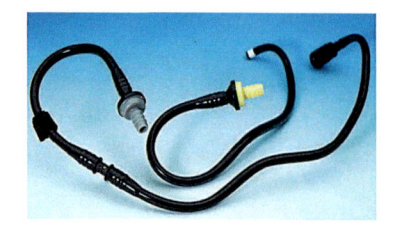

Figure 23.7 Tubes, hoses, and cables: two-dimensional products manufactured by extrusion. [Photos: DuPont]

Three-dimensional products can either be solid or hollow and are of variable complexity.

Hollow products can be produced by several different processing methods:

1. Injection molding:
   - Gas injection
   - Water injection
   - Two parts that are joined together
2. Blow molding
3. Rotational molding

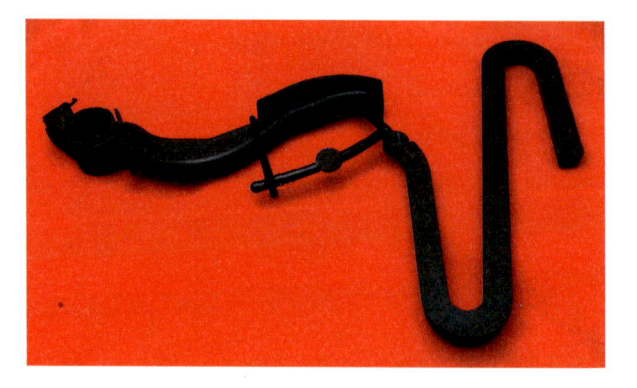

Figure 23.8 This picture shows a hollow accelerator pedal made from glass fiber reinforced polyamide using a gas injection-molding technique. To the right in the picture there is an overflow pocket of the material that was displaced by the gas.

The most common method for producing hollow components is blow molding.

The advantages of this method are:

- Very large products can be produced
- Complex shapes can be produced
- Thin-walled goods can be made

And the limitations are:

- Limited choice of materials
- High equipment costs need to be produced by large batch sizes (> 10,000)
- Poor surface finish compared to injection molding
- Hard to maintain strict tolerances

Figure 23.9 The picture shows a blow-molded ventilation pipe for a Volvo 6-cylinder gasoline engine and is a good example of how even a complex design can be produced by this method. [Source: Hordagruppen]

Rotational molding is less widely used for producing hollow components.

The advantages of this method are:

- Short development time
- Very large products can be produced (up to 20 m^3)
- Low tool costs and good profitability for batch sizes > 100 components

And the limitations are:

- Extremely long cycle time (30–60 min)
- Very limited choice of materials
- Hard to maintain strict tolerances

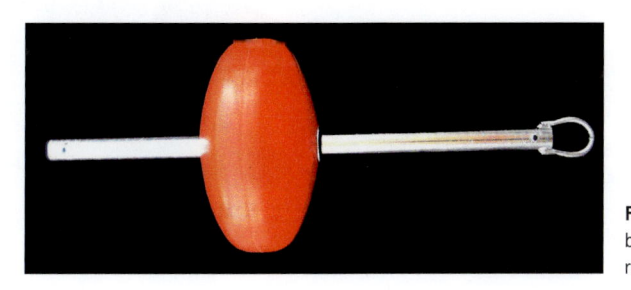

Figure 23.10 Buoys and fenders for boats are often manufactured using rotational molding.

Chapter 23 – Requirements and Specification for Plastic Products

When producing solid three-dimensional components, injection molding is the most common method used. These products can be classified as those made using a simple parting line, and those with a more complex design made using several parting lines and slides.

Figure 23.11 A complete plastic storage box consisting of a box, lid, and snap-lock fastenings, all of which are made in a mold using a simple parting line.

Figure 23.12 A car's intake manifold produced in polyamide 66 is an example of a product made by using a multiple parting line mold and slides in different directions. To make it hollow, you can either make it in two parts and vibration-weld them together, or use advanced melt-core technology. [Photo: DuPont]

When you are considering material replacement from metal to plastic, you normally need to change the dimensions or design of your part in order to withstand the loads. The furniture screws in Figure 23.1 are an exception to this rule.

Figure 23.13 These angle couplings for 32 mm PE pipes are a good example of how the dimensions and design have to change when making the replacement from metal to plastic, in order to withstand water pressure and load when connecting the hose.

Brass → PP

23.6 Assembly Requirements

Many components include several parts. When developing a new plastic product, you should start by trying to integrate as many functions as possible to reduce the number of parts. If this is not done straight away, you are left with the problem of how to join the different parts together in a sensible way. When working with thermoplastics, you have the benefit of numerous assembly methods, as shown in Chapter 25.

Figure 23.14 These worm gears are constructed in two identical halves that can then be lined up and rotated 180° to snap them together. Making the complete gear in one shot is not possible due to the undercut.

23.7 Mechanical Load

For plastic components that will be exposed to a mechanical load or stress, it is important to know:

- The size and direction of the load/stress
- The duration of the load/stress
- The kind of load/stress: e.g. single or repeated impact
- The load/stress frequency
- If the part is under a permanent deformation and can be subjected to relaxation

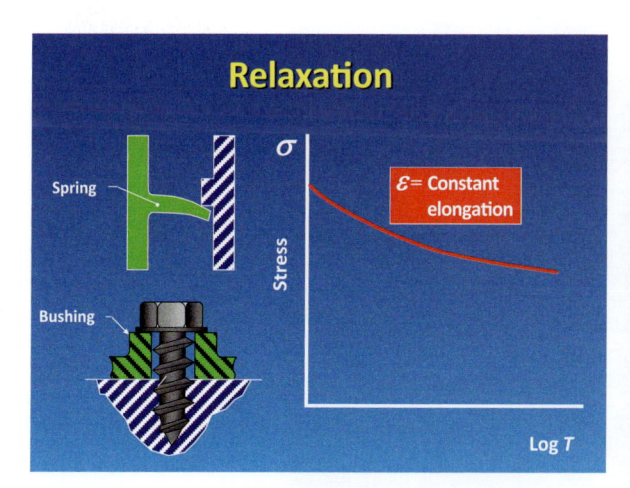

Figure 23.15 A typical relaxation curve for a plastic component subjected to permanent deformation, such as a spring or the bushing of a screw joint. [Source: DuPont]

Figure 23.16 A polycarbonate helmet is a good example of a product that can only be exposed to a single hard impact. The smallest crack can affect its protective ability.

23.8 Chemical Resistance

The capacity to not be affected by different chemicals varies greatly between different types of plastic and can be an important key requirement when selecting materials for a specific application.

You can find chemical resistance tables on the Internet for a large proportion of materials. Simply by typing "chemical resistance of plastics" into Google, you will get many good hits. In addition, the CAMPUS database contains detailed chemical resistance information for many materials. However you find your information, remember to take into account the temperature of the chemical when it comes into contact with the plastic material.

Figure 23.17 Polypropylene is one of the few polymers that can withstand strong acids, including sulfuric acid, making it highly suitable for use in car battery cases.

Figure 23.18 Fluorocarbon polymer (e.g. Teflon) has the best chemical resistance combined with high temperature resistance. The material is also food approved and is often used for coating frying pans. [Photo: DuPont]

Figure 23.19 Water bottles are commonly made of the thermoplastic polymer PET. However, they cannot cope with temperatures above 60 °C without suffering from chemical degradation (hydrolysis reaction). This makes PET unsuitable for use in baby bottles, for example, since these are often boiled to sterilize them.

23.9 Electrical Properties

Most plastics act as electrical insulators, and there are a variety of test methods for establishing the electrical properties of plastics. Usually the material's insulating properties are specified as dielectric strength (through the wall) or leaking currents (on the surface). Good information is available online, e.g. at www.ulttc.com.

The following methods are generally included in datasheets provided by plastic suppliers:

- Dielectric strength
- Volume resistivity
- Arc resistance
- Surface resistivity
- CTI, Comparative Tracking Index

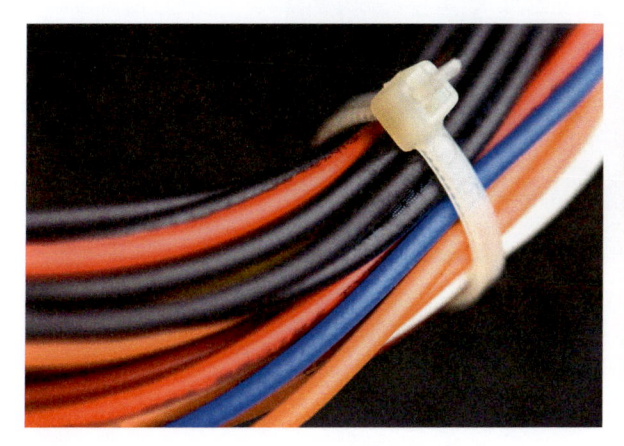

Figure 23.20 Low price combined with good electrical insulation has made PVC the dominant material in cable manufacturing.

Figure 23.21 When you need a material that has good electrical properties and can withstand high service temperatures, polyamide 66 and thermoplastic polyester PBT are both good options and are available in self-extinguishing grades.

23.10 Environmental Impact

When "environmental impact" appears on the requirement specification, it generally refers to the effect of the environment on the plastic and not the other way around, although in many cases that could also be a relevant consideration. All plastics are affected to some extent by the environment they are used in. Over time, they are degraded by different environmental factors, such as:

- UV rays from the sun or other radiation (e.g. microwave ovens or radioactive sterilization)
- Oxygen in the air
- Water or steam
- Temperature fluctuations
- Micro-organisms (e.g. fungi and bacteria)
- Different chemical solutions and pollution

Figure 23.22 The poop-scoop bags provided free by many municipalities are made from a bioplastic that dissolves if it is left outside. This is an example of a product where the requirement specification includes not leaving litter in nature, but also using something that will degrade naturally after a relatively short period of time.

Figure 23.23 For plastic bottles and plates that are used in microwave ovens, the plastic material must be transparent to microwave radiation so that the plastic does not melt or thermally degrade. The symbol to the right of the picture is taken from the bottom of a jar manufactured in polypropylene and indicates that it is safe for use in microwave ovens.
NOTE: mugs and plates made of melamine will be destroyed instantly if used in a microwave!

23.11 Color

One of the major benefits of using plastic for a product is that you can color it directly during the production. However, it should be noted that some materials have a natural color such that they cannot be colored in bright or light colors. If a transparent color is required (e.g. a part with a smoky effect), you must choose an amorphous plastic. In the past, raw material suppliers would often supply custom-colored material in relatively small volumes. Today, the larger suppliers usually require an order of hundreds of tons per year before they will produce a special color. Therefore, most molders are obliged to self-color their materials with color concentrates, usually in the form of masterbatch.

Figure 23.24 These golf tees are made from sawdust-filled bioplastic. The material's natural dark coloring does not allow them to be produced in bright colors without being painted. [Source: Plastinject AB]

When coloring a plastic material with masterbatch, you need to be aware that both the pigment and the so-called carrier (e.g. polyethylene) can alter the mechanical properties of the plastic and/or its ability to be processed. Mold shrinkage can also be affected by different pigments. For example, the coloring of a part made from polyamide will change as the material absorbs moisture from the air. With all coloring, it is very important to choose the right light source (e.g. daylight, fluorescent or bulb) when the color is matched.

You should also be aware that different surface structures affect how the eye perceives a color.

If plastic components are to be assembled with painted metal ones and need to have the same color, it is much better to make the color standard in plastic and match the metal paint after this. It is much easier to match a lacquer than a plastic.

Figure 23.25 This freezer door handle is made in a colored thermoplastic. The door is made of painted steel plate. In order to get the best possible consistency in color, the paint should be matched to the handle rather than the other way around.

23.12 Surface Properties

Just as coloring can provide significant benefits when manufacturing a plastic product, having the freedom to choose the texture can also be a great advantage. This can range from a high-gloss finish (known as a "Class A" surface) to a textured surface (e.g. with a leathery feel).

Materials that are very fluid (low melt viscosity) provide the best high-gloss finish. Thermoplastic polyester has a low melt viscosity and, even with a 30–40% glass fiber content, can provide a very good surface gloss.

Figure 23.26 This car door panel is made in a fully colored black PMMA. The surface has a high-gloss mirror finish, a so-called "Class A" surface.

When a matte finish is required, this is usually accomplished in the mold, either by blasting, EDM (electrical discharge machining), or etching. The sink marks appear much more on dark glossy surfaces, and if they cannot be removed by other means, your only alternative options are to use a matte finish or a lighter color.

Figure 23.27 The picture shows a part made in polyamide. The matte surface has been achieved by EDM, which offers much better control over the surface finish than sand blasting.

Figure 23.28 The leather or textile effect on this plastic armrest is accomplished by etching the required pattern onto the inner surface of the mold.

Scratch and abrasion resistance can also be included in the list of surface characteristics. As a rule, scratch resistance is related to the material's surface hardness. Soft materials with good abrasion resistance include rubber and polyurethane, while PEEK and acetal are good stiff alternatives.

The abrasion properties of many materials (such as acetal) can be improved by blending with a low-friction material such as fluoropolymer or with a lubricant such as molybdenum disulfide.

Figure 23.29 Chain links manufactured in acetal. The addition of fluoropolymer or silicon improves both friction characteristics and wear.

Sometimes, for aesthetic or functional reasons, the plastic surface needs to be metal coated, painted, or printed (e.g. screen or tampon printed).

Figure 23.30 As a rule, the reflectors in car headlights are made in thermoplastic polyester PBT, which can withstand high temperatures and has a good surface gloss, in addition to being easy to metal-plate.
The headlight lens is made of polycarbonate, which has excellent impact strength and transparency. The lens surface is coated with siloxane in order to improve scratch resistance, UV protection, and solvent resistance.

23.13 Other Properties

Other properties in the requirement specification for a plastic product could include:

- Low or high friction
- Thermal conductivity or insulation properties
- Electromagnetic shielding
- Thermal expansion coefficient

Figure 23.31 A red-colored, high-friction material has been chosen to provide a good grip on this pen.

Figure 23.32 When selecting the cover for a cell phone, one of the key requirements is low electromagnetic shielding in order not to restrict the phone's range.

23.14 Regulatory Requirements

Where regulations must be met, the requirements are obviously part of the key/"must have" category and will often include the following:

- Electrical requirements
- Food approval
- Drinking water approval

Figure 23.33 The CE marking on the top left of the packaging shows that the enclosed plug is made of a plastic material that meets the requirements within the European Economic Area for electrical insulation at the specified power (3,680 W).

- Medical requirements
- Flame retardance
- All additives in the plastic material must be cleared for health and safety purposes
- Imposing a returnable deposit on plastic packaging

23.15 Recycling Requirements

Most thermoplastics can be recycled both mechanically and chemically. Mechanical recycling involves the grinding down of either material wasted during manufacturing or used-up products so that they can be recycled and processed again. Chemical recycling breaks the material down to the molecular level, enabling the production of virgin material. As of 2014, the prices of crude oil make chemical recycling unprofitable.

Some municipalities require waste separation to boost recycling rates.

Figure 23.34 "Climate smart" plastic bags. These are made exclusively of collected recycled material.
[Photo: Miljösäck AB]

Improved and increased labeling is required to facilitate the sorting and recycling of plastic products. The symbols shown in Figure 23.35 show how to identify plastic products for recycling.

Figure 23.35 Packaging and disposable products should be labeled with the recycling symbols shown here. If the material does not have its own code, e.g. ABS, it is denoted by the number 7. A better alternative is to mark the product in the same way that technical moldings are marked (see below).

Technical moldings usually put the polymer designation between angle brackets or under the triangular recycling symbol.

Figure 23.36 The top image shows the recommended marking for technical moldings. This particular material is polyamide 66. Additives to the material can also be indicated, e.g. polyester PBT with a 30% glass fiber content would be labeled as > PBT GF30 <. The lower image shows a less common label for a polyamide product.

One way to encourage recycling is to include a returnable deposit on a product, which is now common for plastic drink bottles made of thermoplastic polyester PET.

Figure 23.37 Including a returnable deposit on packaging reduces the risk of littering and stimulates recycling. We will probably see an increase of this requirement in the future.

23.16 Cost Requirements

This requirement is ever present when developing a new product, since the manufacturing costs directly affect the selling price and ability to compete in the marketplace. You cannot make a reasonable estimate of the price of your plastic product until you have selected the material, the production method (including post-production), the manufacturer, and the estimated mold and other equipment costs.

A complete cost estimate for an injection-molded part, such as those shown in Chapter 19, is very comprehensive and includes:

- Estimated annual production and number of units per batch
- Net weight of the component
- Waste material and scrap value
- Material price

- Handling time per component
- Staffing levels for the machinery
- Salaries and related employee expenses
- Manufacturer overheads
- Percentage of color masterbatch
- Masterbatch price
- Actual cycle time
- Number of cavities in the production mold
- Utilization levels for the machinery
- Hourly running costs for the machinery
- Machinery depreciation, plus utilities and maintenance costs
- Setup time per batch/machine run
- Setup costs per batch/machine run
- Post-production processing (per component)
- Post-production handling time
- Post-production overheads

Most plastic manufacturers use computerized costing calculations and can usually get a good estimate of what the component will cost in different materials.

23.17 Requirement Specification – Checklist

The checklist below, while not claiming to be complete for every product, may still be helpful in the preparation of the requirement specification for many plastic components.

23.17.1 Background information

- Have we developed a similar product?
- What new features will the product have?
- Is this just a new size (upscaling/downscaling) of an existing product?
- Can we modify the geometry of an existing product to create this new one?
- Does the new product require a radical change of materials?
- How do the competitors' products work?
- What tests, studies, or reports already exist regarding this type of product?

23.17.2 Batch Size

- How many of the components are likely to be produced each year?
- Which manufacturing methods are most appropriate?

23.17.3 Part Size

- Is the part's size a limiting factor in the choice of material, production method, or producer?

23.17.4 Tolerance Requirements

- Which dimensions are critical, and what are the required tolerances?
- Would redesigning the product help to reduce the tolerance restrictions?

23.17.5 Part Design

- Can the product be made in one piece?
- Which processing methods may be appropriate?
- What additional functions can be integrated?

23.17.6 Assembly Requirements

- Can the assembly simplify or improve the functionality?
- Can assembly lower the manufacturing costs?
- Which assembly methods are appropriate for the proposed material?

23.17.7 Mechanical Load

- Is the part design optimal (in terms of e.g. gate location) with regard to the anticipated mechanical load?
- Will the part be under constant load?
- What are the likely normal and peak loads?
- How long will the product be under peak load conditions?

23.17.8 Chemical Resistance

- What chemicals will the component be exposed to under normal and extreme conditions?
- What are the likely chemical concentrations?
- At which temperatures will the component be exposed to chemicals?

23.17.9 Electrical Properties

- Does the material need to have electrical conductivity or insulating properties?
- Does the component need to comply with electrical regulations?

23.17.10 Environmental Impact

- Will the part be used outdoors or exposed to UV rays?
- Will it be subjected to any other kind of radiation?
- Is the material affected by atmospheric oxygen (oxidation)?
- Will the part be exposed to water or steam?
- What will be the range of the service temperatures (normal/minimum/maximum)?
- How long/how often is the part likely to be exposed to peak temperatures?
- Will the part be exposed to micro-organisms or attack by animals or insects (e.g. rodents, termites)?

23.17.11 Color

- Does the proposed material have an effect on the choice of custom colors?
- Does the material need to be UV stabilized to protect the color?
- Does the part need to be painted?
- Does the part need to be assembled and color-matched to other colored or painted parts?

23.17.12 Surface Properties

- What surface texture should the part have?
- Is the proposed material suitable for the intended surface finish (e.g. high gloss)?
- Is there a risk of sink marks (e.g. ribbing) that would need to be hidden with a matte finish?
- Does the surface need to be scratch resistant?
- Does the surface need to be metal coated?
- Does the surface need to be paintable?
- Will the plastic surface be printed (e.g. hot stamp, pad, or screen printing)?

23.17.13 Other Properties

- Does the proposed material need to have high or low friction properties?
- Will the material be able to insulate from or conduct heat?
- Is the functionality of the component affected by electromagnetic radiation?
- Will the part's functionality be influenced by the material's linear expansion coefficient?

23.17.14 Regulatory Requirements

- Will the part be CE-approved (electrical regulations)?
- Will the part be FDA-approved or meet other food approval standards?
- Will the part meet medical requirements:
  - Be compatible with human tissue?
  - Able to be sterilized?
- Does the part need to be flame retardant to a specific classification?
- Does the material contain any ingredients or additives that are prohibited in the intended marketplace?

23.17.15 Recycling

- What are the recycling requirements (e.g. return deposit)?

23.17.16 Costs

- What will the component cost be:
  - To produce?
  - To the client or consumer in its use?

CHAPTER 24
Design Rules for Thermoplastic Moldings

Designing in plastic is a science in itself, and a lot has been written on the subject. This chapter aims to show some of the most important rules that a designer should bear in mind when developing a new product in plastic. These rules are divided into the following 10 sections:

1. Remember that plastics are not metals
2. Consider the specific characteristics of plastics
3. Design with regard to future recycling
4. Integrate several functions into one component
5. Maintain an even wall thickness
6. Avoid sharp corners
7. Use ribs to increase stiffness
8. Be careful with gate location and dimensions
9. Avoid tight tolerances
10. Choose a suitable assembly method

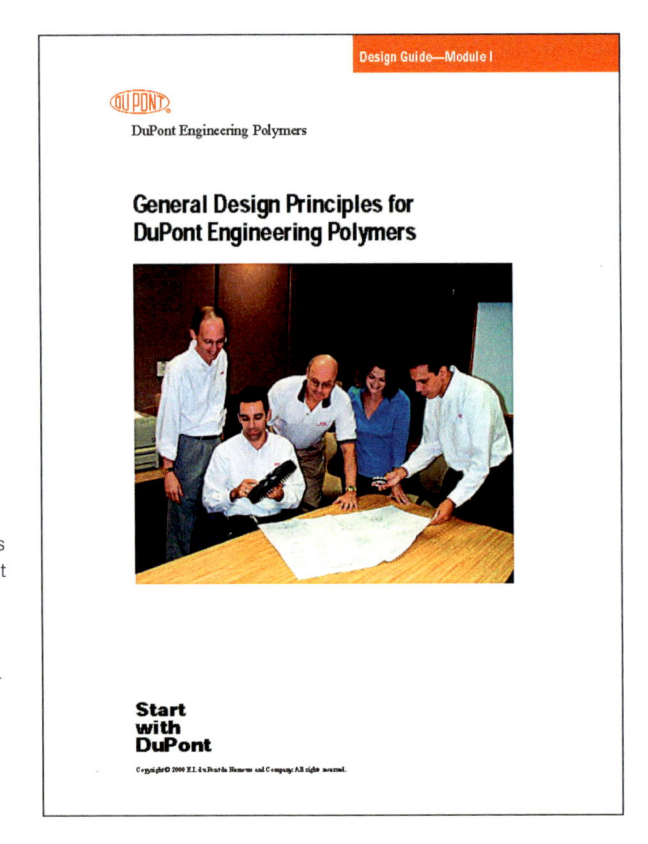

Figure 24.1 There is some good design literature online available for free download. One such document is "General Design Principles for DuPont Engineering Polymers" containing 136 pages of useful information. See: http://www.dupont.com/content/dam/dupont/products-and-services/plastics-polymers-and-resins/thermoplastics/documents/General%20Design%20Principles/General%20Design%20Principles%20for%20Engineering%20Polymers.pdf

24.1 Rule 1 – Remember That Plastics Are Not Metals

Some engineers still design plastic components as if they were made of metal. If you can succeed in maintaining the strength, the product will be lighter and often much cheaper. However, if the main purpose is to reduce production costs, it is by default necessary to make a total redesign when plastic is intended to replace metal.

If a direct comparison is made, the metal will have a higher:

- Density
- Maximum service temperature
- Stiffness and strength
- Electric conductivity

While the plastic material has an increased:

- Mechanical damping
- Heat expansion
- Elongation and toughness

Table 24.1 This table shows that thermoplastics have certain advantages over metals, such as decreased weight, corrosion durability, thermal and electric insulation, freedom of design, and recycling potential.
However, they are clearly disadvantaged in terms of stiffness, strength, and sensitivity to high temperature. Thermosets follow the same patterns as thermoplastics but are much harder to recycle.

	Metals	Thermoplastics	Thermosets
Weight	⇔	⇑⇑	⇑
Corrosion	⇔	⇑	⇑
Stiffness	⇔	⇓⇓	⇓
Strength	⇔	⇓⇓	⇓
Temperature	⇔	⇓⇓	⇓
Thermal insulation	⇔	⇑	⇑
Electric insulation	⇔	⇑	⇑
Freedom of design	⇔	⇑⇑	⇑
Recycling	⇔	⇑	⇓

⇔ Reference / equal ⇑ Better ⇓ Worse

24.2 Rule 2 – Consider the Specific Characteristics of Plastics

To be able to achieve an optimal design, the specific characteristics of plastics have to be considered.

On the one hand, thermoplastics have:

1. Nonlinear stress-strain curves, which give a more complicated strength calculation
2. Anisotropic behavior, where you must be more careful about the gate location
3. Temperature-dependent behavior, which makes it necessary to know the maximum service temperature and the duration the plastic component must resist it
4. Time-dependent stress-strain curve (creep and relaxation)
5. Speed-dependent characteristics, which demand careful studies of stress, strain, and impact rate, etc.
6. Environmentally dependent characteristics, which demand knowledge of the humidity and any chemicals and/or radiation (e.g. UV light) that the product will be exposed to

On the other hand, thermoplastics are:

7. Easy to design and process with various different cost-effective methods
8. Easy to color, which eliminates the need for surface treatments
9. Easy to assemble (e.g. self-threading screws, snap-fits, or welding)
10. Easy to recycle

In the graph shown in Figure 24.2, one can see that thermoplastics have a nonlinear load curve. When calculating the loads for steel structures, Hooke's Law can be applied to the proportional (linear) segment, which is a simple mathematical formula. However, the calculations for plastic structures are much more complicated since the stress-strain curve is nonlinear.

Extensive computing power is required to make these more detailed calculations.

24.2 Rule 2 – Consider the Specific Characteristics of Plastics

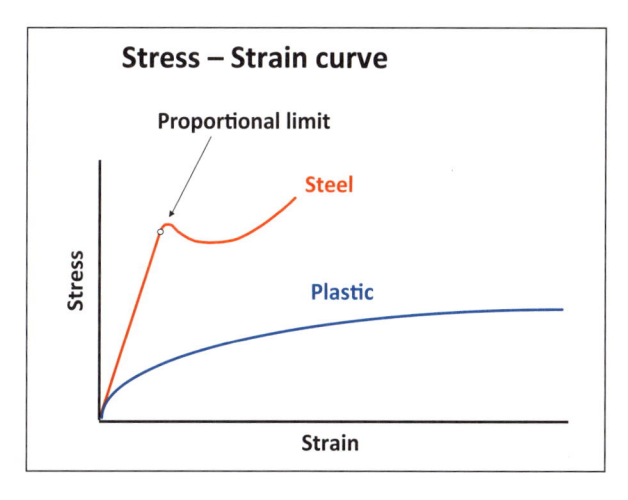

Figure 24.2 Stress-strain curves for steel and thermo plastics. [Source: DuPont]

24.2.1 Anisotropic Behavior

Anisotropic behavior means that the characteristics of the material vary in different directions. This can depend on the orientation of the polymer chains or the fiber direction in reinforced materials. A big difference in the mechanical characteristics along or across the flow direction in a cavity can often be seen. That is why the gate location is important in a component that will be subjected to loading.

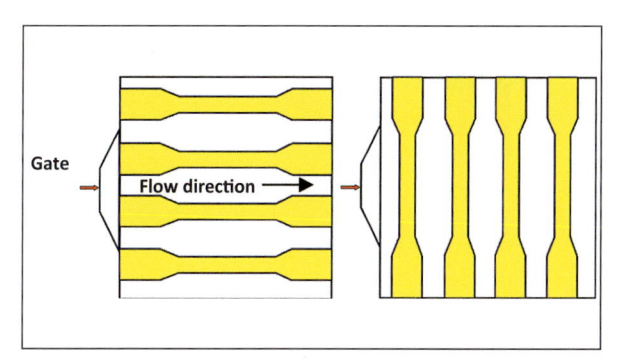

Figure 24.3 Test bars punched from plates in the flow and across the flow direction.

Figure 24.4 Both the tensile and compression load are considerably stronger in the material's flow direction. [Source: DuPont]

24.2.2 Temperature-Dependent Behavior

The mechanical properties of most thermoplastics vary extensively when they are exposed to different environmental temperatures. The graph shown in Figure 24.5 shows a great difference in the stress-strain behavior between room temperature and 93 °C.

241

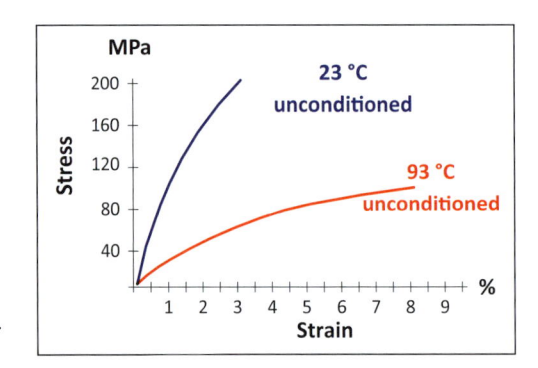

Chapter 24 – Design Rules for Thermoplastic Moldings

Figure 24.5 Load curves for a 33% glass fiber reinforced unconditioned PA66 at 23 °C and 93 °C respectively. [Source: DuPont]

24.2.3 Time-Dependent Stress-Strain Curve

24.2.3.1 Creep

Plastic components when exposed to a constant stress will be deformed over time. This phenomenon is called creep.

If a straight plastic cylinder is loaded with a weight, it will, over time, develop into the shape of a barrel.

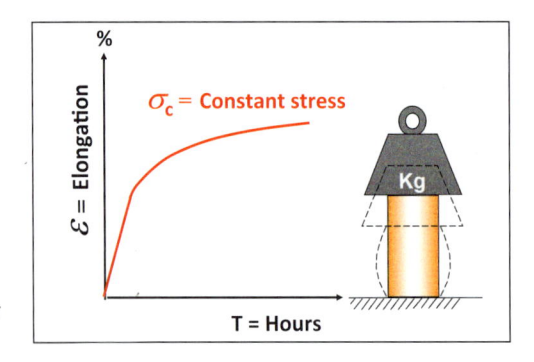

Figure 24.6 A typical creep curve for thermoplastic. [Source: DuPont]

24.2.3.2 Relaxation

If a plastic component is subjected to a constant deformation, stresses will emerge. Over time, these stresses will decrease, and eventually the component will achieve a stress-free state. This phenomenon is called relaxation.

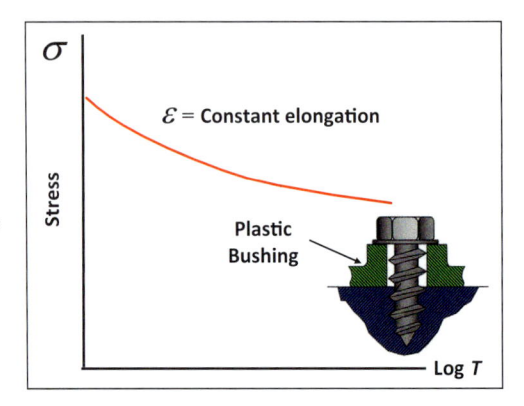

Figure 24.7 A typical relaxation curve for thermoplastic. For a screw joint with a plastic bushing, there is a great risk that the screw will become loose when the stresses in the bushing fade and the bushing relaxes. [Source: DuPont]

24.2.4 Speed-Dependent Characteristics

The mechanical properties of thermoplastics are dependent on the characteristics of the load (static, dynamic, or impact). Figure 24.8–taken from a series of articles by DuPont entitled "Top Ten Design Tips"–shows a range of cases with differing loads, types of deformation, and the calculation method that is most applicable.

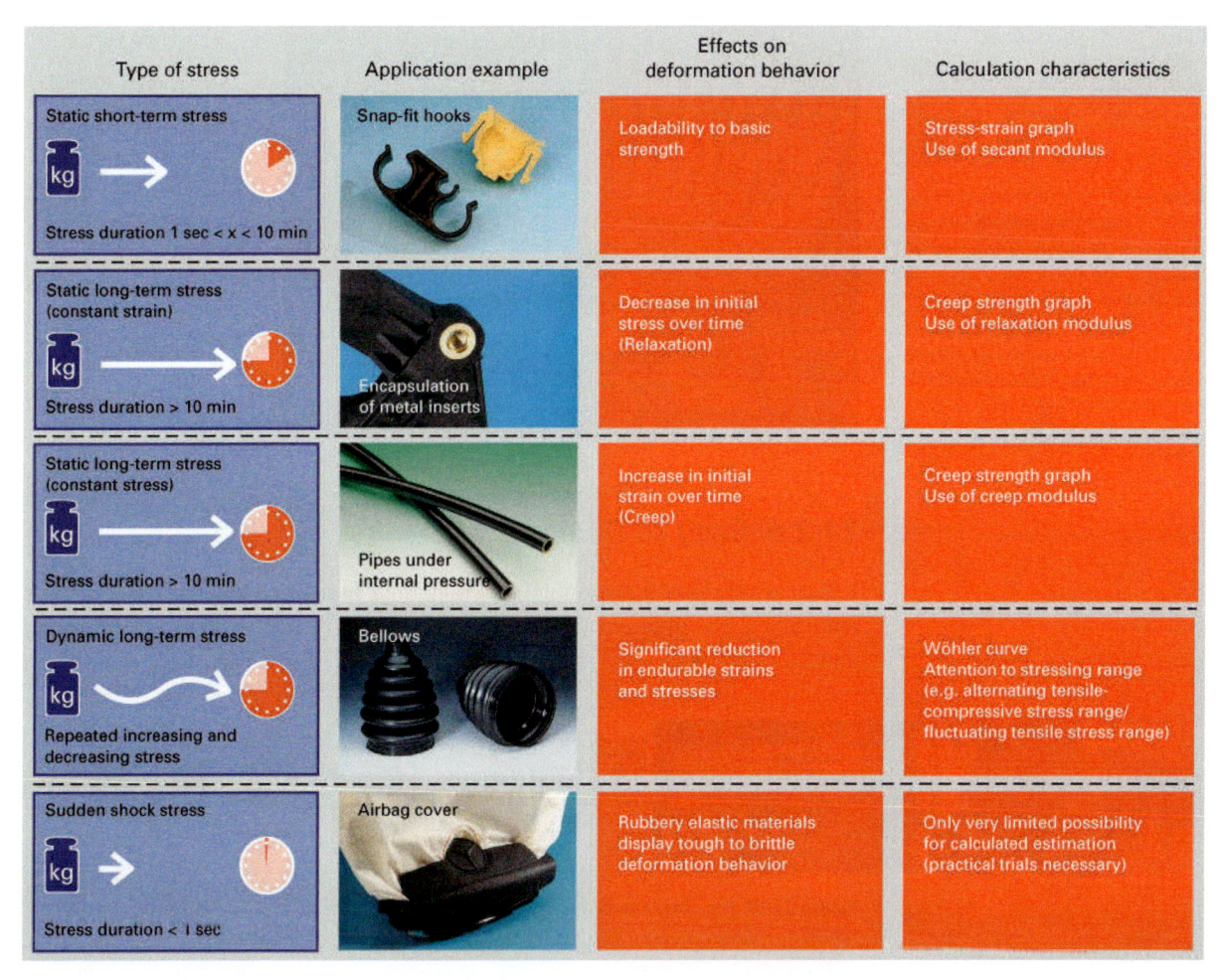

Type of stress	Application example	Effects on deformation behavior	Calculation characteristics
Static short-term stress Stress duration 1 sec < x < 10 min	Snap-fit hooks	Loadability to basic strength	Stress-strain graph Use of secant modulus
Static long-term stress (constant strain) Stress duration > 10 min	Encapsulation of metal inserts	Decrease in initial stress over time (Relaxation)	Creep strength graph Use of relaxation modulus
Static long-term stress (constant stress) Stress duration > 10 min	Pipes under internal pressure	Increase in initial strain over time (Creep)	Creep strength graph Use of creep modulus
Dynamic long-term stress Repeated increasing and decreasing stress	Bellows	Significant reduction in endurable strains and stresses	Wöhler curve Attention to stressing range (e.g. alternating tensile-compressive stress range/fluctuating tensile stress range)
Sudden shock stress Stress duration < 1 sec	Airbag cover	Rubbery elastic materials display tough to brittle deformation behavior	Only very limited possibility for calculated estimation (practical trials necessary)

Figure 24.8 This image reflects five different load situations (case studies) and the type of deformation and calculation methods involved. The last example shows an airbag cover on the steering wheel of a car. The airbag is released in fractions of a second, and the load speed increases so fast that the rubber-like thermoplastic elastomer in the cover becomes stiff and brittle. It is difficult to make computerized simulations of this type of load, so it is better to make prototype tools and conduct practical tests. [Source: DuPont]

Chapter 24 – Design Rules for Thermoplastic Moldings

24.2.5 Environmentally Dependent Characteristics

Thermoplastics do not corrode in the same way as metals, but they can be affected by the humidity in the air, or degraded by different chemicals, micro-organisms, or radiation.

Figure 24.9 This rake is produced in acetal. It has been outdoors for five years. Even though a UV stabilizer was applied to the acetal, the original color (shown as an orange dot) has faded. Tiny cracks can be seen in the grayish area. With each passing year, these cracks in the faded surface layer will become thicker, and in the end the rake will break when it is used.

24.2.6 Easy to Design

It is relatively easy and economically beneficial to manufacture products in thermoplastics. The different processing methods are described in Chapters 12, 20, and 21.

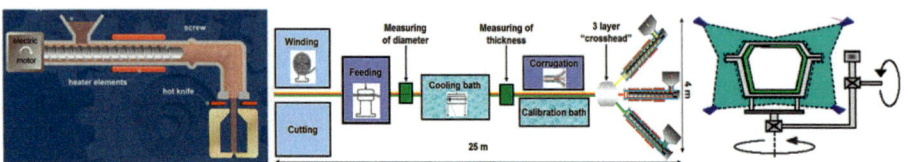

Figure 24.10 The images show some of the processing methods that have been described previously.

24.2.7 Easy to Color

Most thermoplastics have a light natural color and are, by default, easy to color. If the raw material supplier does not offer custom colors, the material will usually be colored during the processing using a masterbatch. Other possibilities include coloring with powder or liquid, and some plastics, such as polyamide, can also be dipped in liquid dye solutions.

24.2 Rule 2 – Consider the Specific Characteristics of Plastics

Figure 24.11 If you want smaller batches but in many colors, like the zippers shown here, it can be advantageous to use liquid dyeing. Preparations for this begin in the production planning stage: when setting up "the color cycle". It begins with the lightest shades and finishes with the darkest, cleaning the equipment thoroughly before starting the next batch run.

24.2.8 Easy to Assemble

The next chapter describes several different assembly methods for thermoplastics.

Figure 24.12 Laser welding is an excellent method to permanently enclose electronics in a device like this car key.

24.2.9 Recycling

In Chapter 7 (Plastic and the Environment) different recycling methods for thermoplastics are described.

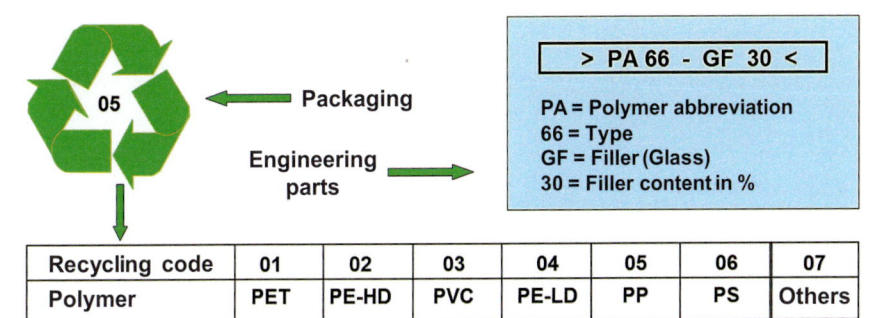

> PA 66 - GF 30 <

PA = Polymer abbreviation
66 = Type
GF = Filler (Glass)
30 = Filler content in %

Recycling code	01	02	03	04	05	06	07
Polymer	PET	PE-HD	PVC	PE-LD	PP	PS	Others

Figure 24.13 Recycling codes

245

24.3 Rule 3 – Design with Regard to Future Recycling

The advantage of most thermoplastics is that they are easily recycled:

- They can be collected and melted over and over again to be used in new products
- They can be chemically recycled and be a source in production of new "virgin" materials
- They can be incinerated and generate a high energy output

For plastic components to be easily recycled, it is important that they are:

1. Easy to dismantle, i.e.:
   - Access and collect
   - Assembled with recoverable methods
   - Designed in a way that any inserts are easy to remove
   - Prepared for robotic dismantling where top mounting is preferred
2. Made in as few plastic materials as possible and preferably only in standard materials
3. Coded so that the material(s) can be identified (see Section 24.2.9)
4. Designed so it will be easy to clean

24.3.1 Dismantling

Minimize the number of included components, such as joints, and avoid permanent assembly techniques, such as welding.

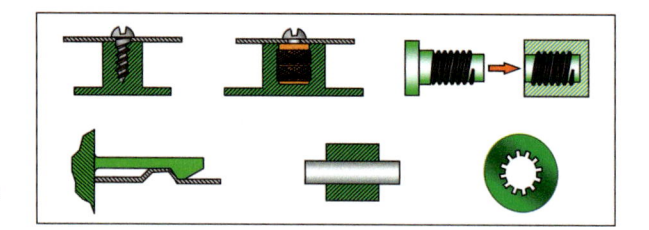

Figure 24.14 Recoverable assembly methods such as screws, snap-fits, press-fits, and splines facilitate future dismantling.

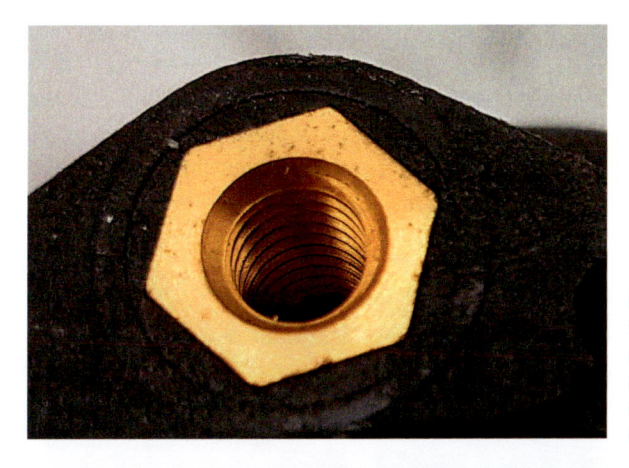

Figure 24.15 Metal inserts should not be over-molded or assembled by ultrasonic welding if they are meant to be easily removed. The inserts should be able to be pressed, pulled, or screwed apart.

Figure 24.16 Metal inserts should not be over-molded or assembled by ultrasonic welding if they are meant to be easily removed. The inserts should be able to be pressed, pulled, or screwed apart.

Chapter 24 – Design Rules for Thermoplastic Moldings

24.3.2 Reused Materials

In order to be able to recycle plastic materials from scrapped products, it is an advantage if the same type of polymer is used in all the assembled components so that they can be recycled without the need to be dismantled first. Wherever possible, it is also advantageous to use standard-grade materials (i.e. without any additives).

Figure 24.17 If it is possible to produce the cap and bottle in the same polymer, there is no need to pick them apart and sort them as different materials. The metal cap is a bad alternative regarding recycling.

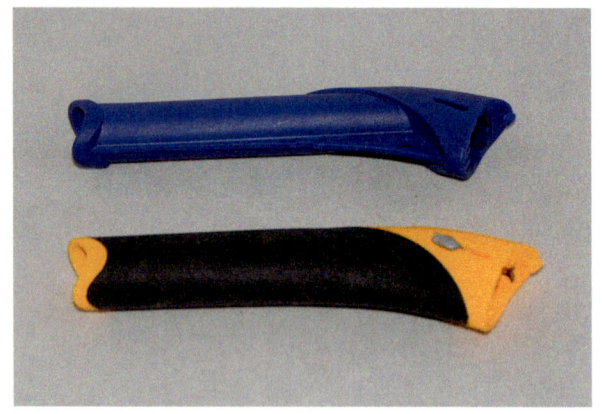

Figure 24.18 For multi-component injection molding, a stiff material is often over-molded with a softer high-friction material. The handle of a ski pole, as shown here, is an example of this. In most cases, incineration for energy production is the only recycling option for this kind of product.

24.3.3 Coding

See Figure 23.35 in Section 23.15 for a recap on recycling codes. If a copolymer is used–such as the impact-resistant PP, which comprises PP + PE–the correct coding is: ⁂.

This is also applicable to multi-component injection molding with different polymers in the product (see Figure 24.18).

Figure 24.19 This freezer box is produced in PP copolymer and thus has a recycling triangle with a seven in it shown on the base. The lid is co-injected with a TPE seal and would be marked with a seven even if it was made of PP homopolymers, since the seal is made of another polymer.

24.3.4 Cleaning

Collected components that are to be ground and recycled by injection molding, extrusion, etc., have to be carefully cleaned beforehand.

Figure 24.20 This chain saw housing has to be carefully cleaned before grinding if any of the nylon material is to be reused.

24.4 Rule 4 – Integrate Several Functions into One Component

One of the great advantages of plastics is that several functions can be integrated into one single component. If this were to be done with metal, different materials would have to be used for the different functions, with additional costs for assembly.

Examples of functions that can be integrated into a plastic component are:

- Snap-fits
- Pipe connectors
- Seals
- Slide bearings
- Threads
- Gear racks
- Stiffeners (ribs)

The following figures show an example of an oil sump with a splash guard (prevents oil splashing up on the pistons in a car engine). Here, the product development has been done in two steps.

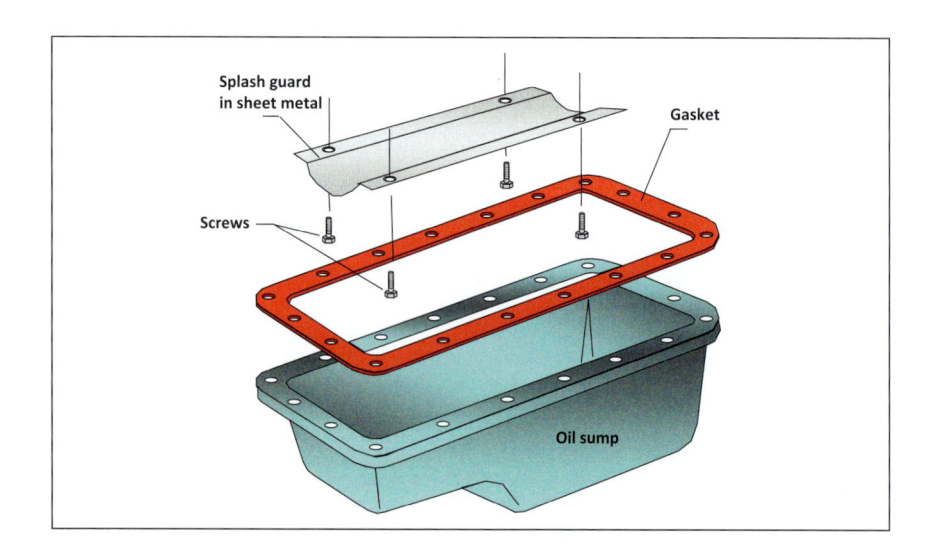

Chapter 24 – Design Rules for Thermoplastic Moldings

Figure 24.21 The original metal design consisted of a total of seven components. [Source: DuPont]

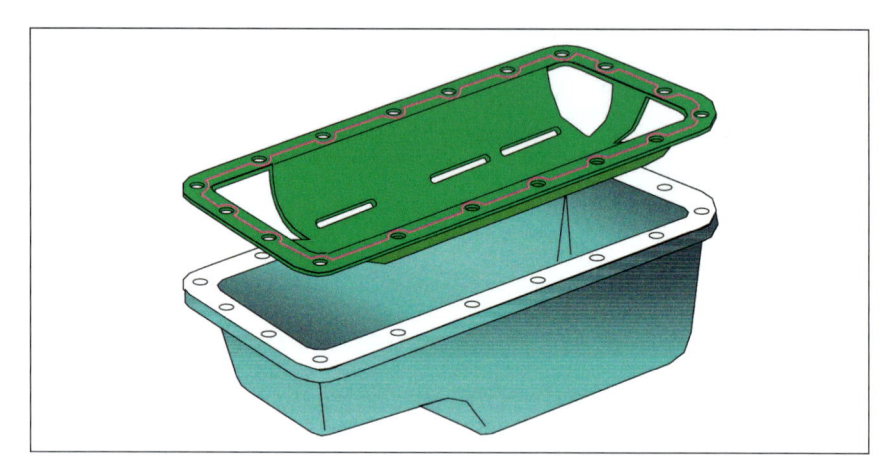

Figure 24.22 By integrating the gasket with the splash guard (and thereby eliminating the four screws) the number of components has been reduced to two. [Source: DuPont]

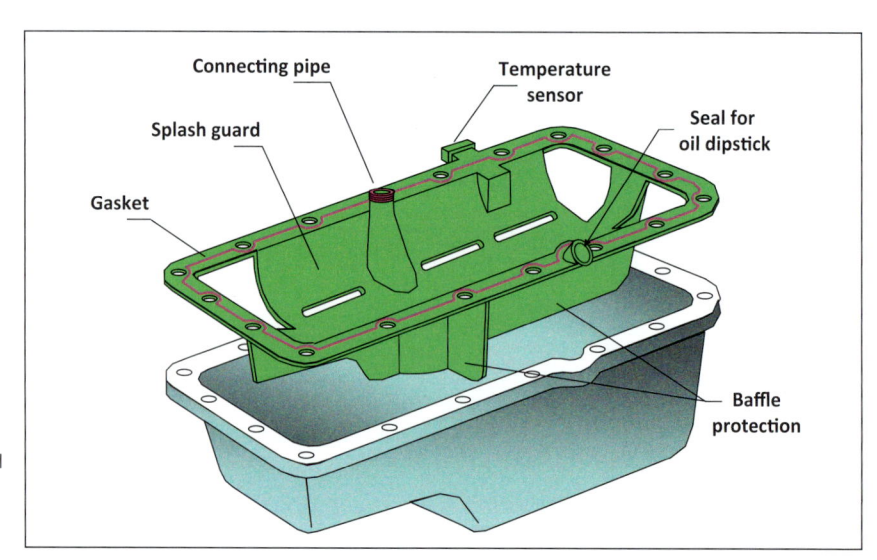

Figure 24.23 By continuing to integrate new functions like the connector pipe, temperature sensor, seal for oil dipstick, and baffle protection, the total costs are considerably reduced compared to multi-component alternatives. [Source: DuPont]

24.5 Rule 5 – Maintain an Even Wall Thickness

When determining the wall thickness of a detail, the load and environmental influences (temperature, humidity, chemicals, sunlight, etc.) that the component will be exposed to must be taken into consideration. The expected service time is also of importance.

As a rule, compromises on thickness are often required, since it should be *thin* enough to:

- Meet the lower weight requirements
- Meet cost demands (a thinner wall gives a shorter hold pressure time and consequently a shorter cycle time)
- Allow quick and effective temperature control during injection molding

But at the same time it should be *thick* enough to:

- Meet functionality and load requirements
- Withstand handling and transportation
- Withstand assembly, mounting, and servicing
- Allow easy filling of the part
- Allow part ejection from the cavities without deformation

It is important that the wall thickness is even (max. ±15% variation), since this affects the mold shrinkage. A thicker wall has higher shrinkage. When you have large wall thickness variation, which means different shrinkage, you will get internal stress between the different parts of the component, often resulting in warpage and distortion.

A "normal" wall thickness for thermoplastic parts is usually specified as between 1.5–4 mm. Anything below 1.5 mm may cause problems with filling of the cavity. If the wall needs to be increased, it should be done in small steps, since a doubling of the wall thickness results in a four times increase of the flow. If more than 4 mm is needed, the cycle time (and production costs) will increase since the hold pressure time versus the wall thickness is not linear.

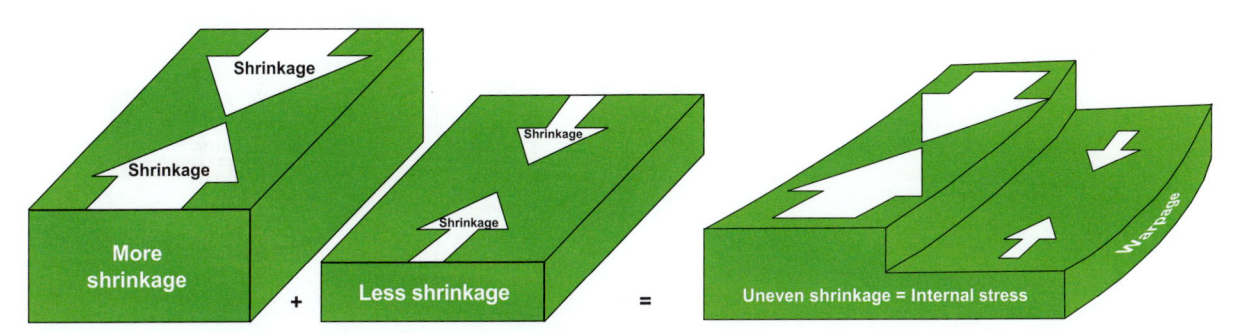

Figure 24.24 If the wall thickness varies in a plastic component, the shrinkage will also vary, which means there is a high risk of stress in the final component, which results in warpage. [Source: DuPont]

Chapter 24 – Design Rules for Thermoplastic Moldings

Figure 24.25 Injection molded flow spirals are used to determine the flowability at different wall thicknesses.

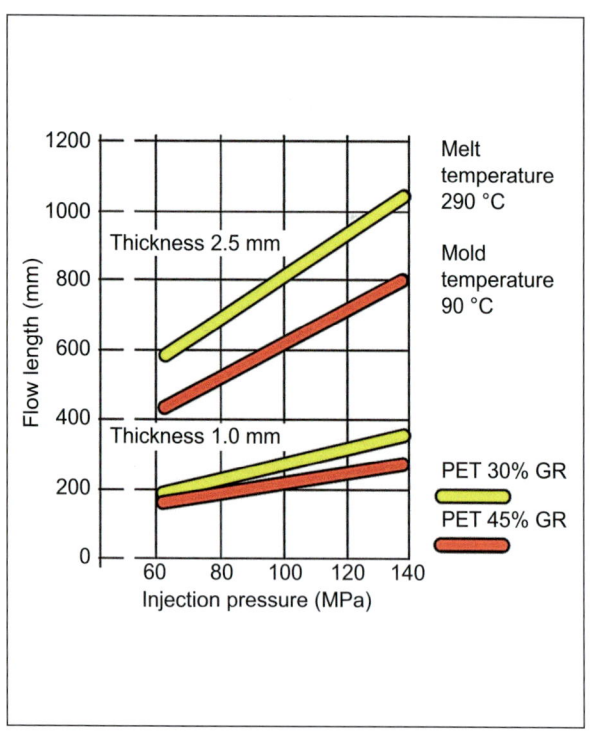

Figure 24.26 In addition to wall thickness, the flow length is also influenced by the melt temperature, the injection pressure, and the mold temperature. [Source: DuPont]

24.6 Rule 6 – Avoid Sharp Corners

Many plastics are sensitive to notches in the form of a too-small corner radius. As a rule of thumb, the corner radius should be at least half the wall thickness. If it is smaller there is a risk that the stress concentration factor becomes too high whereupon the component will break even under moderate load (see Figure 24.27).

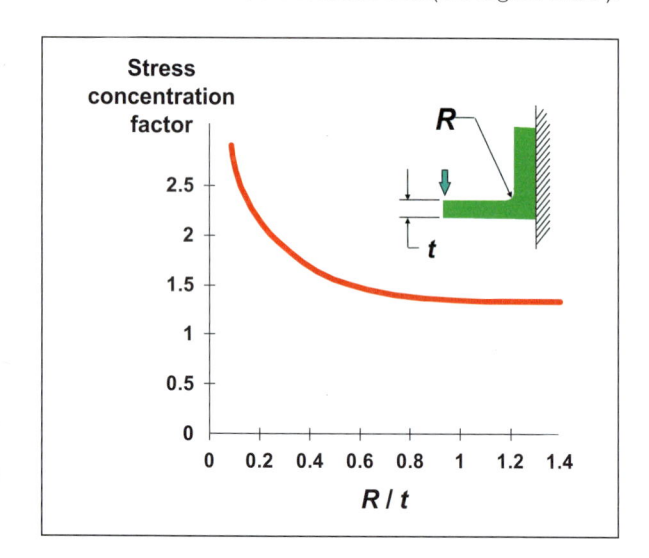

Figure 24.27 It is recommended that the corner radius R should be at least 0.5 times the wall thickness t. If it is below 0.3, the stress concentration significantly increases. Sometimes grinding the corner with an emery cloth can give a surprisingly good result. [Source: DuPont]

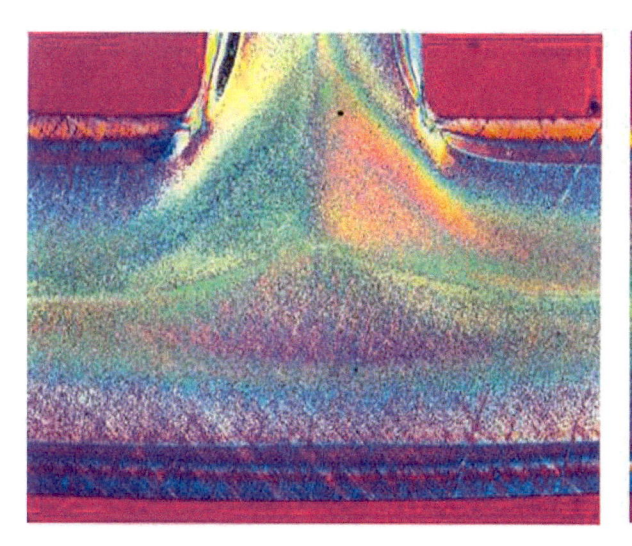

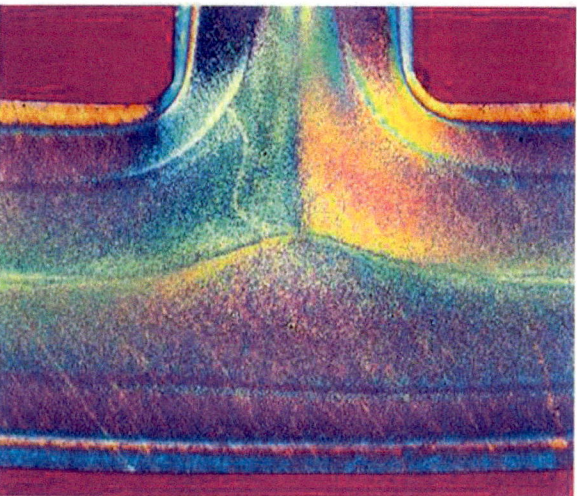

Figure 24.28 Each of the pictures here shows a rib connecting to a wall. The images are taken using a microscope at about 40 times magnification. Polarized light is used to highlight the samples from below. In the left-hand picture, a disruption can be seen in the crystal structure at the root of the rib. In the right-hand picture, this disruption is absent, due to the corners being rounded. [Photo: DuPont]

Notch sensitivity varies greatly between different plastic materials. One way of testing this is to make test bars as shown in Figure 24.29 and measure the breaking force. The areas between the V- and U-shaped notches on the test bars are identical. The power needed to break the test bar in acetal is about nine times higher for the U-notch compared to the V-notch. For a test bar made in a conditioned standard polyamide 66 grade the difference between U- and V-notches can be more than 40 times, whereas it is negligible for a "super tough" impact-modified polyamide 66.

Figure 24.29 Test bar for testing notch sensitivity for thermoplastics.

24.7 Rule 7 – Use Ribs to Increase Stiffness

In general the stiffness of a component can be increased by:

- Increasing the wall thickness
- Increasing the E-modulus of the material, i.e. increasing the reinforcing fiber content
- Adding ribs to the design

In those cases where sufficient stiffness cannot be acquired by modifying the design, use a material with higher stiffness. The most common way is to choose a grade with a higher fiber content (most often glass fiber). With a consistent wall thickness, this will result in a linear increase of stiffness. A significantly more effective way is to increase the stiffness with the help of ribs. Here, the increase in stiffness is a result of the increased moment of inertia.

24.7.1 Limitations when Designing Ribs

By increasing the height and thickness of the rib, a high moment of inertia is obtained. However, when designing with engineering polymers this may cause serious problems such as sink marks, porosity, and warpage. If the height of the rib is too big, there is an additional risk that it will get buckled when exposed to a load. With these negative effects in mind, it is important to keep the rib dimensions within the recommended limits.

In order to facilitate the ejection of the ribbed component from the cavity, it is important to have demolding tapers on the rib.

This angle is dependent not only on the height of the rib but also on the material being used. The mechanical load on a ribbed design is often at its peak at the root of the rib. It is therefore necessary to have an adequate corner radius (see Rule 6).

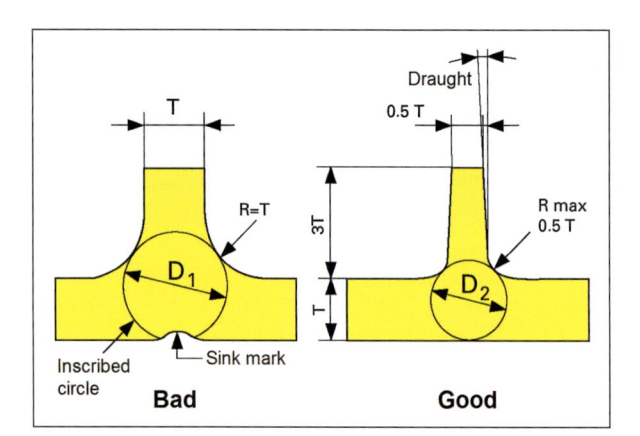

Figure 24.30 Recommendations for rib dimensioning. [Source: DuPont]

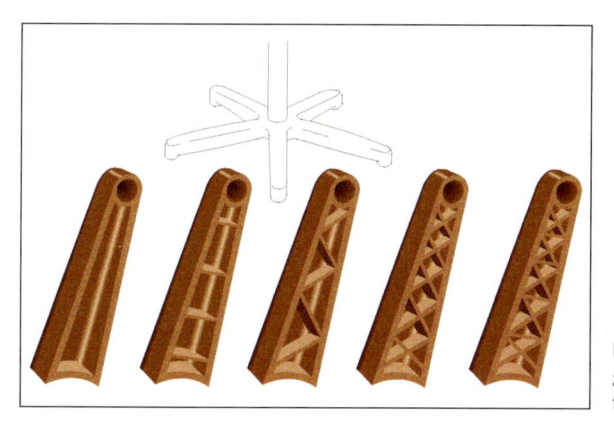

Figure 24.31 The rib design has a great significance for the stiffness of the part design. [Source: DuPont]

24.7.2 Material-Saving Design

It is important that the ribs are located properly to achieve an optimal design. An example of this is the rib structures of the chair base shown in Figure 24.31. Here we can see that the cross-rib design in the base on the far right has a 30 times stiffer design in comparison to the non-ribbed base on the far left. Aluminum chair bases are mainly designed as the beam to the far left, while plastic chair bases most often have ribs according to the beam on the far right. This highlights the importance of making a radical redesign when going from metal to plastic to achieve the most economical product.

24.7.3 Avoid Sink Marks at Rib Joints

Sink marks often appear where the rib meets the outer wall, which can be minimized by:

- Making the rib thin enough, i.e. less than half the wall's thickness
- Avoiding accumulation at rib joints
- Etching the surface or picking a lighter color for the product

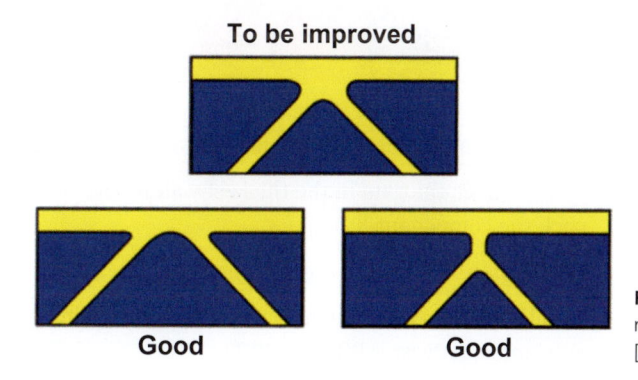

Figure 24.32 Accumulation at the rib joints should be avoided. [Source: DuPont]

24.8 Rule 8 – Be Careful with Gate Location and Dimensions

The choice of gate location and dimensioning in a plastic component is extremely important, as it affects the following characteristics:

- The mold filling process (flow path and length)
- Part dimensions and tolerances (mold shrinkage)
- Warpage (internal stress)
- Mechanical properties
- Surface finish and surface marks

Not all designers take this into account. They often let the tool maker decide on gate location and the runner system without being aware of the full requirement of the part. This sometimes leads to the expected characteristics of the final product not being achieved.

Aside from designing the component and making the strength calculations, you should always ensure that the number of gates is sufficient and that their locations are in accordance with the expected weld lines. Since gates and weld lines always are the weakest spots of a part, it is important to avoid them in highly loaded areas.

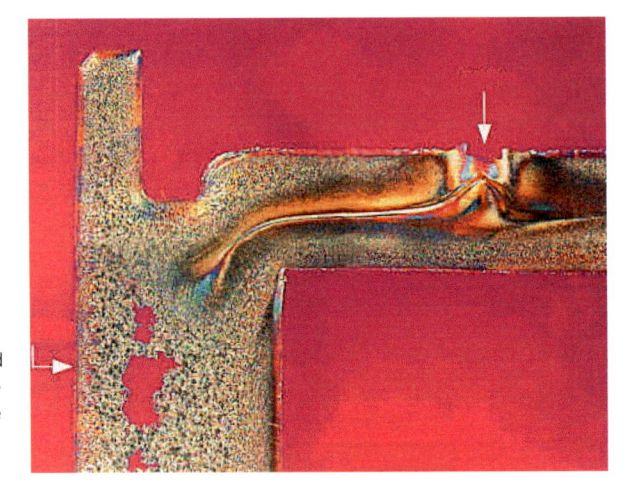

Figure 24.33 Shown here is a photo of a thin layer of an acetal part at 25 times magnification and illuminated from below by polarized light. The gate is in the thinnest wall, and you can see voids in the thickest wall.
[Photo: DuPont]

Some points to consider when determining gate location:

- Always place the gate in the thickest wall if possible, in order to sufficiently pack the component
- Do not place the gate in areas of high load
- Avoid conical gates for semi-crystalline engineering polymers (e.g. acetal, polyamide, polyester PBT and PET)
- A gate that is too small not only hinders the optimal packing of the component, but also increases the risk of shearing problems during the filling process (delamination, surface marks, or glass fiber streaks). This is particularly true at high injection speeds.

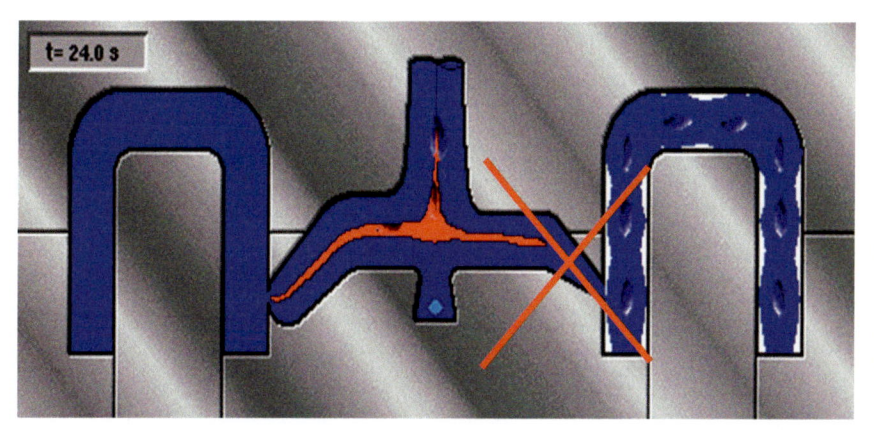

Figure 24.34 Conical gates are not suitable for semi-crystalline engineering polymers. The gate size is identical in both cavities. The left part is completely packed, while the right one shows voids and sink marks. [Source: DuPont]

24.8.1 Weld Lines

Some points to consider when it comes to weld lines:

- A weld line occurs between every gate if there is more than one
- A weld line occurs after every hole
- Eliminate or reduce the number of weld lines if possible
- Do not place the weld lines in areas of high load
- Reinforcement fibers do not have a reinforcing effect in the weld line
- Try to avoid a "butt" weld line

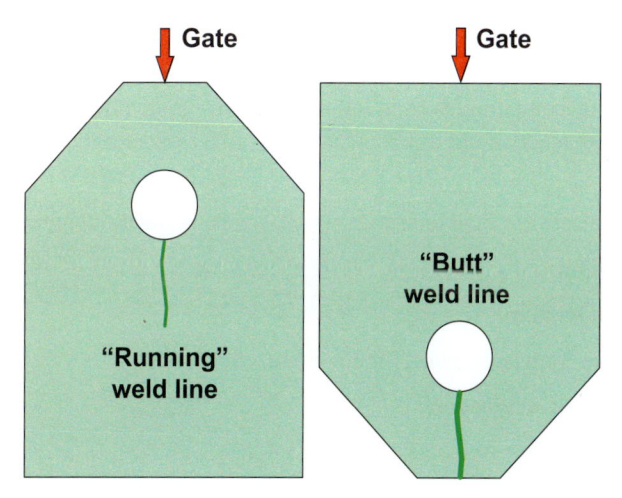

Figure 24.35 If the material is allowed to flow on after the weld line, the line is more or less erased, and the component becomes considerably stronger.

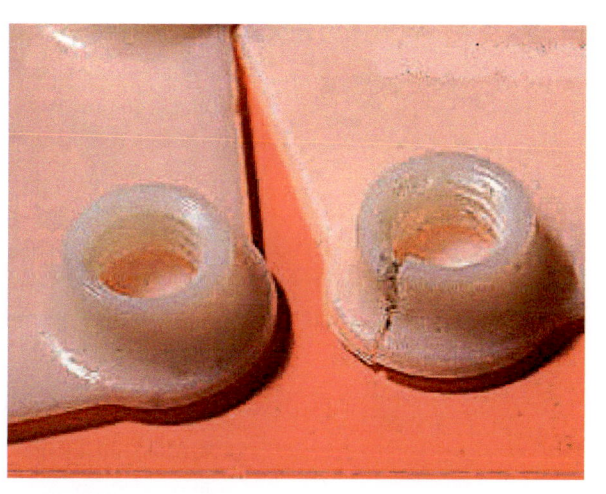

Figure 24.36 A threaded hole that has cracked along the weld line. [Photo: DuPont]

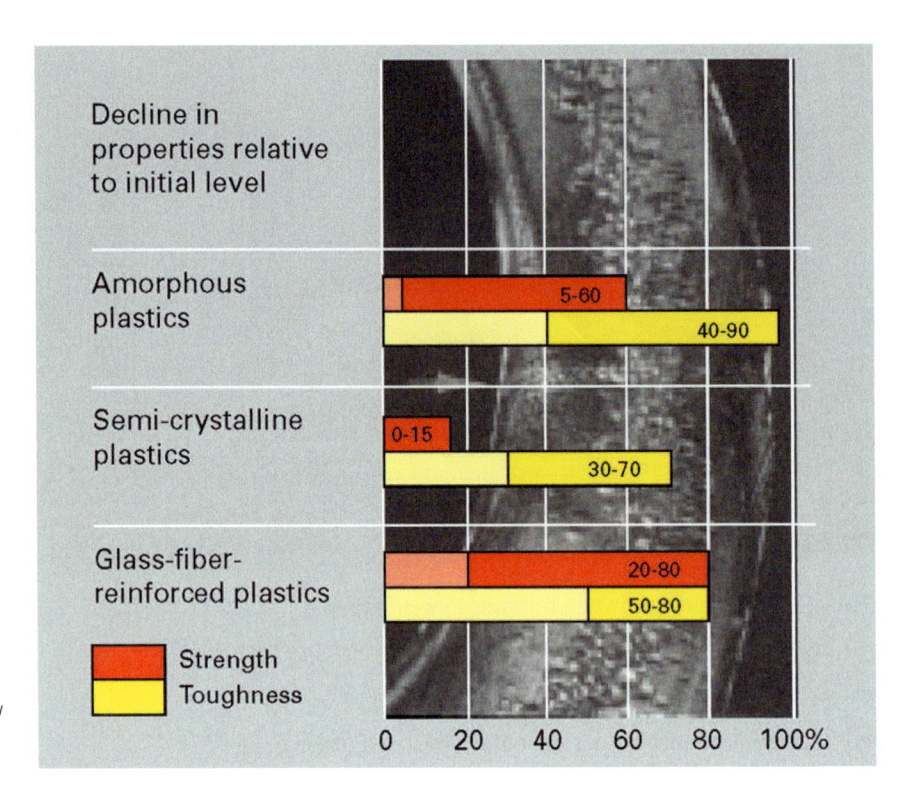

Figure 24.37 The picture shows how strength and toughness are impaired by a weld line. [Source: DuPont]

24.9 Rule 9 – Avoid Tight Tolerances

Injection-molded components cannot be manufactured with the same tolerances as machine-processed metal components. Even though most designers know this, tolerances that cannot be reached or are unnecessarily expensive to manufacture are still specified.

Properties of plastic that affect the final tolerance of a plastic component include:

1. Tolerances for tool making
2. Tolerances for injection molding
3. Tolerances of plastic raw materials (glass fiber content, etc.)
4. Warpage of a component dependent on:
   - Mold shrinkage
   - Post-shrinkage
   - Part design (varying wall thickness, etc.)
   - Flow direction or glass fiber orientation
   - Internal stresses
   - Uneven temperature control in the mold

5. Measurement variations of the finished part dependent on:
   - Moisture absorption
   - Thermal expansion (which can be 10 times higher for plastics versus metals)

The tolerances should not be as tight as possible, but as tight as required for the functionality of the part. Commonly accepted tolerances for a cost-effective production is 0.25–0.3% deviation from the nominal dimension, but this is of course dependent on the application.

Manufacturing a component with a nominal measure of 30 mm requires a tool tolerance of 30 ± 0.01 mm and a tolerance of the injected-molded part of 30 ± 0.03 to 0.04 mm. For parts up to 150 mm in size, ±0.15% is usually specified for high precision and ±0.30% for technical molded components. For components above 150 mm, the recommended values are ±0.25% for high-precision moldings and ±0.40% for technical moldings.

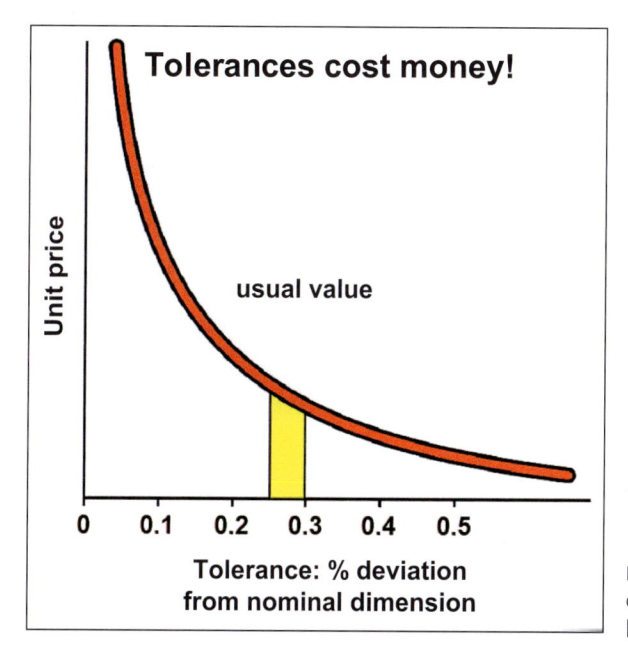

Figure 24.38 Unit price as a function of tolerance requirement. [Source: DuPont]

24.10 Rule 10 – Choose an Appropriate Assembly Method

See the next chapter!

CHAPTER 25
Assembly Methods for Thermoplastics

Most designers seek to make their plastic products as simple as possible while at the same time integrating all the necessary functions. The product should preferably come out of the mold complete and ready, but sometimes—for functionality or cost purposes—it can be necessary to make the product in two or more parts that are assembled at a later stage.

There are several assembly methods for thermoplastic products, and this chapter considers most of them. To begin with, it is common to divide the assembly methods into those where the product can be disassembled and reassembled several times (e.g. using screw joints) and those permanent methods, whereby components are assembled only once (e.g. welding).

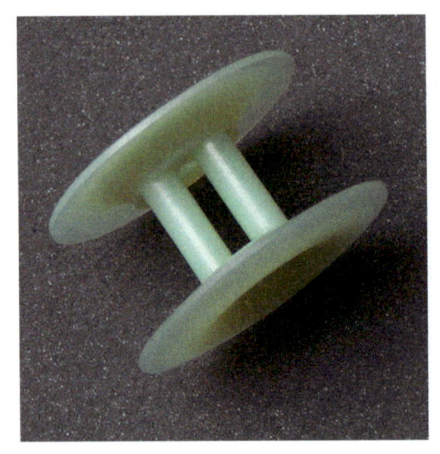

Figure 25.1 This bobbin is made as two identical halves in a mold with a simple parting line. The halves are then rotated 90° in relation to each other and then joined together by press-fitting.

25.1 Assembly Methods That Facilitate Disassembly

Among the dismountable methods, the following methods are usually used when it comes to plastic details:

- Self-tapping screws
- Threaded inserts
- Screw joints (with an integrated thread)
- Snap-fits (specifically designed to allow disassembly)

Figure 25.2 If a good screw joint is required for a self-tapping screw, the plastic material should have a stiffness lower than 2800 MPa (i.e. the same as for POM). For stiffer materials (e.g. glass fiber reinforced), a threaded hole or threaded inserts are recommended. It is also important to use a self-tapping screw that is specifically developed for plastic materials.

Figure 25.3 On the left is a threaded brass bushing in the wall of a pump housing made in glass reinforced polyamide 66. The bushing can either be overmolded or pressed into the plastic wall.
The plastic caps on the plastic bottles in the picture on the right are typical examples of a screw joint with integrated threads.

25.2 Integrated Snap-Fits

Plastic snap-fits can be designed for use with both dismountable and permanent assembly methods.

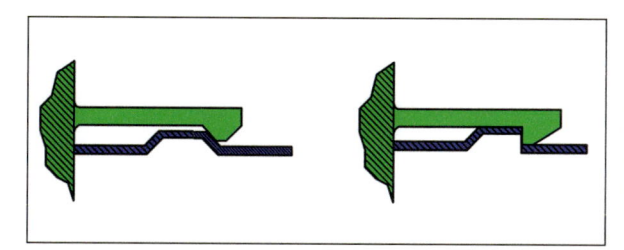

Figure 25.4 The snap-fit on the left can be dismounted as it can climb over the black plate if the tensile load is coming from the left. The snap-fit to the right is designed for permanent assembly, and has a 90° angle which cannot be pulled apart.

Figure 25.5 It is often a great advantage to show the user which direction to use to dismount a snap-fit.

25.3 Permanent Assembly Methods

In addition to the snap-fits described above, the list below shows the most common permanent assembly methods:

- Ultrasonic welding
- Vibration welding
- Rotational welding
- Hot plate welding
- Infrared (IR) welding
- Laser welding
- Riveting
- Gluing

25.3.1 Ultrasonic Welding

Ultrasonic welding is a fast and relatively cheap method that is used for assembling smaller components.

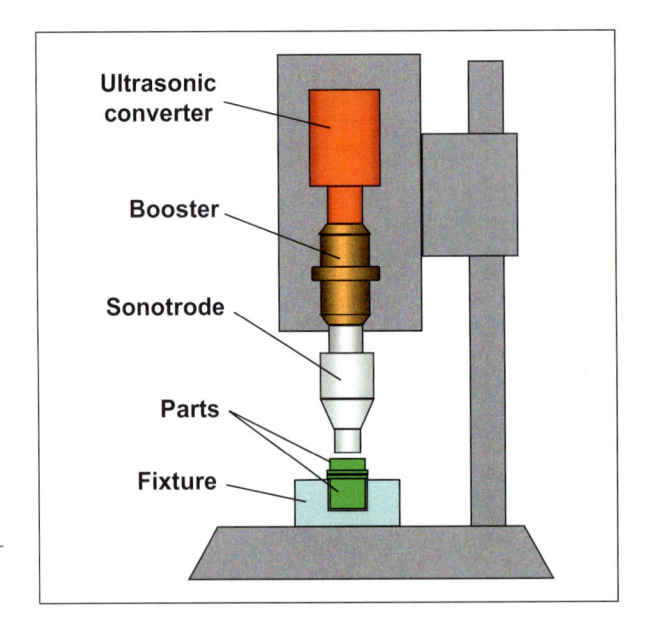

Figure 25.6 Shown here is an ultrasonic welding equipment. An electromagnetic signal is amplified in the booster and then transmitted to the sonotrode that is pressed against the plastic component (shown here in green). The sonotrode vibrates with high frequency and transmits the movement to the upper part, developing frictional heat between the two halves of the plastic component.

Advantages

+ Fast (normal welding time < 1 second)
+ Easy to automate
+ Leak-free seals
+ Economical for mass production

Limitations

- Component size max. 80 mm × 80 mm (> 80 mm requires several sonotrodes)
- Different polymers cannot be welded together (although the same material with different reinforcements can be welded together)
- Hygroscopic materials (like polyamide) must not be allowed to absorb any moisture before welding
- High frequency (unpleasant noise)

Figure 25.7 This BIC lighter is an example of an ultrasonic welded product. The white bottom oval is welded to the green gas container, and there is a high demand on the welded joint not to allow any gas leakage.

25.3.2 Vibration Welding

Vibration welding is a method that is primarily used when welding larger plastic components. As with ultrasonic welding, kinetic energy is transmitted to the surface between the plastic parts, where they are melted by the resulting frictional heat. Welding time is around 1 to 4 seconds.

Figure 25.8 A vibration welding machine. A machine of this size is required to weld the tank halves shown in Figure 25.9. [Photo: Stebro Plast AB]

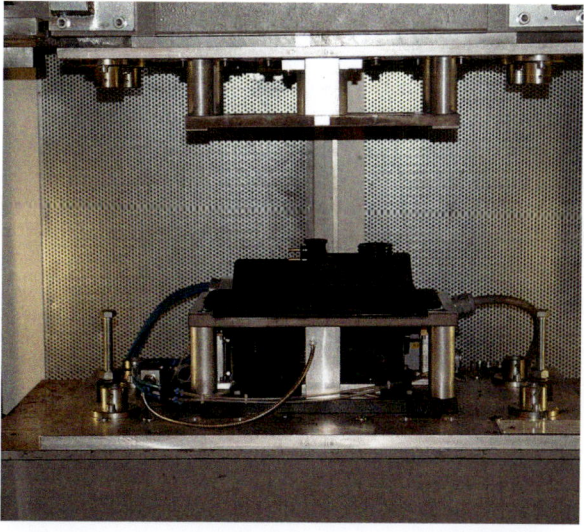

Figure 25.9 This vibration welded tank is made in two halves using impact-resistant polyamide 6. It is important that they are welded directly after the injection molding before they have had time to absorb moisture from the surrounding air. [Photo: Stebro Plast AB]

Advantages

+ Is suitable for very large components
+ Is suitable for components with multiple levels of parting lines
+ Several components can be put in a fixture and be welded at the same time
+ Leak-free joints

Limitations

− Very expensive welding equipment
− Different polymers cannot be welded together (although the same material with different reinforcement levels can be welded together)
− The components must be able to move up to 3.5 mm in relation to each other during the welding sequence.

25.3.3 Rotational Welding

This method is fast and cheap but requires that the welding joint be circular. During this method, the lower plastic part is held in a fixture while the upper part is fixed in a rotating chuck. The rotating chuck is released and pressed down against the lower part. The kinetic energy is thus transformed into frictional heat, and the parts melt together.

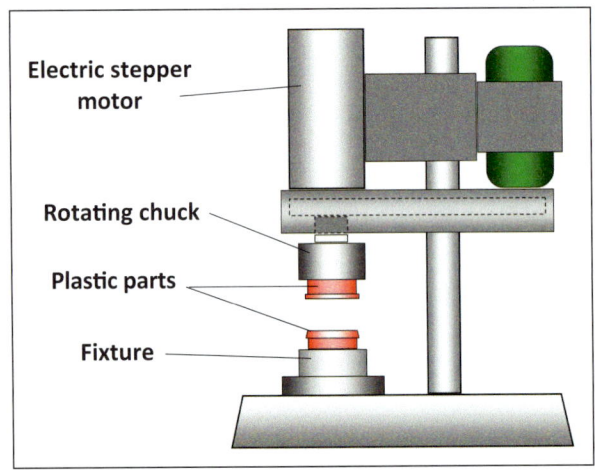

Figure 25.10 Rotational welding equipment.

Figure 25.11 A ball-shaped float made in glass fiber reinforced polyamide with the help of rotational welding.

Advantages

+ Fast method with welding times below 1 second
+ Low investment costs for manufacturing prototypes
+ Leak-free joints

Limitations

− Only circular surfaces can be welded
− Orienting the different parts in relation to each other is quite complicated

25.3.4 Hot Plate Welding

Hot plate welding is another method that is suitable for large components. The method is specifically suitable for amorphous plastics or soft materials. Figure 25.12 illustrates the principle behind this method.

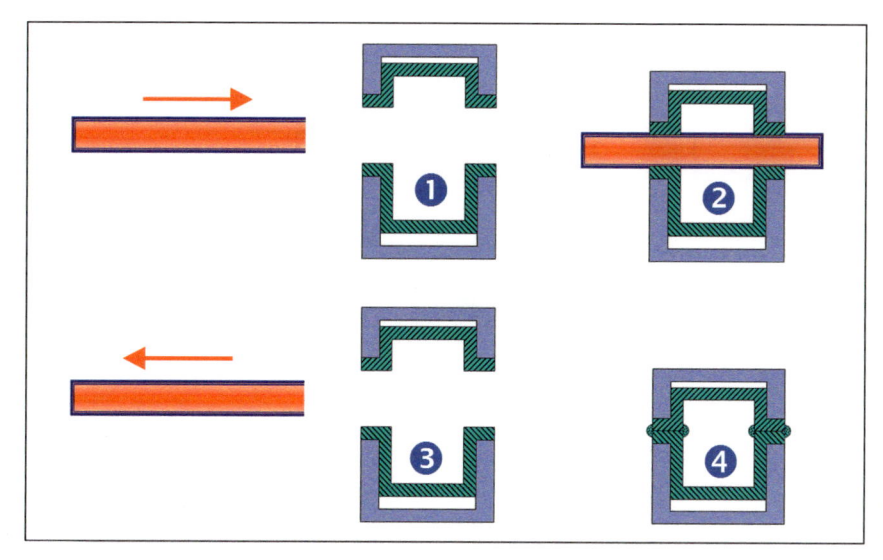

Figure 25.12 The parts that are to be welded are placed in their own fixtures. A heated metal sheet, called a hot plate, is placed between the parts in position ❶.
In position ❷ the parts are pressed against the hot plate, which can be above 300°C if welding polyamide. The surfaces will now begin to melt.
In position ❸ the parts are separated and the hot plate is quickly withdrawn.
In position ❹ the parts are pressed together, thus the surfaces melt together.
The hot plate is often coated with a layer of fluoropolymer (Teflon) to prevent sticking of the plastic.

Advantages

+ Suitable for parts with multiple levels of parting lines
+ Suitable for very large parts
+ Several parts can be welded simultaneously
+ Leak-free joints

Limitations

− Polyamides can be difficult to weld due to oxidation in the surface
− Hot plate welding takes a relatively long time (20–45 sec)
− Only materials of the same kind can be welded
− The polymer can stick to the hot plate
− Strong requirement on flatness for the parts that are to be welded
− High temperatures around the welded joint

Figure 25.13 Here you can see that there will be a big flash around the welded joint during hot plate welding. [Photo: DuPont]

25.3.5 Infrared Welding

This method is reminiscent of hot plate welding but instead of a metal sheet, a strong infrared (IR) source is placed between the plastic halves, causing the surfaces to melt.

The melting temperature is determined by the distance to the elements and the heating time.

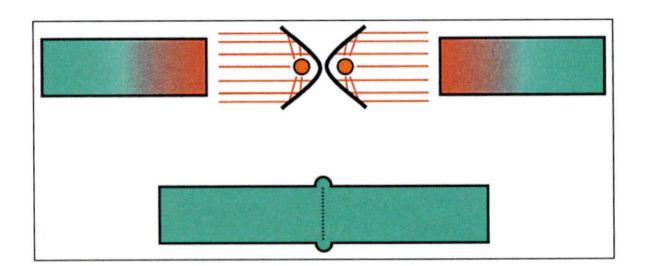

Figure 25.14 Here you can see that the IR source (the red dots in the reflectors) acts as a substitute for the metal sheet used in hot plate welding.

25.3.6 Laser Welding

This is the most advanced welding method, which requires that the plastic material in the upper part (shown in yellow in Figure 25.15) is transparent to the laser beams while the material in the lower part (shown in blue) will absorb the energy of the laser, causing the surfaces to melt together.

Advantages

+ No visible marks or damage on the outside of the weld joint
+ No vibration damage inside the component
+ Minimal heat generation while welding, i.e. minimal temperature influence and warpage
+ Different colored parts can be welded together by the addition of an invisible laser-absorbing pigment
+ No flash is generated

Limitations

− Requires material with different absorption characteristics for the chosen laser wavelength
− Not all polymers are "laser transparent"
− Perfect fit of meeting surfaces is required (i.e. no amount of warpage can be accepted)
− Special laser-transparent fixtures are required in certain cases
− Thin-walled components are required for the best results
− Welding at different parting line levels is very difficult

25.3 Permanent Assembly Methods

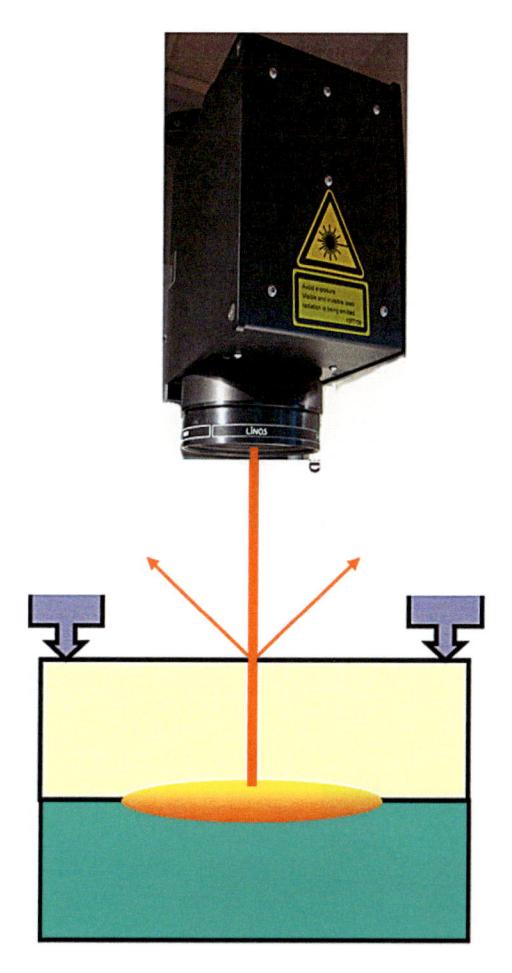

Figure 25.15 The picture shows the principle of laser welding. The blue arrows symbolize that the parts are being pressed against each other, and the yellow oval is the laser beam that is absorbed by the surface of the lower part while causing the surface of the upper part to melt.

Not all plastics are suitable for laser welding as the upper part has to be transparent to the laser beam. Special pigments can be added to make the lower part absorb the energy of the laser.

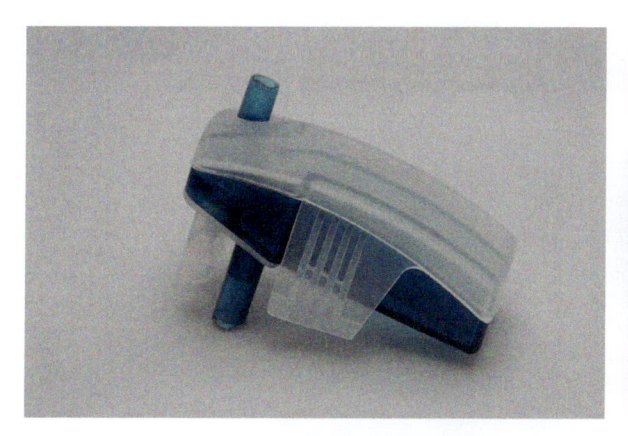

Figure 25.16 Shown is a filter where the laser-transparent upper part has been laser welded to the laser-absorbing blue base. [Source: Arta Plast AB]

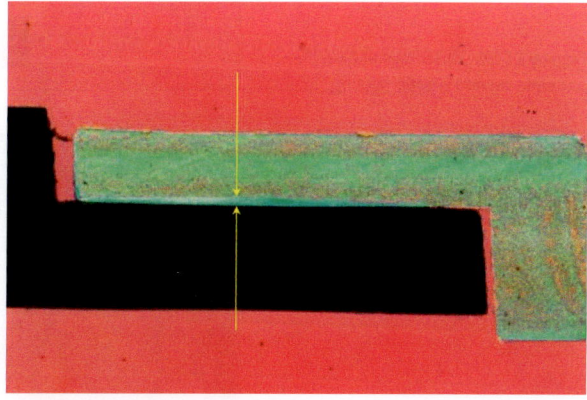

Figure 25.17 Totally flash-free joints can be obtained by laser welding. [Photo: DuPont]

Chapter 25 – Assembly Methods for Thermoplastics

25.3.7 Riveting

This method uses a punch that sprains the rivet in plastic (colored in green in Figure 25.18). The punch can either be cold, warm, or transferring ultrasonic energy.

Advantages

+ A strong permanent assembly method
+ Fast and economical
+ Allows assembly of different materials (for example metal to plastic)

Limitations

- Hygroscopic materials such as polyamides must be conditioned or impact modified

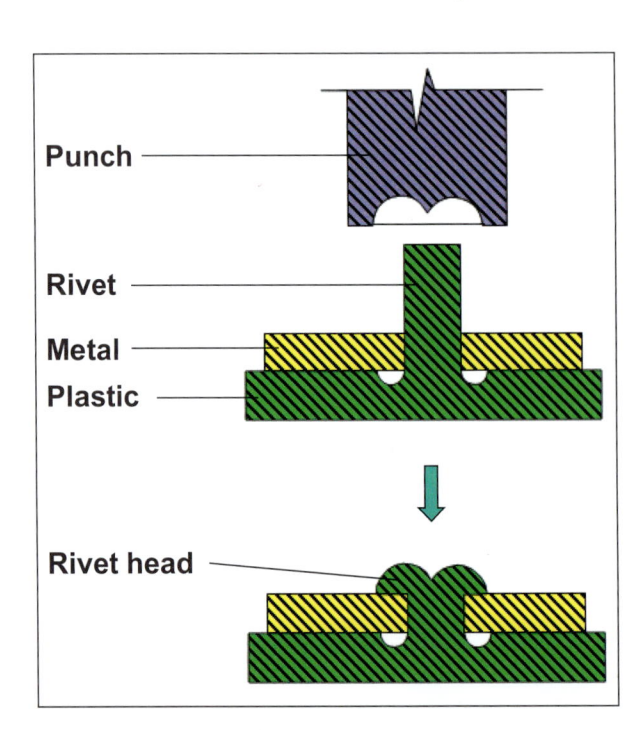

Figure 25.18 The image shows the principle behind plastic riveting.

Figure 25.19 This belt clip for a mobile phone is made in two parts that are then riveted together by heat.

268

Figure 25.20 Chain links in acetal for a conveyor belt. The wear durability has been improved by riveting a steel plate to the upper side of the chain link.

25.3.8 Gluing

This method is not so common for large-scale mass production. It is mostly used for making prototypes. The great advantage is that most materials—such as metal, glass, fabrics, wood, or other plastics—can be glued together with the aid of adhesives.

Advantages

+ Allows assembly of different materials
+ Suitable for big or complicated surfaces
+ Suitable for low volume and prototypes

Limitations

– Labor intensive and expensive
– Not all plastics can be glued together (for example olefins and fluoropolymers)

Figure 25.21 Different types of polymers require different types of adhesives, a fact which most of the larger adhesive suppliers can inform you about.

CHAPTER 26
The Injection-Molding Process

26.1 Molding Processing Analysis

In this chapter, we will go through the main injection-molding parameters that affect the quality of the moldings. We will also emphasize the value of working systematically and having good documentation.

Figure 26.1 shows a document called "Injection Moulding Process Analysis". There is an Excel file that can be downloaded at www.brucon.se. On this sheet we can record most of the parameters that need to be documented to describe the injection-molding process for a molded part.

This document was designed by the author of this book when he was responsible for the technical service at one of the leading plastic suppliers in the Nordic region.

You may think: Why should I spend time to fill it in when I can get all the parameters printed out directly from the computer system in my molding machine?

The answer is that you would probably drown in all the figures and only with difficulty find the cause of the problem. You would also have difficulties in finding the key parameters as the printouts from different machines are completely different.

This document is perfect for use both in problem solving and as a basis for process and cost optimization as well as for documenting a test drive or a start-up of a new job. If you fill in the document when the process is at its best you will have good benchmarks for comparison when there is a disturbance in the process. Therefore, we will closely examine the structure of this document and explain the meaning of the information in each input field. On the last page of this chapter, the document is presented in full-page format (Figure 26.48).

INJECTION MOULDING PROCESS ANALYSIS

Please, use tab button when filling in this sheet

Customer | Location | Date

Contact person | Phone no | Email

Problem / Desire

Material | Alternative usable material

Lot no | Masterbatch | MB content % | Regrind %

Machine | Hold pressure profile possible ☐ | Clamping force kN

Screw type | Shut-off nozzle ☐ | Vented barrel ☐ | Screw diameter mm

Mould / part name | Hot-runner system | No of cavities

Wall thickness at gate mm | Max. wall thickness mm | Min. wall thickn. mm

Sprue dimension mm | Runner dimension mm | Gate dimension mm

Nozzle diameter mm | Parts weight (sum) g | Full shot weight g

Drying | Hot air dryer... ☐ | Dehumidified dryer... ☐ | Direct transport of dried resin to hopper... ☐

Drying temp °C | Drying time hours

Processing

Cylinder temp: Nozzle (front) °C | Zone 4 °C | Zone 3 °C | Zone 2 °C | Zone 1 °C

Melt temperature °C | Mould temp. moving °C | Mould temp. fixed °C | Temp. checked by pyrometer ☐

Injection pressure MPa | Hold pressure >>> Profile? <<< MPa | Hold press. time sec

Injection speed >>> Profile? <<< % ☐ mm/sec ☐ | Fill time sec

Back pressure MPa | Screw rotation RPM | Peripheral screw speed Calculated m/sec

Dosing time sec | Cooling time sec | Total cycle time sec | Hold-up time Calculated min

Dosing length mm ☐ cm³ ☐ | Max. dosing length mm ☐ cm³ ☐ | Suck-back mm ☐ cm³ ☐

Hold pressure switch mm | Cushion mm | Cushion stable ☐

Comments

www.brucon.se / 2016

Figure 26.1 The working tool "Injection moulding process analysis", which is described in this chapter.

26.2 Contact Information

In the top part of the page there are some fields that can be filled in. If you only plan to use the page for internal documentation, it will probably be redundant to fill in these data.

However, you should always fill in the date and contact person. If after several years you need to go back and see how a particular setting was made, it is usually interesting to know who did it and then get additional information.

It is also important to know when the setting was made if several settings have been made over time.

If you use the document to communicate externally with a raw material supplier or a sister company, etc., it also facilitates this if the contact details are filled out.

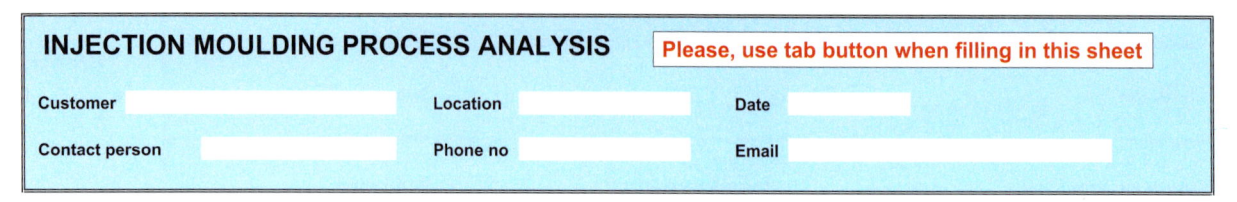

Figure 26.2 Contact information.

26.3 Information Pane

If after a long time has passed you want to make use of the information contained in this document, it is of great importance to understand the context in which it was documented. Was it when you had a problem, when you had solved the problem, after you had quality or cost optimized the part, or was it during a test trial with a different material? This is the type of information you should enter in the pane shown in Figure 26.3. Complete if possible the information with photos describing the problem.

Figure 26.3 The Information pane where you specify the type of problems or optimization desire, etc.

26.4 Material Information

In the pane shown in Figure 26.4 you can fill in the type of information about the plastic material that might be needed afterward to reproduce a successful trial, to investigate a complaint, or to understand a particular result.

Figure 26.4 Useful material information.

In the *Material* pane you can enter the plastic grade such as Rynite FR531 NC010, which in this case means that it is a flame-retardant glass fiber reinforced polyester PET from DuPont. The designation NC010 is a color code that means "natural color". If you use masterbatch coloring, enter the name of the masterbatch supplier and their designation in the field *Masterbatch* and the content in the field *Masterbatch content*.

The field *Alternative usable material* can be used when you have two raw material suppliers or when you run a trial with a new material.

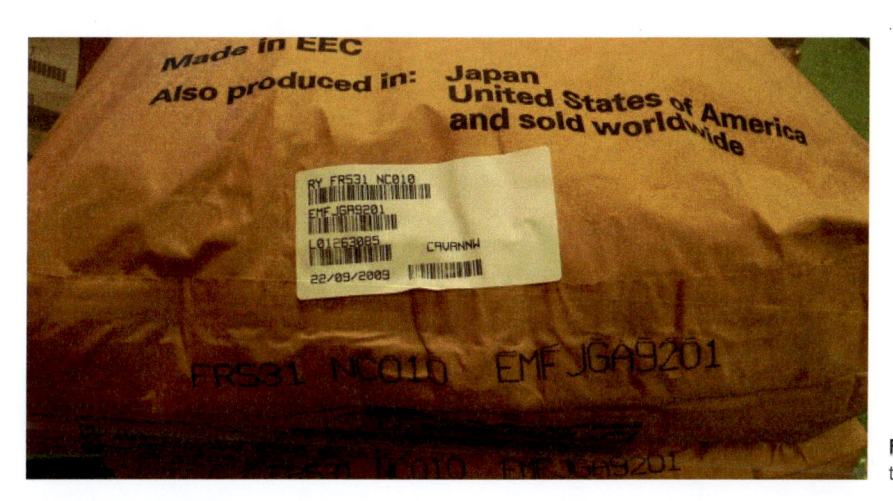

Figure 26.5 The lot no. is specified in the label on the bag.

The field *Lot no* is very important to fill in when there are problems, which can lead to a complaint on the resin. Most plastic producers want the serial or lot number for their own investigation of their production journals to see if something has deviated from the normal. With the help of the lot number, which generally is printed on the bag or container, you can usually get information about the viscosity and glass fiber content, etc. of the material. This is very useful if afterward you want to analyze the difference between different material batches.

In the last field *Regrind* you can fill in the mixing ratio of regrind. Normally up to one-third regrind can be mixed with virgin resin without any noticeable effect on the properties.

However, you should pay attention to:

- Not using parts or runners that are discolored or thermally degraded
- Have the grinder in direct connection to the machine or have such good procedures that foreign material or contaminations cannot accidentally get into the handler
- Have slowly rotating and less noisy grinders to reduce the concentration of small particles that otherwise must be screened away so as not to influence the dosing.

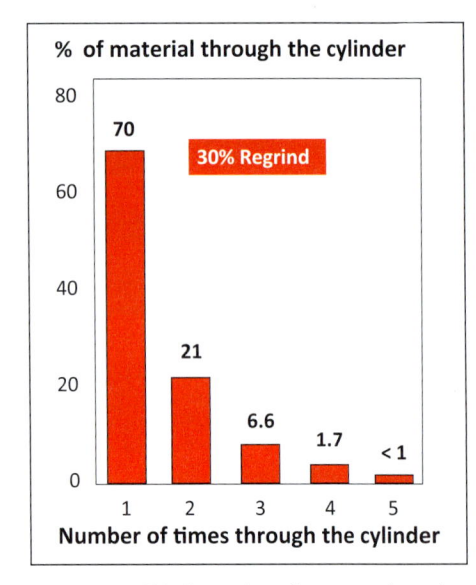

Figure 26.6 This figure shows how many times the material has passed through the cylinder at continuous mixing and thus has been affected thermally. The mixing ratio is 30%, and you can see that at this mixture 70% of the material is virgin and has passed through the cylinder only once. 91% of the material has passed through the cylinder once or twice, and only 1.7% of the material has passed through the cylinder more than three times.
It is only in exceptional cases that you need to use 100% virgin material, e.g. in some medical applications.

Figure 26.7 You should never grind discolored or thermally degraded material and mix it into virgin material. Although you can hide the discoloration with a black masterbatch, the mechanical properties, which have already been affected by the first degradation that caused the discoloration, will be further reduced each time material passes through the cylinder of the injection-molding machine.
On some thermoplastics such as polyester PET, you will not observe any changes other than the lower mechanical properties due to the thermal degradation. You will not see any discoloration or splays in the surface if the material was not dried enough at the first run.
If you mix degraded material, depending on a hydrolysis reaction in the cylinder, you will get a dramatic lowering of the impact strength of the injection-molded parts.

26.5 Information about the Machine

In the pane shown in Figure 26.8, you should enter the information about the injection machine that was used when you documented the injection-molding process.

In the field *Machine*, do not only enter the brand of the machine, e.g. Demag 250-1450, but also enter the internal numbering of machines used by your company. This is because you can get different results in the molding process in two apparently identical machines.

26.5 Information about the Machine

| Machine | | Hold pressure profile possible ☐ | Clamping force | kN |
| Screw type | | Shut-off nozzle ☐ | Vented barrel ☐ | Screw diameter | mm |

Figure 26.8 The machine information pane.

The checkbox *Hold pressure profile possible* was not available for of years, but this setting option is now available on most new injection-molding machines. In some situations and for some materials, it can be the hold pressure profile during the hold pressure time that is the setup parameter that solves problems with insufficient packing.

The *Clamping force* of the machine was earlier provided in tons. The correct unit is kN (kilo-Newton), and to get the correct value in this unit you should multiply the ton value by 10, e. g., a 250 ton machine has a clamping force of 2,500 kN.

Screw type: Most machines are as a standard equipped with general-purpose screws, abbreviated to "GP screws", that fit most thermoplastics. Other types of screws that occasionally will be used are high or low compression screws, barrier screws with multiple flanks, or screws specially designed for glass fiber reinforced material called armor or bimetallic screws. If you use masterbatch-colored materials in the production, a screw with a mixing head gives the best result. It is important that you check the box *Shut-off nozzle* if the cylinder is equipped with this type of nozzle.

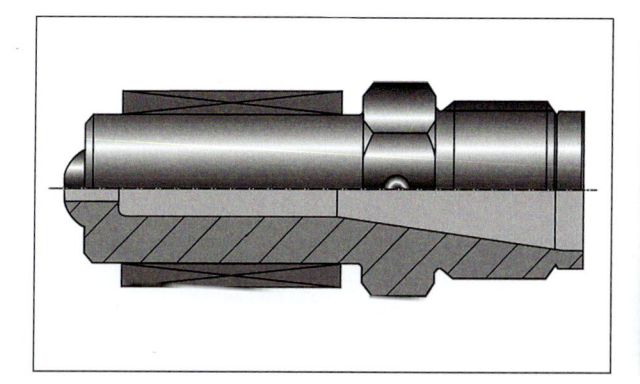

Figure 26.9 In most cases it is recommended to use an open nozzle of the type shown here. If you have extremely short cycle times and need to dose up the next shot under the opening, ejection, and closing phase or use gas injection molding, a shut-off nozzle is a must. After having finished such a job you should change the nozzle back to an open one. [Image: DuPont]

Figure 26.10 In general, you can see from the outside of the machine if the nozzle is of an open or a shut-off type. Many shut-off nozzles are equipped with a moving bar that opens or closes the flow. Others, such as in this picture, are equipped with a hydraulic or pneumatic valve. However, you should pay attention to whether there are spring-operated shut-off nozzles that cannot be detected from the outside.

Regarding the checkbox *Vented barrel*, it is very rare to see this ticked nowadays. Earlier, before dehumidifying dryers made their breakthrough on the market, you would sometimes see this type of cylinder, which was then used to dry the plastic material directly in the cylinder. Figure 26.11 shows the principle of a vented barrel.

Figure 26.11 This shows a venting barrel with the exhaust hole in the center. Through this hole water vapor can evaporate (symbolized by blue arrows). The idea is good, but unfortunately there are often problems as many materials instead degrade faster in this kind of cylinder.

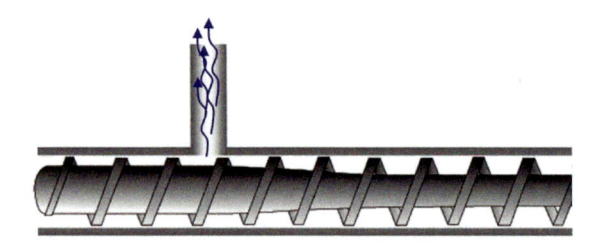

The last field in the machine information is *Screw diameter*. This information is important because you use it to find out if the peripheral speed (shear rate between screw flight and cylinder) is OK for your material. Between the cylinder and the screw flight there are some tenths of a millimeter, and that is where the material flows backward and can be sheared. If the shearing becomes too high, there will also be an increase in the temperature in the cylinder (frictional heat).

26.6 Information about the Mold

In the pane shown in Figure 26.12, you can enter information about the mold needed to understand possible problems in the injection-molding process that are mold-related. Many times problems in the molded parts depend on thermal degradation of the material in unfavorable hot runner systems (see e. g. Figure 26.13 and Figure 26.14). It is therefore of interest to fill in the field *Hot runner system* with brand and type because some systems are not suitable for all materials.

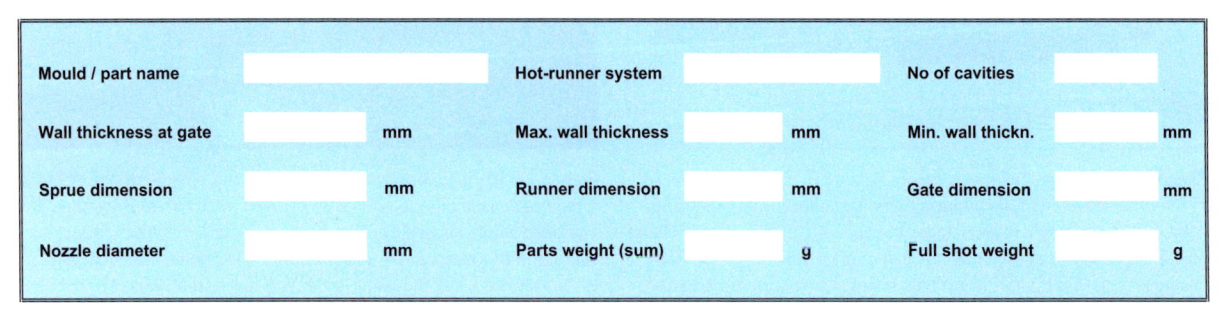

Mould / part name		Hot-runner system		No of cavities	
Wall thickness at gate	mm	Max. wall thickness	mm	Min. wall thickn.	mm
Sprue dimension	mm	Runner dimension	mm	Gate dimension	mm
Nozzle diameter	mm	Parts weight (sum)	g	Full shot weight	g

Figure 26.12 The mold information pane.

26.6 Information about the Mold

Figure 26.13 If you get discolored parts when producing in a hot runner mold, this is often a sign that you have hold-up pockets somewhere in the hot runner that are causing degradation of your resin. Then when the degraded material loosens you will see a discoloration of the parts. Sometimes you may get 10 to 20 perfect parts before a discolored one shows up. The only way to solve this kind of problem is to correct the error in the mold.

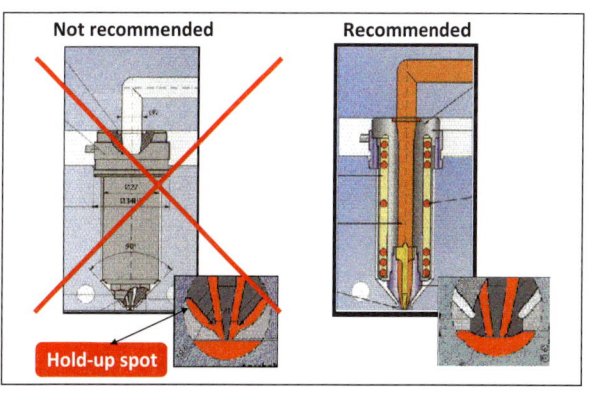

Figure 26.14 To the left there is a hot runner nozzle isolated thermally from the cavity through air gaps. When these are filled with plastic (red color in the image) a degradation of the plastic will start after a short time. The nozzle to the right is an example of a better system where the cavity is thermally isolated by closed air isolation where plastic cannot enter and got trapped.

The field *No of cavities* may give clues to the question as to whether all of the parts of the shot are faulty or just one single part from a particular cavity. Often there is a number engraved in the surface of the part if there is more than one cavity.

In order to get sufficient packing of parts in semi-crystalline plastics, you need to have the correct relation between the *Wall thickness at gate* and *Max. wall thickness*.

The reason why you want to know *Min. wall thickness* is because if this deviates too much from the maximum wall thickness this may explain why the part warps.

Just as it is important to know the relationship between the wall thickness at the gate and the maximum wall thickness, it is also important to know the relationship between the wall thickness at the gate, *Sprue dimension*, *Runner dimension*, and *Gate dimension* to be able to judge if sufficient packing and shrinkage compensation can be done.

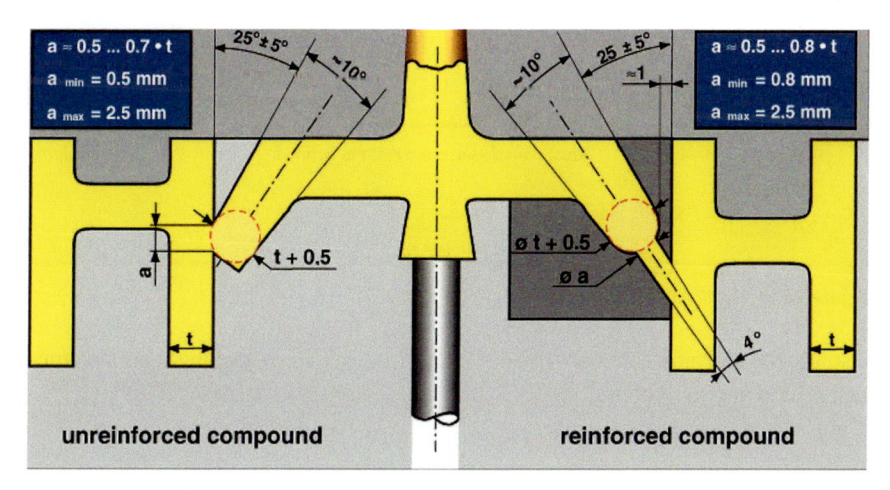

Figure 26.15 The image shows the recommended relations between the wall thickness "t", the sprue, the gate size, and the runner for both unreinforced and reinforced materials. [Source: DuPont]

The value in the *Nozzle diameter* can be entered in relation to the sprue dimension. Normally, the nozzle diameter should be 1 millimeter smaller than the smallest diameter of the sprue. If the nozzle diameter is less than 1 mm, there is a risk that you cannot pack up and shrink-compensate the part well enough.

Figure 26.16 This picture shows the same sprue produced by two different nozzle diameters. The left one had a nozzle with a hole diameter of 4 mm less than the smallest diameter of the sprue. The one to the right is made with a nozzle having a hole diameter that is about 1 mm less than sprue. The parts are rail insulators, and those that were made with the left sprue were about 20% weaker than the parts made with the right one.

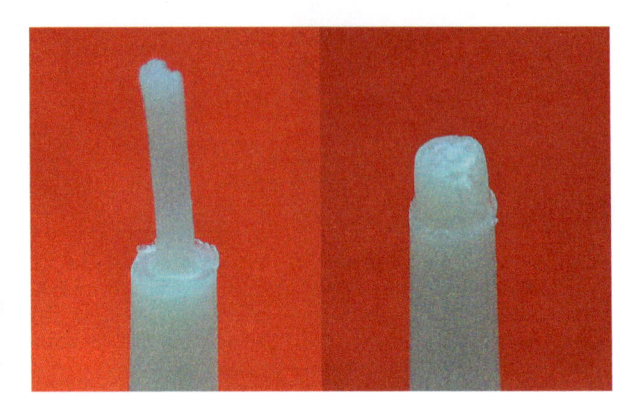

The last two fields in the mold information pane are *Parts weight (sum)* and *Full shot weight.* The difference between these weights is that the weight of the sprue and runners are included in the full shot weight. If you have very small parts it can sometimes be a good idea to take the average weight of several shots if your balance does not have enough accuracy. Using the difference between parts weight (sum) and full shot weight you can calculate the percentage of regrind material generated during the process.

26.7 Drying

Many thermoplastics, both amorphous and semi-crystalline, must be dried down to a maximum moisture content for optimum results. Drying time and drying temperature vary among different polymers, so it is good if you prepare for a trial run the day before, as the material may need to be dried for up to 8 hours before the trial can start.

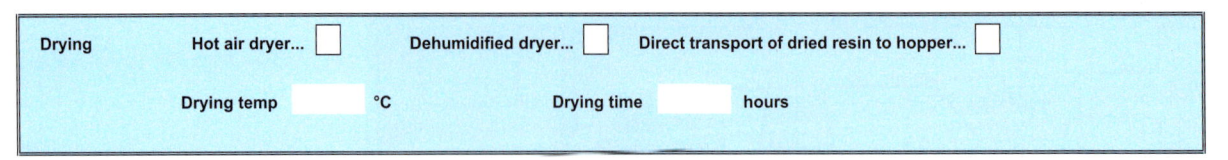

Figure 26.17 The drying information pane.

Usually the thermoplastics are divided in different groups regarding their moisture sensitivity during processing:

- Non-hygroscopic (e.g. PE, PP, PS, PVC, and POM)
- Hygroscopic (e.g. ABS, PA, PC, and PMMA)
- Hydrolysis sensitive (e.g. PBT, PET, PPA, LCP, and PUR)

The non-hygroscopic materials do not absorb moisture and need only in exceptional cases to be dried. Such exceptions are when the granules have condensed on the surface, which can occur if you bring material from a cold storage room into a warm production hall in winter time. POM, which is non-hygroscopic, may also need to be dried if you want to have a very high-gloss surface on a polished part or if the acetal is impact modified (usually by PUR). It is only the non-hygroscopic materials that can be dried during all months of the year in hot air dryers.

It is therefore important to specify in the checkbox what type of dryer has been used:

Hot air dryer, Dry air dryer, and if there is *Direct transport* of the material to the machine.

Figure 26.18 Dryer and direct transport unit.
Here you can see in the left picture a dry air dryer with sucker and direct transport to the injection-molding machine. Unlike a hot air dryer, where the air is taken directly from the room and heated, the heated air will pass through a moisture-absorbing filter in the dry air dryer. In general these dryers have two filters (often with silica) where one is active and the other is regenerated (dried).
It is also important to know if the material has been transported in a closed system directly to the machine. Dried material that comes straight from the dryer can become useless after a few minutes if exposed to the moisture of the room. For example, polyester PET is a material that will absorb enough moisture in about 10 minutes from the air in the room to become extremely brittle.
Thus PET is hydrolysis sensitive, and it can only be processed with good results when its moisture content is below 0.02%. Polyamide is a hygroscopic material and can withstand 10 times higher moisture content than PET during processing, i.e. 0.2%. If the materials are wet during the processing, the mechanical strength will be poor and you can get splays on the surface e.g. polyamide.

In the processing recommendations from the raw material suppliers you can get information on the drying temperature and drying time. In some cases the temperature can be lowered when using a longer drying time. This can be useful if you will run a trial the next day and must dry the material overnight. The temperature setting on the dryer and the actual drying time should be entered in the fields *Drying temp* and *Drying time*.

26.8 Processing Information

The processing information pane (see Figure 26.19) is the most comprehensive one, and for the sake of clarity it will be split into three segments.

Processing

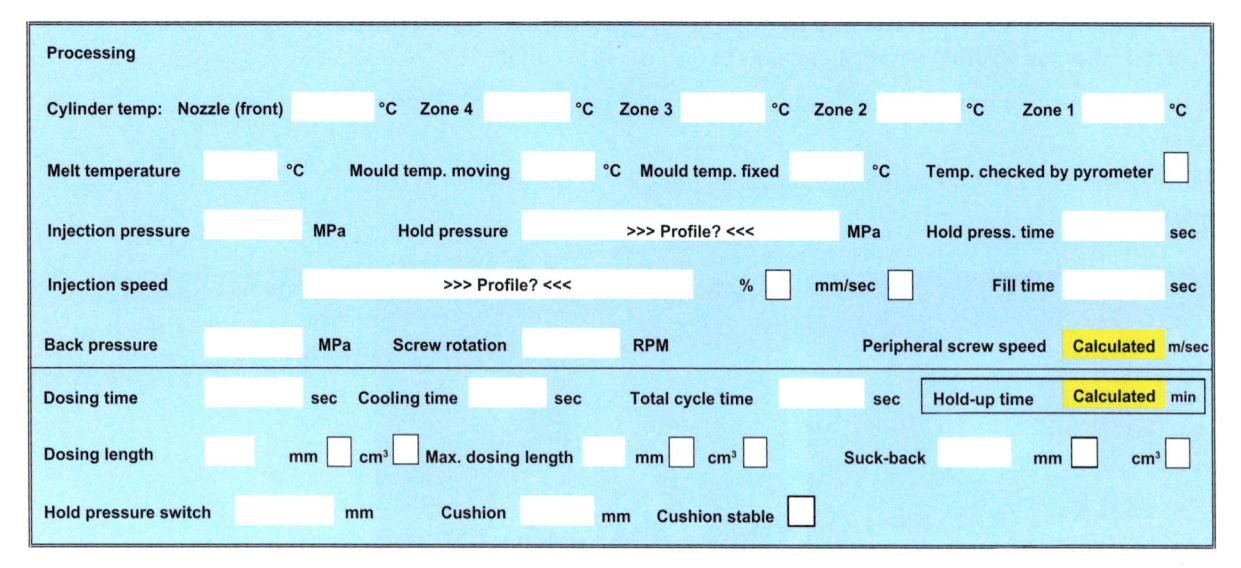

Cylinder temp:	Nozzle (front)	°C	Zone 4	°C	Zone 3	°C	Zone 2	°C	Zone 1	°C
Melt temperature	°C	Mould temp. moving	°C	Mould temp. fixed	°C	Temp. checked by pyrometer	☐
Injection pressure	MPa	Hold pressure	>>> Profile? <<<	MPa	Hold press. time	sec
Injection speed	>>> Profile? <<<	%	mm/sec	Fill time	sec
Back pressure	MPa	Screw rotation	RPM	Peripheral screw speed	**Calculated** m/sec
Dosing time	sec	Cooling time	sec	Total cycle time	sec	Hold-up time	**Calculated** min
Dosing length	mm	cm³	Max. dosing length	mm	cm³	Suck-back	mm	cm³
Hold pressure switch	mm	Cushion	mm	Cushion stable	☐

Figure 26.19 The processing information pane

In the pane above you will find the five most important processing parameters that always affect the quality of injection-molded parts.

We will systematically go through these parameters and also explain the meaning of some other parameters that also may affect the quality of the parts:

The five most important parameters	Other important parameters
Cylinder temperature profile	Back pressure
Melt temperature	Screw rotation or peripheral speed
Mold temperature	Cooling time
Hold pressure	Injection speed
Hold pressure time	Hold pressure switch point

26.9 Temperatures

The first two rows in the pane shown in Figure 26.19 contain the temperature settings.

You should enter the temperatures of all zones of the machine cylinder and the nozzle. This is to control if the temperature profile (see Figure 26.20) is correct.

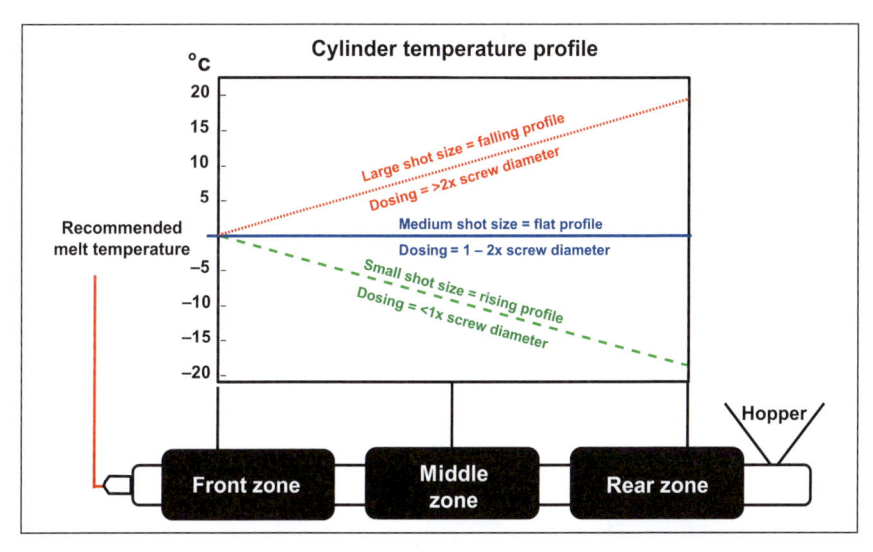

Figure 26.20 The green dashed curve in this figure symbolizes a rising temperature profile, i.e. the zone nearest the hopper has the lowest temperature and the zone closest to the nozzle has the highest. This is the most used setting. You should use a rising profile when the dosing length (in mm) is smaller than the screw diameter.
Especially when you produce in semi-crystalline plastics that require a high specific heat (see Figure 26.21), it is important to have the right temperature profile. If you have a larger dose than the length of the screw diameter you should either have a flat profile, i.e. the same temperature in all zones, or a falling profile, i.e. the highest temperature at the hopper and the lowest at the nozzle. A flat profile (blue) is recommended when the dosing length is between one and two times the screw diameter and a descending (red) one when the dosing length is greater than twice the screw diameter.
[Source: DuPont]

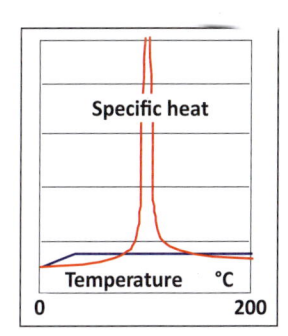

Figure 26.21 Specific heat.

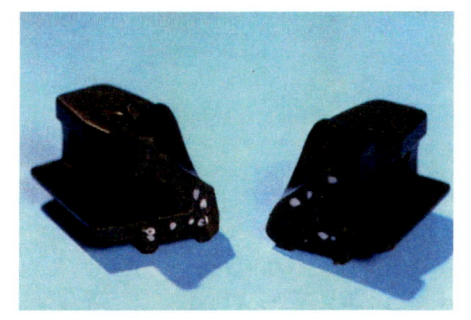

Figure 26.22 Unmelts in an acetal part. [Photo: DuPont]

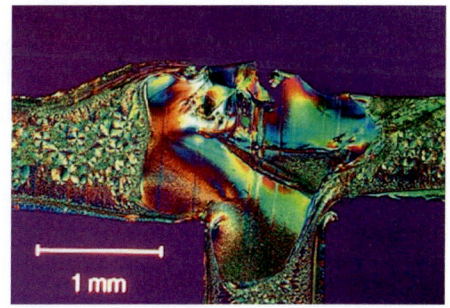

Figure 26.23 Unmelts seen by microscope. [Photo: DuPont]

The blue curve in Figure 26.21 represents the specific heat of an amorphous thermoplastic. Above the glass transition temperature, where the curve becomes constant, you require exactly the same amount of energy to raise the temperature one degree regardless of where in the temperature range you are.

Chapter 26 – The Injection-Molding Process

The red curve represents a semi-crystalline material. Here we can see that within a small area on the temperature curve you require a significant energy boost to increase the temperature one degree. This is the area around the material's melting point, and it is called the melting heat, i.e. the amount of energy required for the material to change from a solid to a liquid phase.

In Figure 26.22 there is an acetal part that has been sawn apart. Thanks to having mixed natural-colored granules with black regrind, you can see that some of the granules have passed the cylinder without melting. If you only had been using natural or black material, you would not have been able to detect the problem visually. But you could have used structural analysis, where you take a thin section cut by a microtome and study it in a microscope illuminated from below by polarized light, as shown in Figure 26.23. If you have unmelts in your part, the strength will be significantly reduced.

Your *Melt temperature* and the *Cylinder temperature profile* are two of the five most important process parameters when it comes to the quality of molded parts. This melt temperature is so important that you need to have it confirmed in the field *Temp. checked by a pyrometer* (Figure 26.24) that it really has been measured and not just copied from the machine's display.

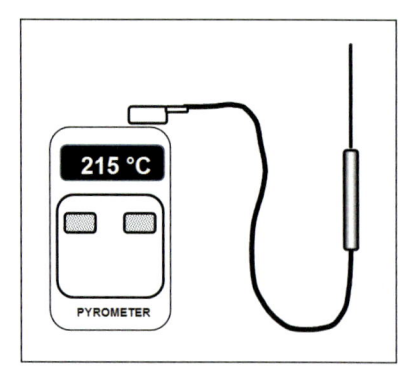

Bild 26.24 Pyrometer.

Figure 26.25 Melt temperature control. [Photo: DuPont]

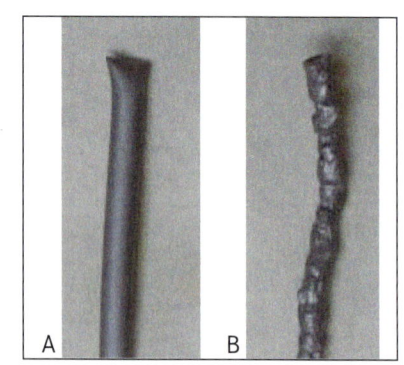

Figure 26.26 Purged PA66 strands.

Another reason why you should measure the melt temperature with a pyrometer is that you should look at the melt visually. In Figure 26.26 are shown the strands of material that came out of the nozzle when silver-colored PA66 was purged out of the cylinder. In both cases the same temperature was measured by a pyrometer that showed 295 °C.

The Figure 26.26 "A" strand looks like a perfect melt should look, i.e. smooth and shiny. In Figure 26.26 "B", however, you can clearly see unmelted granules in the strand, and this is because of a faulty temperature profile.

26.9 Temperatures

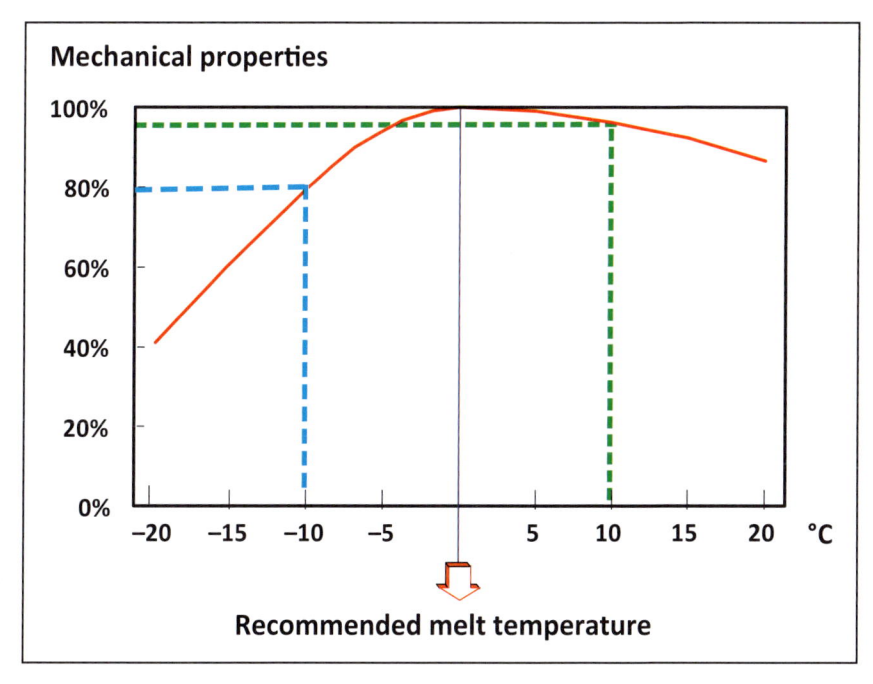

Figure 26.27 Here you can see the tensile strength as function of the melt temperature for a semi-crystalline plastic. The highest value, which gives 100%, is the value that the supplier of the material recommends in its processing manual.
If you process the material 10 °C below the recommended, the strength is reduced, in this case by 20%. If you increase the temperature 10 °C above the recommended, the strength is reduced by only a few percent. The reason the strength of the material decreases significantly more when the temperature is lowered compared to the increased one is the risk of unmelted granules in the melt. This is important to know because variants of the same polymer can have different melting temperatures. The melt temperature difference between the acetal copolymer and homopolymer is 10 °C, i.e. 205 °C and 215 °C, respectively. This results in significantly lower mechanical properties for the homopolymer when running at the copolymer temperature setting.
NOTE: Between PA6 and PA66, the difference is 30 °C, which can result in a lot of scrap and a waste of time if not taken in account.

The last temperature in the pane is the *Mould temperature*, which also belongs to the five most important parameters. This temperature should also be measured by the pyrometer with a special sensor for surface measurements.

You can never trust the setting on the mold temperature control unit as the temperatures can vary greatly between the different halves and even within them. It is not rare that cores can have more than 100 °C higher temperature than the cavity itself.

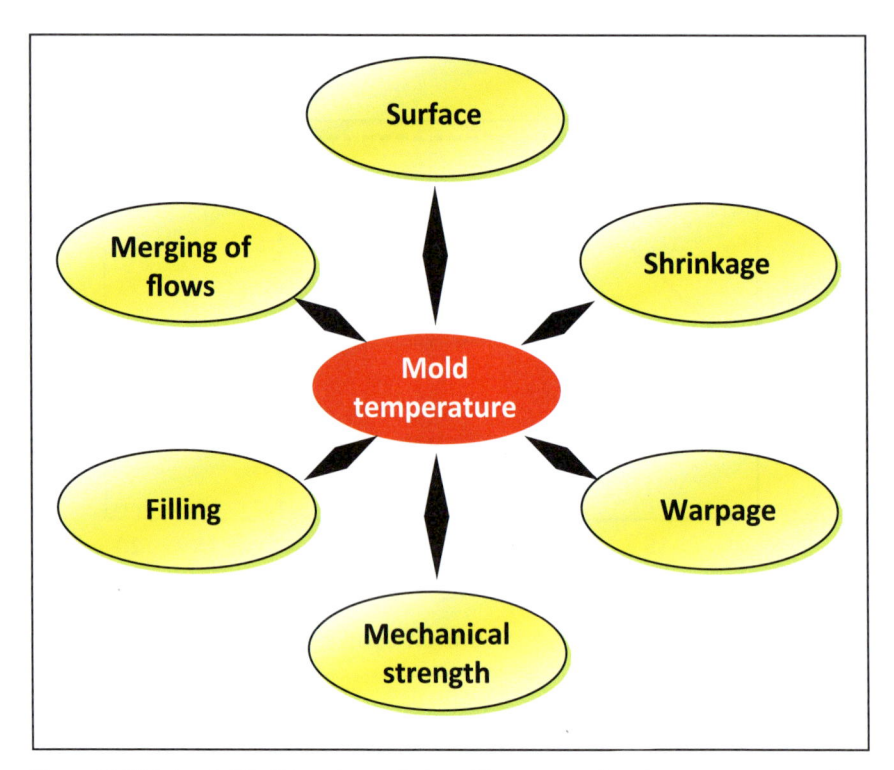

Figure 26.28 It is not difficult to understand why mold temperature is an important parameter when you can see, as here, how many properties are affected by it:
- Higher temperature results in better surface gloss. The mold temperature also affects the ability for early ejection. If the surface temperature is too high, you can get ejector marks or warpage on your part. If you compensate for this with a longer cooling time, this will have a negative effect on your economics.
- Varying temperatures in different parts of the mold cause various shrinkages, with internal stresses and warping as a result.
- The mechanical strength of semi-crystalline plastics is influenced by the crystallinity, where higher temperature results in higher crystallinity and higher strength.
- If you increase the temperature, the material does not solidify as fast in the mold, and you get easier filling.
- If the material is cooled too fast, before the melt flows meet, you will get a poor weld-line.
The mold temperature is crucial on the shrinkage of the part shrink i.e. dimensions.

On most part drawings there are requirements for dimensions and tolerances. The reason that the injection-molded parts do not have the same dimensions as in the mold cavities depends on shrinkage. This causes some problems for the molders, unless a foaming agent is used. Shrinkage is a physical process that goes on for a long time, and it can be divided in two parts: mold shrinkage measured after 16-24 hours and post-shrinkage, which is a temperature and long-term dependent process. The size of these shrinkages depends on the wall thickness of the part, polymer type, fillers, or reinforcing agents and the process parameters where the mold temperature is of crucial importance.

If you have high mold temperature, the mold shrinkage will be high while the post-shrinkage and the total shrinkage, i.e. mold shrinkage + post-shrinkage, will be lower than if you have a low mold temperature.

You can accelerate the post-shrinkage and thus the total shrinkage of semi-crystalline plastics by annealing the parts in an oven.

26.9 Temperatures

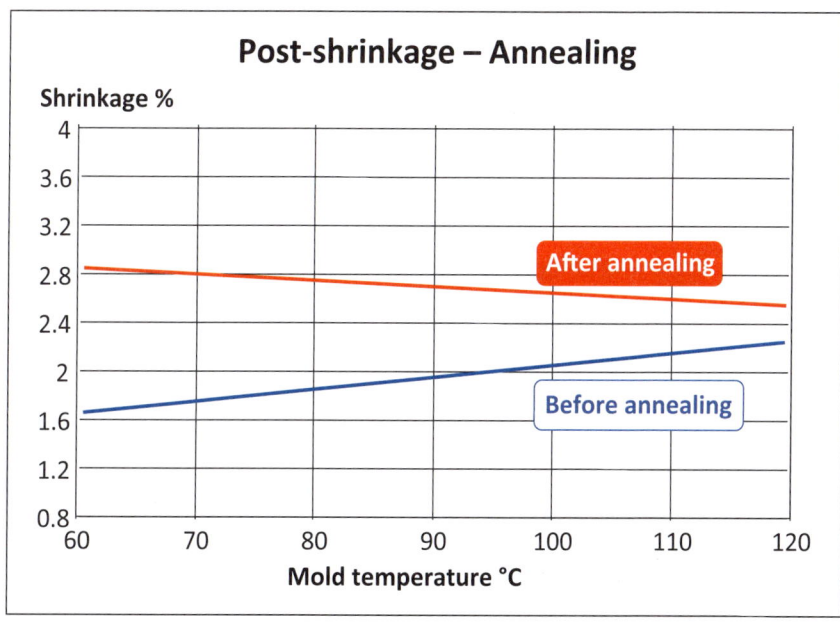

Figure 26.29 In this figure you can see the total shrinkage for 3.2 mm thick test bars in Delrin 500 acetal homopolymer by DuPont as a function of the mold temperature before and after annealing. Annealing means that you have placed the samples in an oven at 160 °C for about one hour.
The blue curve shows the mold shrinkage measured after about 24 hours.
The red curve shows the total or final shrinkage after annealing. This shrinkage you will see after the test bar has shrunk for several months at room temperature if it has not been annealed.
If you produce the test bars at low temperature you will get a mold shrinkage of 1.6% and a total shrinkage of 2.9%. If you produce the test bars at the plastic producer's recommended mold temperature (90 °C), the mold shrinkage will be slightly higher (1.9%) while the total shrinkage drops (2.6%). If you increase the temperature further, the total shrinkage continues to decrease. The blue and the red curves will meet in the melting point (175 °C) of the material. [Source: DuPont]

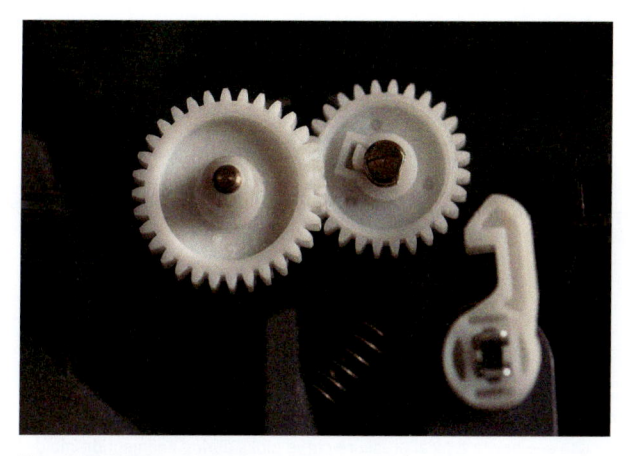

Figure 26.30 Gears are often made in acetal.
This semi-crystalline thermoplastic has very high crystallinity. In order to avoid surprises with dimensional tooth problems after a certain time, it is common to anneal the gears to the right (final) dimension before assembly. The picture shows a gearbox of a laser printer.

26.10 Pressure, Injection Speed, and Screw Rotation Speed

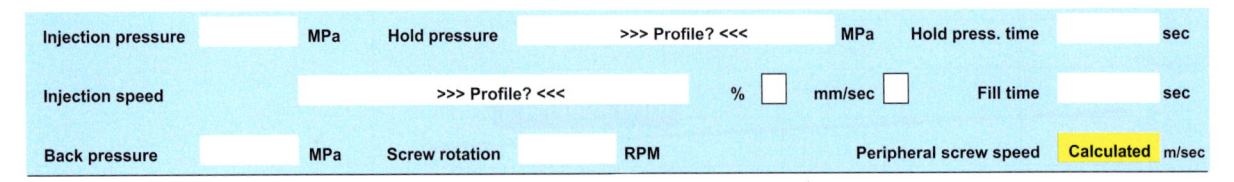

Injection pressure		MPa	Hold pressure	>>> Profile? <<<		MPa	Hold press. time		sec
Injection speed			>>> Profile? <<<		% ☐	mm/sec ☐	Fill time		sec
Back pressure		MPa	Screw rotation		RPM		Peripheral screw speed	Calculated	m/sec

Figure 26.31 The pressure pane and screw speed pane.

In this pane we will go through the most important pressure settings: hold pressure time, screw speed, and fill time.

We will start with the field *Injection pressure*. In many injection machines there is a direct relation between the injection pressure and the *Injection speed*, which means if you want to achieve maximum screw speed you must also have maximum injection pressure.

A common practice is to set the hold pressure switch (which will be dealt with in the next pane; see Figure 26.41) in such a way that the cavities will not be completely full during the injection phase. Thus, you can choose the maximum injection pressure during the setup trials and then limit it in order not to damage the mold. When the injection phase (mold-filling phase) comes to an end, you can then switch from the injection to the hold pressure.

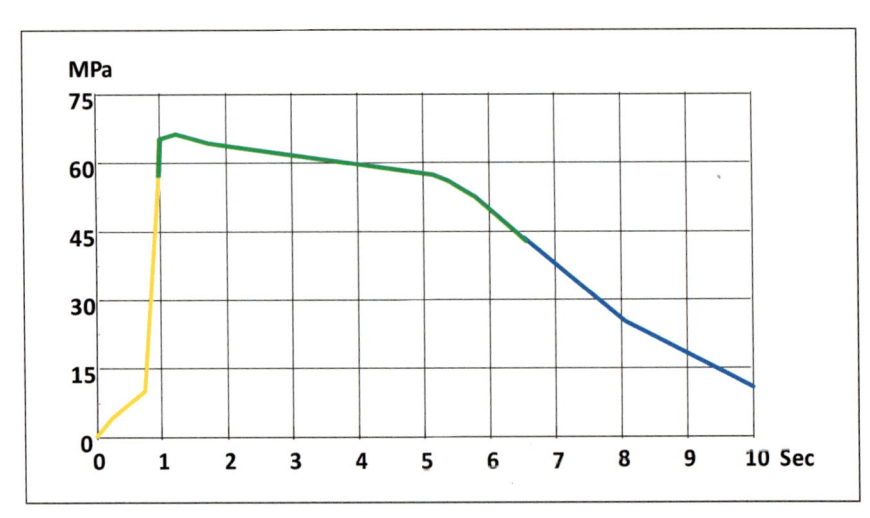

Figure 26.32 In this figure you can see how a typical pressure curve looks during an injection-molding cycle when using a pressure sensor inside the cavity. The yellow part of the curve is the injection phase, which in this case lasts one second before it switches over to the holding pressure phase (the green curve), which in turn lasts about 5.5 seconds.
When the green curve shifts to the blue, the dosing phase for the next shot has started and the pressure in the cavity drops continuously. The part is cooling due to an interaction of both the decreasing thermal expansion and the mold shrinkage.
After 10 seconds the part will be ejected, and now it is important that the remaining pressure (about 10 MPa) is not too high, as there could be problems with sticking in the cavity, deformation, or cracking.

26.11 Hold Pressure

Hold pressure is one of the five important injection-molding parameters, and normally the plastic raw material suppliers have specified an interval for the hold pressure in their processing recommendations.

Regarding amorphous plastics, you normally want to have as low a hold pressure as possible to avoid the risk of parts sticking in the cavity. At the same time, it must be high enough to avoid sink marks. In order to achieve this compromise you might have to profile the pressure (let it drop in steps) during the hold pressure time.

When doing this, you can enter these steps with a dash in between in the field *Hold pressure* in the pressure pane.

Regarding semi-crystalline plastics, you should try to use the highest possible hold pressure.

The reason for this is that higher pressures during the packing phase give higher crystallinity, lower shrinkage, better dimensional stability, and higher strength.

How high can we go in hold pressure? The answer is:

1. Until you get flashes in the parting line of the mold.
2. Until you get ejection problems due to the part sticking in the mold. Normally you can hear a crackling sound, when ejecting, before the parts stick.

You should be careful and only increase the pressure in small steps because some parts may sit so hard in the cavity once they are stuck that you need to take down the mold and spend unnecessary time to get them loose.

Once you have found the maximum hold pressure, which you can use in mass production, you should optimize the hold pressure time too. The semi-crystalline suppliers recommend that you use a constant hold pressure during the whole hold pressure time. But sometimes you are forced to profile it to get the best results.

On the question "What hold pressure time should we choose?" the answer will be:

1. Regarding amorphous plastics, as short as possible without getting sink marks.
2. Regarding semi-crystalline plastics, you must think differently.

In some plastic suppliers' processing manuals you can find a time specified in relation to the maximum wall thickness of the part, e.g., for unreinforced PA66, 4 sec/mm; for glass-reinforced PA66, 2–3 sec/mm; and for acetal, 7–8 sec/mm for wall thicknesses up to 3 mm. If you exceed 3 mm you need to increase the time per millimeter further (see Figure 26.34). These values are for guidance only, and you should determine the optimum hold pressure time yourself by practical testing (see Figure 26.35) and enter it into the field *Hold pressure time*.

Chapter 26 – The Injection-Molding Process

Figure 26.33 This picture shows a polyamide nut with a severe flash in the parting line.
When this happens you should decrease the pressure until you feel no flash with your thumb.

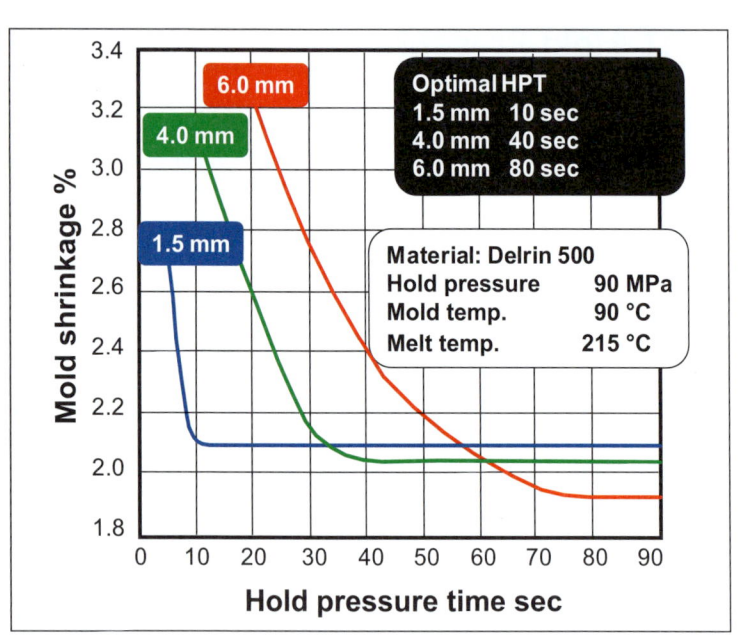

Figure 26.34 In this figure, you can see shrinkage as function of hold pressure time for a standard acetal homopolymer grade. You can usually specify a recommended hold pressure time of 7–8 sec/mm wall thickness up to 3 mm, which will correspond to the blue curve. If you have 4 mm thickness corresponding to the green curve you will not achieve the optimum hold pressure time until after 40 seconds (i. e. 10 sec/mm wall thickness).
For even thicker walls such as 6 mm in the red curve it will take 80 seconds (i. e. 13 sec/mm wall thickness). [Source: DuPont]

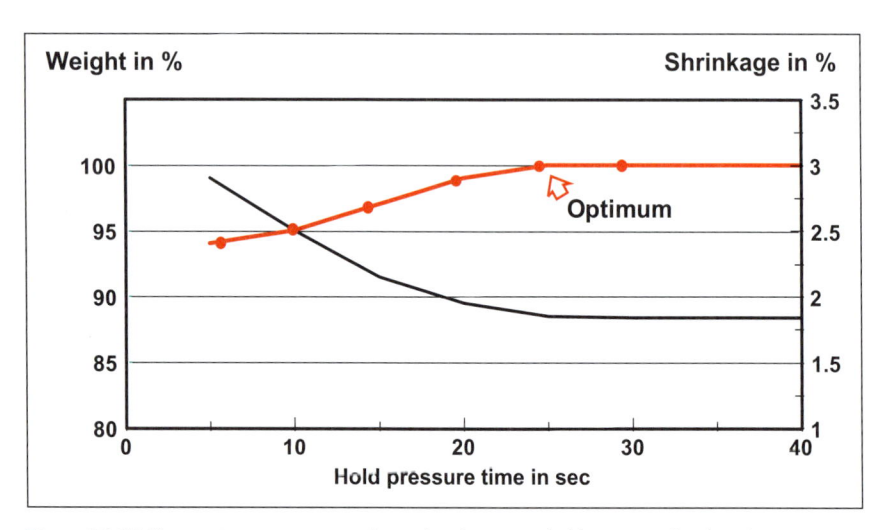

Figure 26.35 The most accurate way to determine the correct hold pressure time is to have a pressure sensor in the cavity and a special computer program that calculates the optimum of it. In Figure 26.32 you can see such a curve where the optimal hold pressure time is 6.6 seconds (red dot) where the green convex pressure curve merges into the blue concave curve.
But the most common way is to make a weight curve. You should weigh the part or parts, if you are using a multi-cavity mold, but you should exclude the sprue and the runners.
If the part/parts will increase in weight when you increase the hold pressure time, you should continue to increase the hold pressure time until the weight does not increase anymore. In the red weight curve shown, this occurs after 25 seconds, which is the optimal hold pressure time with the best strength and dimensional stability. The black curve shows that the shrinkage decreases continuously until you achieve the optimum hold pressure time.

288

Just as the level of the hold pressure affects the strength and dimension of the part, the effect of the hold pressure time will be the same up to the optimum hold pressure time. The time that the raw material suppliers may recommend in their processing manuals can only be achieved if the gate is correctly sized and positioned (see Figure 26.36). You must also ensure that there is enough material in front of the screw during the whole hold pressure phase, i.e. you must have a material cushion that should not be less than 5 mm or 10% of the screw diameter.

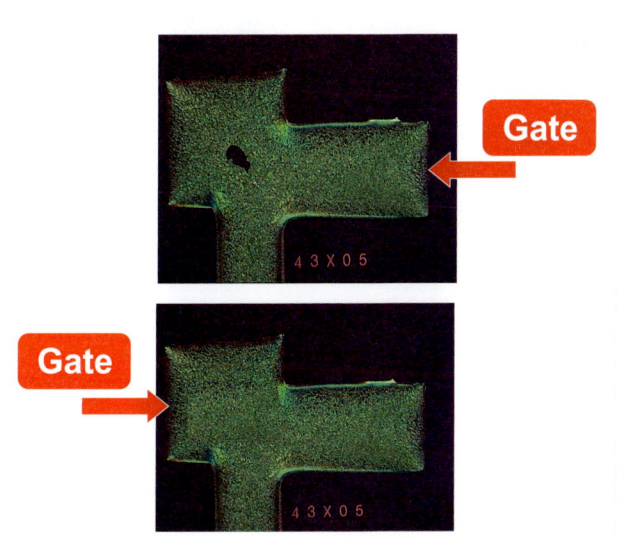

Figure 26.36 In the upper picture the gate is not located in the thickest wall, which means that you cannot pack the thickest wall sufficiently. You will therefore get voids when the specific volume decreases during the cooling process. In the lower picture the gate is correctly located in the thickest wall of the part. With an optimum hold pressure time the part will get a smooth and pore-free structure. [Photos: DuPont]

26.12 Injection

The injection speed is the speed at which the screw moves linearly during the injection phase. Depending on the geometry of the part or venting problems it can sometimes be an advantage to profile the injection speed in several steps. On the question "Which injection speed should be set?" the answer is: You have to test it by trial and error. By increasing the injection speed you may get a better surface but simultaneously increased risk of shear in the gate, if it is too small or poorly rounded. You can also get problems with the venting causing burn marks on the surface.

Enter the *Injection speed* in the field of the processing pane. If the speed is profiled in several steps, you can specify this with a dash between the values. Since the speed will be measured in *mm/sec* in some machines and in % of maximum speed in others you must also tick one of the checkboxes to specify the correct setting.

Fill time is not a parameter to be set but the result of the injection volume, the length of the runners, and the injection speed.

Back pressure is the parameter that influences even dosing. This pressure is applied when the screw rotates during the dosing phase to prevent the backward linear movement of the screw from being too fast. Hereby we also gain a better mixing, dispersion of color pigments, and homogenization of the melt. The back pressure makes it easier for the ring in the back flow valve to move forward and open so that the molten plastic can flow to the front of the screw tip (see Figure 26.37).

Normally you should use as low a back pressure as possible to ensure even dosing (a small variation of dosing time and size of the cushion). The higher the back pressure is, the longer the dosing time will be, which can lengthen the cycle time and impair the economy.

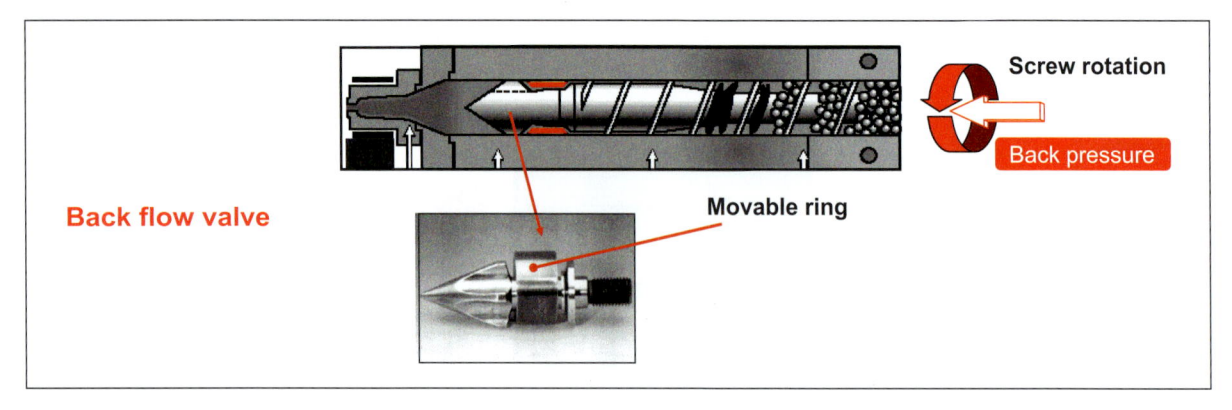

Figure 26.37 The white arrow here shows the direction of the back pressure during the dosing phase. At the left of the screw there is the screw tip with back flow valve. The movable ring moves forward and opens the valve during dosing. Upon injection, it moves backward and seals the screw so it will act as a piston.
It is important to check that the back flow valve is not worn to prevent backward leakage of the material during the injection. The ring has a lower hardness than other parts in the back flow valve and will wear faster and should therefore be replaced periodically.
If the valve leaks you will not get a stable cushion; due to that, the screw will continue to creep forward. [Images: DuPont]

26.13 Screw Rotation Speed

Another setting that affects the dosing and the quality of the melt is the *Screw rotation*. A higher screw rotation speed will reduce the dosing time while the shearing of the material between the screw flight and the cylinder wall will increase. If the *Peripheral screw speed* is too high, the molecular chains in the material will shear apart and generate frictional heat and thermal degradation. The result will be deposit on the screw (see Figure 26.39) and in the mold. The deposit on the screw can also give black specks on the surface of the part. The only way to get rid of screw deposit is to bring out the screw from the cylinder and clean it manually with a wire brush.

26.13 Screw Rotation Speed

Figure 26.38 In the picture you can see a white part with black specks all over the surface, giving a dirty impression. If you have a multi-cavity mold and all the parts show the same problem, the cause is usually to be found in the machine's nozzle, cylinder, or screw.
The first thing you should do is measure the melt temperature and carefully study the appearance of the strand that is purged out. If you see black specks here the normal procedure is to stop the production and purge the cylinder with a cleaning resin. If this does not help you must bring out the screw and clean it.
When some polymers are degraded they will form corrosive gases that can cause pitting on the screw (see the flanks of the screw in this picture). If that happens, you will never get rid of the problem unless the screw is repaired or replaced.

Figure 26.39 The picture here shows a screw that has a deposit due to too-high screw rotation speed in the production of parts in red acetal. When acetal copolymer is degraded it can form formic acid, which is extremely corrosive and will etch out craters in the steel surface (see the screw flanks in the picture). Once that has happened the problem will accelerate.

On the question "How high a screw rotation can be used for a particular polymer?" there is no direct answer. First you must find out the screw diameter in your machine and then calculate the peripheral screw speed. The raw material suppliers used to have recommendations of the maximum peripheral speed of their different grades. The peripheral speed is the rotational speed of a point on the surface of the screw flight when the screw rotates.

If you study the minute hand of your watch you can hardly see that the tip of it moves as peripheral speed is so low. If you look instead on a large wall clock you can often see that the hand moves as the peripheral speed is much higher. In both cases the speed of the minute hand is the same, i.e. one rotation per hour. In the formula shown in Figure 26.40 there is a relationship between the maximum allowed screw rotation speed and the maximum allowed peripheral speed.

$$\text{Max screw rotation} = \frac{\text{Max peripheral speed} \times 1000 \times 60}{\text{Screw diameter} \times \pi}$$

Figure 26.40 The screw rotation is specified in the unit "revolutions per minute" (RPM). The peripheral speed is specified in meters per second and the screw diameter in millimeters, which is why "1000 × 60" is included in the formula in order to get the right units. By the formula, you will understand that if you have a specified maximum peripheral speed and will increase the screw diameter, it is necessary to reduce the screw rotation.

If you enter *Screw diameter* and *Screw rotation* in the fields in the pane (see Figure 26.41) the peripheral speed will automatically be calculated and shown in the field *Peripheral screw speed*.

26.14 Time and Length Settings

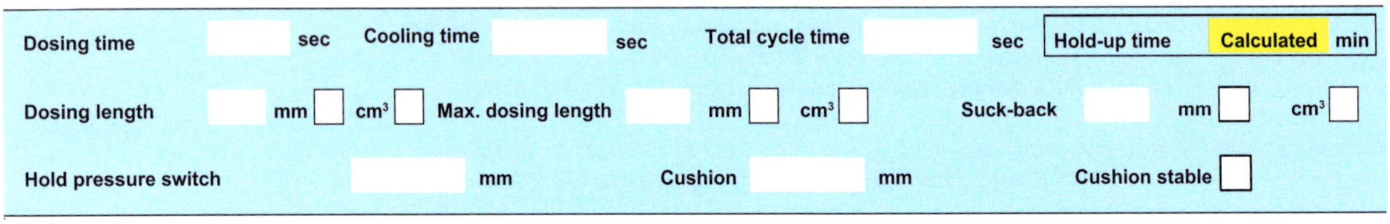

Dosing time	sec	Cooling time	sec	Total cycle time	sec	Hold-up time	Calculated min
Dosing length	mm / cm³	Max. dosing length	mm / cm³			Suck-back	mm / cm³
Hold pressure switch	mm	Cushion	mm			Cushion stable	

Figure 26.41 The pane showing the processing times and length settings.

The *Dosing time* is not a setting parameter but the result of several factors. It is affected by the parts volume, the cylinder size, the actual screw speed, and the back pressure.

The value is normally shown in the display of the machine.

The *Cooling time* is however an input parameter. To get a short cycle time (and good economy), you should have the cooling time as short as possible without getting the parts so soft that they will be deformed during the ejection. If you have an open nozzle, which is the normal case, the cooling time must be a little bit longer than the dosing time. A rule is to add 0.5 to 1 second to the dosing time to get the cooling time. This additional margin can be seen as a safety margin to prevent production stop, if the dosing time varies.

If you have a shut-off nozzle, you can have a shorter cooling time than dosing time since in this case you can continue to dose throughout the mold opening, ejection, and closing phases. The *Total cycle time* is the sum of all part times sections (see Figure 26.43).

You can read this time on the display on most injection-molding machines, and when you have this time and know the *Max dosing length*, actual *Dosing length*, and *Cushion*, you will be able to calculate the *Hold-up time* for the material in the cylinder during continuous production. The formula shown in Figure 26.42 is not completely accurate, but it gives a good guideline.

26.14 Time and Length Settings

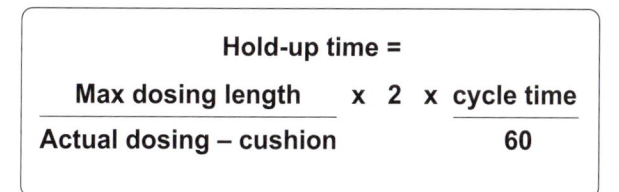

$$\text{Hold-up time} = \frac{\text{Max dosing length}}{\text{Actual dosing} - \text{cushion}} \times 2 \times \frac{\text{cycle time}}{60}$$

Figure 26.42 A simple formula to calculate the hold-up time.

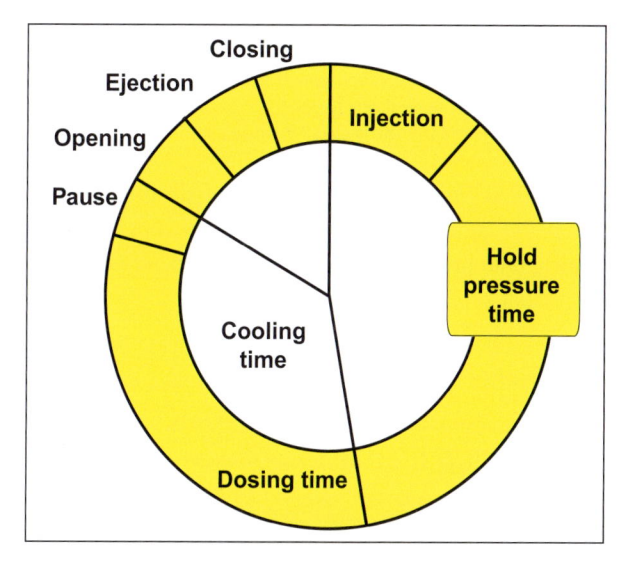

Figure 26.43 The sum of all part times will give you the "Total cycle time".

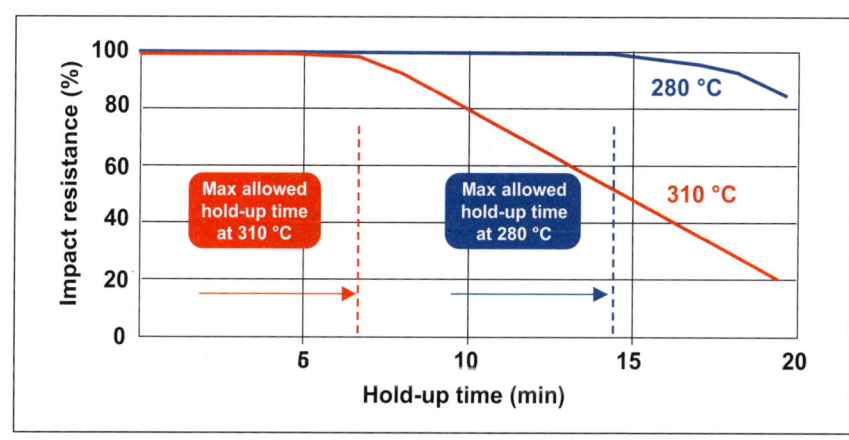

Figure 26.44 The figure here shows that the impact strength of a toughened PA66 is affected by the hold-up time in the cylinder. At a melt temperature of 280 °C (blue curve) the material resists a hold-up time of about 15 minutes before the degradation starts. At 310 °C (red curve) the material resists a hold-up time of only 7 minutes before the molecular chains start to degrade. The most sensitive resins regarding the hold-up time used to be the flame-retardant grades. [Source: DuPont]

On many injection-molding machines you can see a needle that mechanically follows the screw's linear motion on a scale. Figure 26.45 shows such a scale, in which the position of the needle during the injection-molding cycle has been marked with colored triangles.

Figure 26.45 The photo of the scale shown here is from a new injection-molding machine.
When the needle is in the starting position on the far left the maximum dose length is 170 mm. The blue triangle marks the selected dosing length, and the yellow is the position after suck-back, which is the final position before the injection can start. During the injection the needle moves rapidly to the position for hold pressure switch, which is marked by the green triangle. Then under the hold pressure phase the needle creeps slowly to reach its final position before starting to dose again. This location marked with the red triangle is called the cushion.

In the field *Dosing length* you should set the distance that corresponds to the volume (+ cushion) that the full shot needs to dose. On most machines, dosing length, hold pressure switch, and cushion are specified in mm, but there are also those that specify this in cm^3.

Normally, however, that indicates the position in mm. Decompression or *Suck-back* is when after dosing you let the screw move linearly a few mm backwards.

Decompression is often used on semi-crystalline materials to suck back material into the cylinder from the nozzle to prevent the material from freezing and blocking the nozzle.

If the suck-back is too big there will be a risk of sucking air into the melt, which then causes air bubbles inside the part (see Figure 26.46).

Figure 26.46 The picture shows a cylinder to a measuring syringe made of PA66.
By mistake, a too large suck-back was set, which caused big bubbles inside the wall. This could be seen with the naked eye in uncolored resin. Due to a quality control by weight when the cylinders were set up, the first time the machine operator observed that something was wrong was when the weight dropped significantly.
If the parts had been made in black material, the weight control would have been the only way to detect this problem, which otherwise would have lead to a substantial weakening of the cylinder wall, with complaints as a result.

During the injection the screw acts as a piston and moves rapidly from the rear position to the point where it switches over from injection pressure to hold pressure. This point on the scale is called the hold pressure switch point. There are several ways to switch from injection pressure to hold pressure:

- On pressure with an internal pressure sensor inside the cavities, which is the most accurate way but quite rare in the plastic industry.

- On distance or volume, which are the absolute dominant ways, and in the field *Hold pressure switch* you tick one of the boxes *mm* or *cm^3* to get the correct unit.

- On time, which is the less accurate way.

On the question "Where shall we set the hold pressure switch point?" the answer will be that you have to locate it by trial and error:

1. Start to run "short shots" (unfilled parts), i.e. set the hold pressure switch at approximately 30% of the dosing length.

2. Remove the hold pressure, i.e. set the value to 0 MPa. When you then inject the parts, the injection pressure goes down to 0 MPa when it reaches the hold pressure switch point. The linear motion of the screw stops then immediately. The cavities are normally not completely full at this setting. But if that is the case:

3. Move back the hold pressure switch point further. The perfect setting is the position where 90–95% of the cavities are filled (see Figure 26.47).

Figure 26.47 Here you can see one-half of a snap lock in acetal. The parts are molded in a two-cavity mold while setting the hold pressure switch by the procedure described above.
The two cavities do not fill completely the same, which may be due to very small differences in the dimensions of the gates. Nevertheless, we can see that the parts are not completely filled. We can also see that about 90–95% of the full volume has been reached, which is the recommendation, when we look for the correct position of the switch point.
When you then adjust the hold pressure from 0 MPa to your normal setting, the remainder of the lower part will fill and both parts will be packed and get correct strength and dimensions.

The last fields of the processing pane are *Cushion* and the checkbox *Cushion stable*. Earlier in this chapter we mentioned that the cushion should be at least 5 mm, and ideally you should reach the same value all the time plus or minus a few tenths of a millimeter for small parts and plus or minus one millimeter for big parts. If the cushion is steadily declining during hold pressure time, this is normally an indication of leakage of the back flow valve. The valve should in this case be corrected immediately, otherwise the quality of the molded parts will be compromised.

The last pane in the "Injection moulding process analysis" is called *Comments*. Here you can write observations, the results of your trial, or other information that may be valuable to save.

In Figure 26.48 you will find “Injection moulding process analysis” in full format, and if you want to have it as an Excel spreadsheet it is available to download free at www.brucon.se

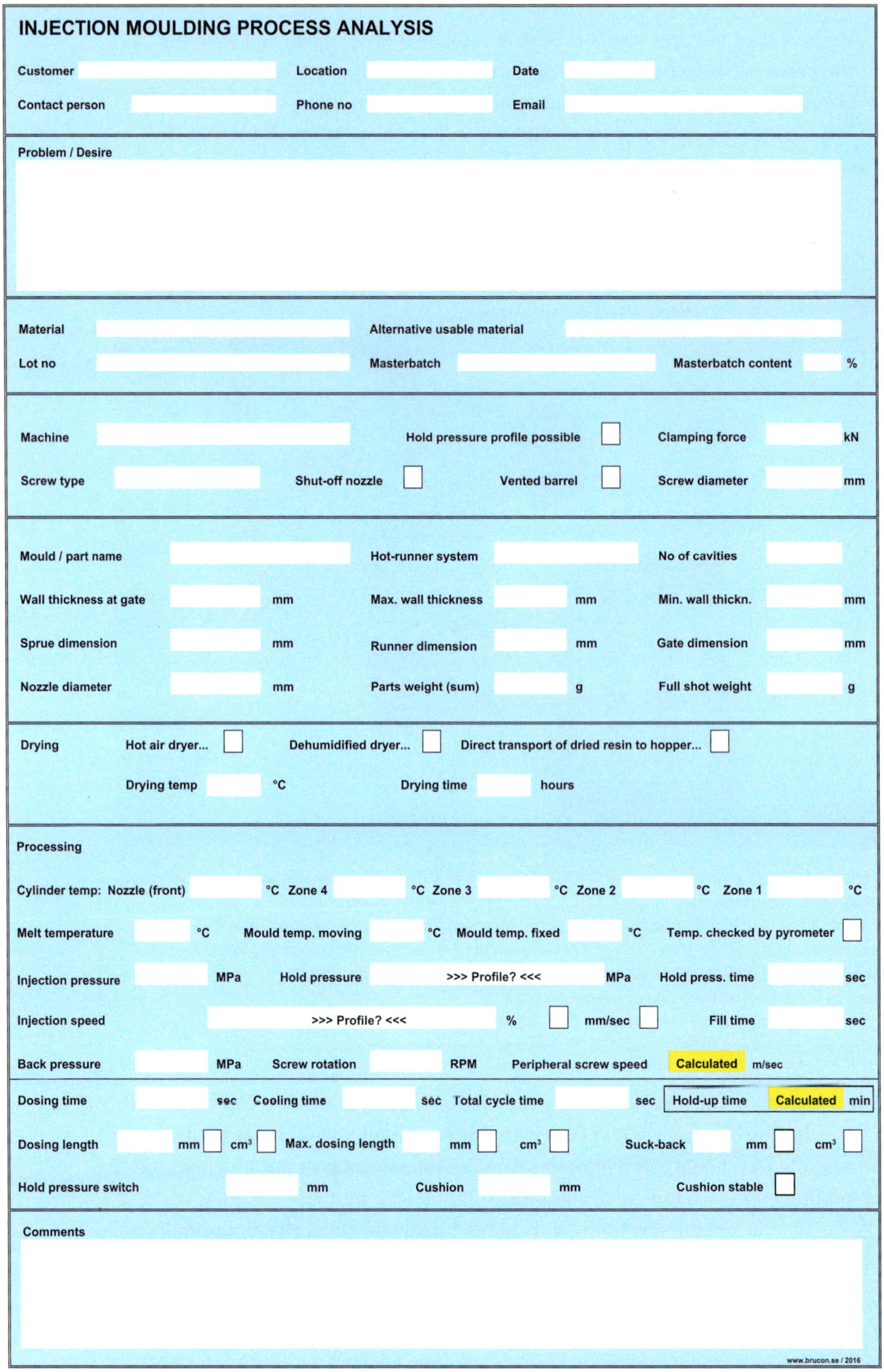

INJECTION MOULDING PROCESS ANALYSIS

Customer | Location | Date

Contact person | Phone no | Email

Problem / Desire

Material | Alternative usable material

Lot no | Masterbatch | Masterbatch content %

Machine | Hold pressure profile possible ☐ | Clamping force kN

Screw type | Shut-off nozzle ☐ | Vented barrel ☐ | Screw diameter mm

Mould / part name | Hot-runner system | No of cavities

Wall thickness at gate mm | Max. wall thickness mm | Min. wall thickn. mm

Sprue dimension mm | Runner dimension mm | Gate dimension mm

Nozzle diameter mm | Parts weight (sum) g | Full shot weight g

Drying | Hot air dryer... ☐ | Dehumidified dryer... ☐ | Direct transport of dried resin to hopper... ☐

Drying temp °C | Drying time hours

Processing

Cylinder temp: Nozzle (front) °C | Zone 4 °C | Zone 3 °C | Zone 2 °C | Zone 1 °C

Melt temperature °C | Mould temp. moving °C | Mould temp. fixed °C | Temp. checked by pyrometer ☐

Injection pressure MPa | Hold pressure >>> Profile? <<< MPa | Hold press. time sec

Injection speed >>> Profile? <<< % ☐ mm/sec ☐ | Fill time sec

Back pressure MPa | Screw rotation RPM | Peripheral screw speed Calculated m/sec

Dosing time sec | Cooling time sec | Total cycle time sec | Hold-up time Calculated min

Dosing length mm ☐ cm³ ☐ | Max. dosing length mm ☐ cm³ ☐ | Suck-back mm ☐ cm³ ☐

Hold pressure switch mm | Cushion mm | Cushion stable ☐

Comments

www.brucon.se / 2016

Figure 26.48 “Injection Moulding Process Analysis”: the Excel document can be downloaded by readers for free.

CHAPTER 27
Injection Molding Process Parameters

In this chapter, we will present the main injection-molding parameters for a number of thermoplastics.

When setting an injection-molding machine with a new resin, you should always use the recommended process data from the raw material producer if available. If you do not have them, look on the producer's website or search for them on the Internet.

NOTE: The values shown in Table 27.1 are typical for an unmodified standard grade of the polymer in question and serve only as a rough guide. Contact your plastic raw material supplier for accurate information about your specific grade!

The **Melt temperature** is one of the most important parameters. When processing semi-crystalline plastics, you should always consider the risk that you may get unmelted granules in the melt. To eliminate this risk, you should use the cylinder temperature profile that depends on the capacity utilization of the cylinder (see Section 26.8). You should also be aware that additives such as flame retardants or impact modifiers often require a lower temperature than the standard grade. Glass fiber reinforced grades should have, as a rule, the same temperature settings as unreinforced grades.

The **Mold temperature** is also one of the most important parameters for achieving the best quality. For semi-crystalline plastics you need a certain temperature to ensure that the material's crystal structure will be correct and thus provide the best strength and dimensional stability (less post-shrinkage). See Section 26.9.

Drying is needed for plastics that are either hygroscopic (absorb moisture) or sensitive for hydrolysis (degraded chemically by moisture). See more in Section 26.7.

We recommend that molders use dehumidifying (dry air) dryers in their production. Therefore we publish both the *temperature* and *drying time* needed to be below the maximum allowed moisture content for the material, provided that the dry air dryer is working with a sufficiently low *dew point*.

Note also that if you dry the material longer than the indicated time in the table you should reduce the temperature 10–20 °C because some materials can oxidize or degrade thermally. Material where *"Does not normally need to be dried"* is given in the table may still need to be dried if condensation will occur on the surface of the granules. If this is the case, a drying temperature of 80 °C and a drying time of 1–2 hours usually works well.

The reason that the maximum **Peripheral speed** is published in the tables is that many molders in good faith are dosing up the next shot with too high of a screw speed and thus unnecessarily degrading the polymer chains in the cylinder by high shear and friction, resulting in poorer quality. In Section 26.13 (Figure 26.40) you will find a formula where you can calculate the maximum allowed peripheral speed to maximum allowed rotation speed depending on the screw diameter. If you cannot find the recommended maximum peripheral speed for your resin, you should take into account that high-viscosity grades sometimes require 30% lower rotation speed compared to a less viscous standard grade. For example, impact-modified acetal with a melt index of 1–2 g/10 min has a recommended maximum peripheral speed of 0.2 m/s, compared to 0.3 m/s for a standard grade with melt index of 5–10 g/10 min. For glass fiber reinforced grades you will usually find the recommended maximum peripheral speed to be 30–50% of the speed for the unreinforced grade. Also, impact modified, flame retardant grades used to be more sensitive to shear than standard grades.

Having sufficiently high **Hold pressure** is especially important for semi-crystalline plastics. Usually it is recommended to have as high a pressure as possible without getting flashes in the parting line or having ejection problems. We provide hold pressures because many molders sometimes in good faith set far too low a hold pressure, resulting in poorer quality.

Other important parameters such as hold pressure time, hold pressure switch, back pressure, injection speed, and decompression are more dependent on the part design and machine conditions. We therefore cannot give any general values of these parameters, but refer you instead to Chapter 26.

Table 27.1 Typical processing data for unmodified standard grades of common thermoplastics.

Semi-crystalline commodities

Material	Type	Melt temperature Nominal	Melt temperature Range	Mold temp.	Drying Temp.	Drying Time	Drying Max moisture	Drying Dew point	Hold pressure	Max peripheral speed
Unit		°C	°C	°C	°C	Hours	%	°C	MPa	m/s
Polyethylene	PEHD	200	200–280	25–60	Does not normally need to be dried				25–35	1.3
Polyethylene	PELD	200	180–240	20–60	Does not normally need to be dried				25–35	0.9
Polyethylene	PELLD	200	180–240	20–60	Does not normally need to be dried				25–35	0.9
Polyethylene	PEMD	200	200–260	25–60	Does not normally need to be dried				25–35	1.1
Polypropylene	PP	240	220–280	20–60	Does not normally need to be dried				35–45	1.1

Amorphous commodities

Material	Type	Melt temperature Nominal	Melt temperature Range	Mold temp.	Drying Temp.	Drying Time	Drying Max moisture	Drying Dew point	Hold pressure	Max peripheral speed
Unit		°C	°C	°C	°C	Hours	%	°C	MPa	m/s
Polystyrene	PS	230	210–280	10–70	Does not normally need to be dried				45–50	0.9
HIPS	PS/SB	230	220–270	30–70	Does not normally need to be dried				45–50	0.6
SAN		240	220–290	40–80	Does not normally need to be dried				45–50	0.6
ABS		240	220–280	40–80	80	3	0.1	−18	45–50	0.5
ASA		250	220–280	40–80	90	3–4	0.1	−18	40–45	0.5
PVC	Soft	170	160–220	30–50	Does not normally need to be dried				40–45	0.5
PVC	Hard	190	180–215	30–60	Does not normally need to be dried				50–55	0.2
PMMA		230	190–260	30–80	8	4	0.05	−18	60–80	0.6

27 Injection Molding Process Parameters

Semi-crystalline engineering polymers

Material	Type	Melt temperature Nominal	Range	Mold temp.	Drying Temp.	Time	Max moisture	Dew point	Hold pressure	Max peripheral speed
Unit		°C	°C	°C	°C	Hours	%	°C	MPa	m/s
Acetal	POM Homo	215	210–220	90–120	Does not normally need to be dried*				60–80	0.3
Acetal	POM Copo	205	200–220	60–120	Does not normally need to be dried*				60–80	0.4
Polyamide 6	PA6	270	260–280	50–90	80	2–4	0.2	−18	55–60	0.8
Polyamide 66	PA66	290	280–300	50–90	80	2–4	0.2	−18	55–60	0.8
Polyester	PBT	250	240–260	30–130	120	2–4	0.04	−29	50–55	0.4
Polyester	PET+ GF	285	280–300	80–120	120	4	0.02	−40	50–55	0.2

Amorphous engineering polymers

Material	Type	Melt temperature Nominal	Range	Mold temp.	Drying Temp.	Time	Max moisture	Dew point	Hold pressure	Max peripheral speed
Unit		°C	°C	°C	°C	Hours	%	°C	MPa	m/s
Polycarbonate	PC	290	280–330	80–120	120	2–4	0.02	−29	60–80	0.4
Polycarbonate	PC/ABS	250	230–280	70–100	110	2–4	0.02	−29	40–45	0.3
Polycarbonate	PC/PBT	260	255–270	40–80	120	2–4	0.02	−29	60–80	0.4
Polycarbonate	PC/ASA	250	240–280	40–80	110	4	0.1	−18	40–45	0.3
Mod. PPO		290	280–310	80–120	110	3–4	0.01	−29	35–70	0.3

Semi-crystalline advanced thermoplastics

Material	Type	Melt temperature Nominal	Range	Mold temp.	Drying Temp.	Time	Max moisture	Dew point	Hold pressure	Max peripheral speed
Unit		°C	°C	°C	°C	Hours	%	°C	MPa	m/s
Fluoroplastic	FEP/PFA	350	300–380	150	Does not normally need to be dried**				Low***	****
Aromatic polyamide	PA6T/66	325	320–330	85–105	100	6–8	0.1	−18	35–140	0.2
	PA6T/XT	325	320–330	140–160	100	6–8	0.1	−18	35–140	0.2
LCP		355	350–360	60–120	150	3	0.01	−29	20–60	Max
PPS		330	300–345	70–180	150	3–6	0.04	−29	45–50	0.2
PEEK		370	360–430	160–200	160	2–3	0.1	−18	50–65	0.2

Chapter 27 – Injection Molding Process Parameters

Amorphous advanced thermoplastics										
Material	Type	Melt temperature		Mold temp.	Drying				Hold pressure	Max peripheral speed
		Nominal	Range		Temp.	Time	Max moisture	Dew point		
Unit		°C	°C	°C	°C	Hours	%	°C	MPa	m/s
Polyetherimide	PEI	380	370–400	140–180	150	4–6	0.02	−29	70–75	0.5
Polysulfone	PSU	340	330–360	120–160	150	4	0.02	−29	50–70	0.4
PPSU		370	350–390	140–180	150	4	0.02	−29	50–70	0.4
PES		360	340–390	140–180	140	4	0.02	−29	60–80	0.2

* If the granules have been exposed to moisture or condensation you should dry approximately 2 hours at 100 °C

** If the granules have been exposed to moisture or condensation you should dry 3–4 hours at 150 °C

*** Hold pressure should be as low as possible

**** Stainless special screws are required

CHAPTER 28
Problem Solving and Quality Management

28.1 Increased Quality Demands

The accelerating developments in both processing technology and thermoplastics have led to new uses, such as metal replacement, electronics, and medical technology. At the same time, the demands on plastic components have increased when it comes to performance, appearance, and other characteristics. The slightest deviation from the requirements and specifications must immediately be addressed, and thus the goal for many molders is to deliver error-free products (zero tolerance) while keeping their own rejects below 0.5% at high utilization of their machines. We can no longer accept the previous "hysterical" troubleshooting methods, where changes in the process parameters (sometimes several at the same time) immediately were made without any detailed analysis of the problem as soon as there was an unacceptable deviation from the specifications. To meet the increasing and intense competition, you have to work with both statistical problem-solving methods and process control. In this chapter we will describe some of these:

- ATS: Analytical troubleshooting
- DOE: Design of experiments
- FMEA: Failure mode effect analysis

In the next chapter, we will describe a large number of errors that can occur during injection molding of thermoplastics and how to solve them.

28.2 Analytical Troubleshooting – ATS

The word "problem" is often used with different meanings, such as production problems, decisions, and plans to be implemented. This diversity can create a lot of confusion when it comes to communication with others.

Working on problems in a systematic and organized way within the area of "analytical troubleshooting" (ATS) requires very specific definitions of the word "problem". This leads to an improvement in communication and understanding among the parties involved.

28.2.1 Definition of the Problem

A problem always consists of a *cause* and an undesirable *deviation*.

Figure 28.1 shows an example of this.

Figure 28.1 The black specks that you can see on the red button of the safety belt lock are an unacceptable deviation. It is normally defined as a surface defect.

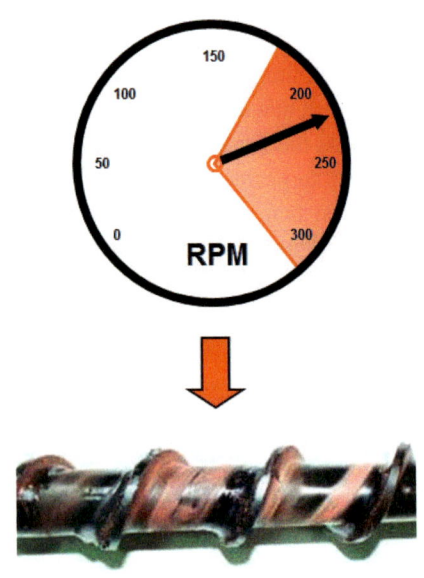

Figure 28.2 The cause of these black specks is usually the use of too-high screw speed in the injection-molding machine. This leads to a screw deposit on the surface of the screw that degrades thermally and causes the specks on the surface of the button.

28.2.2 Deviation Definition

A deviation is defined as a "failure" or a "problem" due to the difference between the *setpoint* and the *actual value*. This definition of the problem involves defining a problem's setpoint without knowing why the actual value differs from the setpoint.

This is represented by the following diagram:

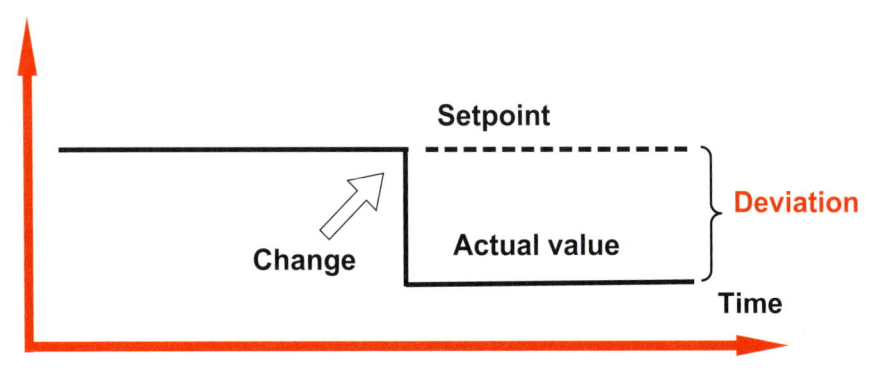

Figure 28.3 Definition of deviation.

When an "acute" problem arises in most cases you will not have the right focus or perspective on the situation. This might mean that the problem ultimately becomes both difficult to solve and unnecessarily expensive.

In order to quickly identify and analyze a problem, you need to have a clear definition of what is to be expected *(the setpoint)* and an equally clear definition of what has actually happened or is generating the problem *(the actual value)*. Setpoints and actual values can be used for various areas such as personnel, equipment, materials, processes, products, and markets.

It is crucial that both setpoints and actual values be clearly and unambiguously defined in order to verify whether the expected goals are met.

The following common defects can be used for an actual value:

Too general:	The goal, also called the norm value, is only available for the entire activity, and not just for the clearly identifiable components that contribute to the overall final result. The observed deviations will thus only be of a general character.
Traditions/ assumptions:	Setpoints are determined as a result of assumptions or traditions and are reconsidered only if significant changes have occurred.
Changed norms:	Demands or results that were acceptable may change due to changes in the business environment. Changed attitudes to the environment or climate changes, energy use, product liability, or gender may generate forced new setpoints.

The setpoint must be measurable, realistic, and accepted by those who must comply with it to get it achieved. This information must also be available for everyone involved. The actual value must contain the same specific information and should be measured in the same way as the setpoint. It is also important that the actual value is measured often enough and is documented.

If there is a deviation from the expected value, positive or negative, you can by describing the scope, timing, and trend decide how to prioritize this problem in comparison to other problems that must be addressed.

28.3 Defining a Problem

When a problem arises, it is important to first get a general view of its extent. This initial survey should result in where all parties involved can agree on the **where** and **how** of the problem. An agreed-upon definition of the problem should be created.

Where does the problem arise?

- Only one machine or part shows the problem?
- Several similar machines or parts show the problem?
- Several different machines or different parts show the problem?
- Several similar machines or similar parts show the problem but to a different extent or frequency?

- Several different machines or parts show the problem but with a different extent or frequency?

How does the problem occur?

- The machine or the part is new or recently taken into production?
- The machine or the part is old or has been in production previously?
- The machine or the part does not reach the goal or desired performance?
- The problem arises suddenly and then persists to the same extent?
- Problems "come and go" but always to the same extent?
- The problem is "coming and going" in a varying extent?
- The problem or error increases in scope or frequency, creating an increasing trend?

28.3.1 Classification of Problems

Problems can range from small and relatively insignificant to very extensive and with major impact on life. Here is a list of how problems can be classified into different categories:

1. Small problems
2. Recurring problems
3. Start-up problems
4. Complex problems
5. Potential problems

1. Small Problems

- Only insignificant values are involved
- The pressure to solve the problems is not acute
- The problems have recently been discovered
- Information can be obtained without too much difficulty

An example of a small problem may be that the tape used to close cardboard boxes full of molded parts has run out and the boxes cannot be sealed off due to this lack of adhesive tape.

2. Recurring Problems

- The deviation comes and goes randomly
- They disappear suddenly without any actions taken but occur again later on
- They can be very expensive, time consuming, and frustrating to solve
- Often DOE and FMEA will be of great value when trying to solve these kinds of problems

An example of such a recurring problem is an injection-molding process where the reject level increases from 5% to 20%. After 3 hours it returns to normal. The problem occurs once a week.

3. Start-up Issues

- Problems in achieving the specified setpoint at the initial start-up
- This kind of problem can often be prevented by using mold-filling simulation software

An example of such a problem is when starting up a brand new mold and getting too high shrinkage so that the parts will not fit the tolerance range.

4. Complex Problems

- The problems are caused by multiple factors that will interact
- The individual causes of the problem may be small one by one but when added together they can be amplified and create a much larger problem
- It can sometimes be very difficult to understand how the individual causes of the complex problem interact
- Often DOE and FMEA can be of great value when trying to solve these kinds of problems

5. Potential Problems

- Small problems that do not require immediate attention but can eventually grow into larger problems if not taken care of

An example of such a problem is shown in Figure 28.4.

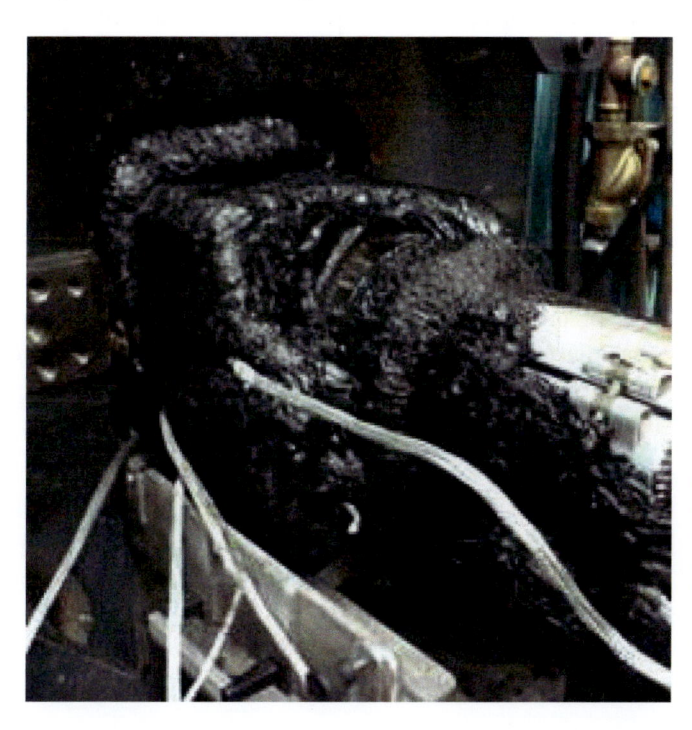

Figure 28.4 This picture shows material that has leaked between the nozzle and the mold. As this part of the machine was hidden behind a cover, the problem was not observed immediately. A small problem thus grew into a larger one which will be difficult to handle with a stop in production for several hours.

28.3.2 Problem Analysis

When a problem is analyzed analytically, the aim is to find the cause for the deviation as well as prevent the problem from occurring in the future. In order to achieve these goals, steps can be taken according to Figure 28.5.

	Action steps	Auxiliary tools
Discover	Identify the problem Classify the problem	Setpoint/Actual value analysis Impact/Time/Trend
Investigate	Define the problem Specify *Is/Is not*	Analyze in steps the cause: What, Where, When, How Evaluate the scope
Find and prove	Characteristics/Changes Probable causes Verify the cause	Search for more information Test against the specification Prove by practical tests
Actions to do	Eliminate short/long term Evaluate the long term consequences of the action	Impact assessment: Current deviation Future deviations

Figure 28.5 The basic steps to discover and investigate problems.

First you have to discover and classify the problem. Problems that have been defined and named have great potential to be quickly solved as they can be searched for on the Internet.

Once the problem has been defined, you usually get the answer to the question: **What** kind of deviation has occurred? You can then begin to investigate the nature of the problem.

When investigating the problem an obvious question is: **Where** has the problem occurred?

It is often important to know **when** the deviation has occurred in order to be able to prioritize the problem. You must understand **how** comprehensive it is and what impact it will have if it is not taken care of immediately. Figure 28.6 shows a graphic presentation of problem analysis.

28.3 Defining a Problem

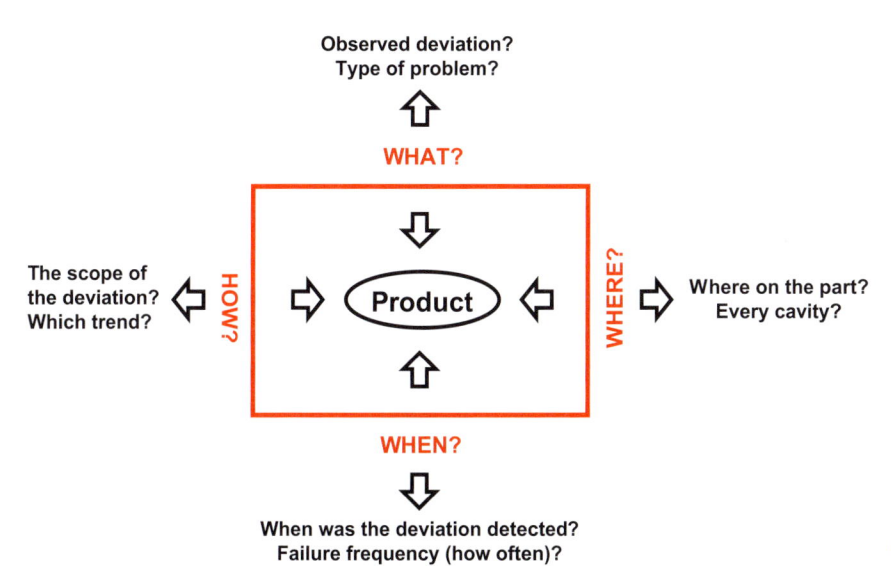

Figure 28.6 This figure shows the questions to be answered in order to understand the nature of the problem.

Once you think that you have an understanding of the nature of the problem, you can start to find the cause of it in order to be able to deploy a countermeasure. If you do not know the cause, you can in many cases find information on the problem with the help of troubleshooting guides found on the Internet. Figure 28.7 shows the proposed remedies to be tested if you get black specks on a part made of polypropylene. This information can be obtained free of charge at the European distributor Distrupol's website: www.distrupol.com.

Troubleshooter

Polypropylene
Black Specks

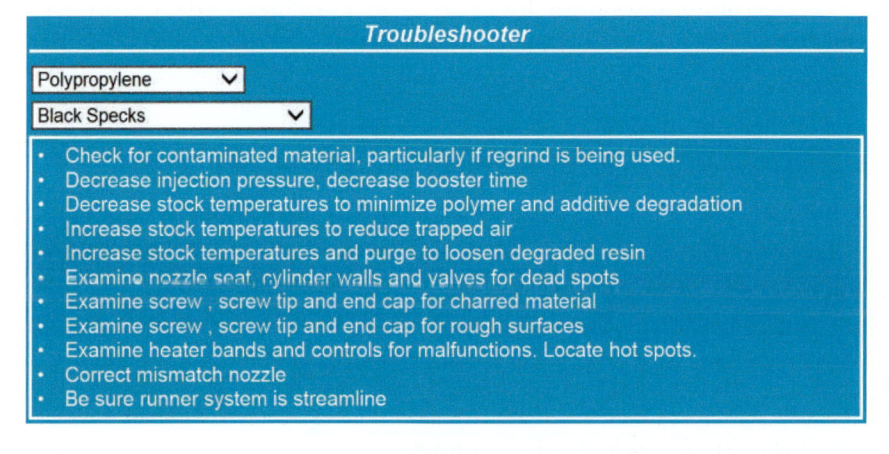

- Check for contaminated material, particularly if regrind is being used.
- Decrease injection pressure, decrease booster time
- Decrease stock temperatures to minimize polymer and additive degradation
- Increase stock temperatures to reduce trapped air
- Increase stock temperatures and purge to loosen degraded resin
- Examine nozzle seat, cylinder walls and valves for dead spots
- Examine screw, screw tip and end cap for charred material
- Examine screw, screw tip and end cap for rough surfaces
- Examine heater bands and controls for malfunctions. Locate hot spots.
- Correct mismatch nozzle
- Be sure runner system is streamline

Figure 28.7 Troubleshooting guide by Distrupol.

28.3.3 Brainstorming

There is always at least one cause of each problem. The difficulty is often in finding the cause. By taking advantage of the collective experience of your colleagues you will see that the time required to solve a problem is reduced significantly. This can be achieved in several ways. Brainstorming can be one way to find a number of possible causes of a problem.

These causes can then be highlighted and evaluated by addressing the characteristics of the problems. At the same time you will assess the probability of your assumptions in order to explain the assumed sequence of events.

Figure 28.8 Brainstorming: a team working with respect.

There is a lot of information about the effectiveness of brainstorming to be found on the Internet or in the literature. Often the information refers to the following four rules:

- Focus on the quantity of ideas: the more ideas, the better the chance to find a solution to the problem
- Ideas must not be criticized before they are completed
- Unusual ideas should be encouraged: the starting point is that all ideas are good ideas. An idea that at first does not seem to be great can actually develop into a brilliant idea
- Combine and develop ideas: all participants should be involved in developing and improving each other's ideas. Many times you get the 1 + 1 = 3 "rule".

28.3.4 Verification of Causes

When you think you have found the most likely cause of the problem, this should be verified by trial and error. If you suspect that a process parameter causes the problem, the value should be increased and then decreased in order to verify that the problem follows the same pattern. If you suspect that several process parameters interact, you should first test them one by one and then create a series of experiments to test them simultaneously. Learn more about this in the section on statistical design of experiments, DOE (Section 28.4).

28.3.5 Planning of Actions to Take

Once the cause(s) are established and verified through practical testing, you can start the process to eliminate them. You must immediately determine when the action should be performed and may need to ask the following questions:

- Should the process be stopped immediately or can unacceptable parts be sorted out manually and the final actions of corrections be put on hold until the ongoing process is completed and the delivery quantity reached?
- Should we do the final solution directly or just do a temporary one?
- Can the product pass the quality control even though it is slightly out of range?
- Should the same action be made on other machines that are in production as well?
- Might the action generate other problems?
- Can we be 100% sure that the planned action will be sufficient?

28.4 Statistical Design of Experiments – DOE

If you have difficulties in understanding and verifying the cause of a problem, a good source of help may be found in using an experimental method called "design of experiments" or DOE.

The goal of this section is not to provide a complete description of this method, but only to inform about its advantages over random tests. For more information about this method there is much to be found on the Internet or in specific literature.

The benefits that you normally get when you use the method of design of experiments are:

- To save time and money as troubleshooting will be made easier and faster
- To find a robust process setting, where the variations that can occur in the process still will give products that are within the specifications of the tolerance range
- To reduce wear on machinery and molds in the processing of some special materials (such as glass fiber reinforced ones)

If you ask if this method is very complicated, requiring great knowledge of statistics, the answer is that you can make a comparison to driving a car. You do not need to have a deep knowledge of how the car engine works in order to drive it. Regarding DOE you can set up both simple experiments such as those in Table 28.1 or use highly sophisticated computer programs that with a limited number of tests still will be able to find the relationships between the different process parameters.

28.4.1 Factorial Experiments

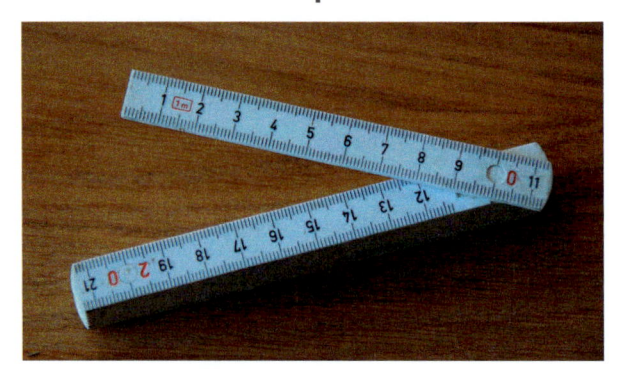

Figure 28.9 Ruler in glass fiber reinforced PA66.

We will now describe an example of problem solving by using factorial experiments. Here a molder was producing folding rulers in a glass fiber reinforced PA66. When performing a cost evaluation six weeks after the production, it was concluded that the profitability was a disaster with a reject rate of 30%. Each ruler had to be measured as the tolerance range did not allow a greater deviation than 1 m ± 2 mm in total. Many of the rulers where either too short or too long.

The molder also discovered that wear of the screw in the injection-molding machine was unusually high. Without understanding the cause of both problems it was decided to find a robust injection-molding process, with a reject level below 1%, as quickly as possible. At the same time reducing the wear of the screw was also an issue. Reducing the reject level got a higher priority than reducing wear.

When trying to figure out the causes for the variations in length of the rulers, the following questions were asked:

- Can variations in the plastic raw material, such as glass fiber content, affect shrinkage, and might this be a notable influence on the length of the ruler?
- Which of the process parameters will have an influence on the shrinkage and how will these vary?
- Are there other external causes, e.g. humidity in the air that can affect the shrinkage?

The same questions were asked when it came to the screw wear, followed by these other obvious questions:

- Can any of the process parameters cause both problems?
- Can any of the parameters interact and thus amplify the problems?

When performing the analysis, the following six machine parameters were concluded:

- The glass fiber content of the material, which can vary from 23 to 27%
- The melt temperature of the material in the cylinder of the injection-molding machine
- The screw rotation speed and the back pressure at the dosing of the next shot
- The hold pressure when packing the material in the cavities
- The hold pressure time
- The mold temperature

A decision was made to set up an experimental matrix to test the different parameters in the injection-molding machine. As the number of trials are limited, three of the parameters were chosen for testing at low and high settings. The mathematical formula for the number of individual trials to be performed in order to complete an experimental series is:

$$T = I^V \tag{28.1}$$

Where T is number of tests, I is number of settings for each variable, and V is the number of variables, i.e. process parameters.

If you want to test two parameters at two settings (low and high), 2^2 tests are required = 4, which is easy to implement. Three parameters with two settings (low and high) require 2^3 tests = 8, which also is realistic to make in one day.

If four parameters with three settings are desired to be tested (low, medium, and high) this will require 3^4 tests = 81, which in general is unrealistic. In such cases you need to use a special computer program that can reduce the number of tests but still find which parameters interact when problems occur. Test series should be run nonstop and in random order in order to get the most accurate results. Regarding the folding ruler, the following matrix was implemented, in which the probabilities of the individual parameters that might affect the problems were taken into consideration:

Table 28.1 Matrix with six possible causes judged in three probability levels for two different types of errors.

Parameters	Influence on wrong length			Influence on severe wear		
	Very big	Big	Small	Very big	Big	Small
Glass fiber content		X		X		
Melt temperature		X			X	
Screw rotation	X			X		
Hold pressure	X					X
Hold pressure time	X					X
Mold temperature	X					X

In order to reduce the number of tests to eight, it was decided to choose the parameters glass fiber content, screw speed, and hold pressure time and set a low and a high setting for each of them. The test was set up according to Table 28.2:

Chapter 28 – Problem Solving and Quality Management

Table 28.2 Matrix with eight tests.
Following parameters where selected: Glass fiber content: low 23% and high 27% (the supplier's delivery tolerances).
Rot. screw speed: 200 and 250 rpm. Hold pressure time: 3 and 5 sec.
A screw rotation of 250 rpm and hold pressure time of 3 sec were used in the six-week production series.
A hold pressure time of 5 sec will give a maximum weight of the folding ruler.

Trial no.	Glass content (%)	Screw rotation (rpm)	Hold pressure time (sec)	Reject level
1	23	200	3	
2	27	200	3	
3	23	250	3	
4	27	250	3	
5	23	200	5	
6	27	200	5	
7	23	250	5	
8	27	250	5	

In Table 28.3 you can see the results of the experimental series. However, the goal of 1% rejects was not quite reached. It was therefore decided to change the parameters to a screw speed of 200 rpm, hold pressure time of 5 sec, and a glass fiber content of 25%. This was performed in a new series of experiments where you can see the influence on the length of the other three parameters: melt temperature, hold pressure time, and mold temperature.

Table 28.3 Here are the results of the experiments. The lengths of the rulers were measured after 24 hours in order to get the full effect of the mold shrinkage. As PA66 takes up moisture and swells slightly, it was estimated that the swelling would be compensated for by the post-shrinking process that normally occurs upon injection molding of semi-crystalline thermoplastics. The levels of wear on the screw could not be determined until after six weeks.

Trial no.	Glass content (%)	Screw rotation (rpm)	Hold pressure time (sec)	Reject level
1	23	200	3	5%
2	27	200	3	14%
3	23	250	3	9%
4	27	250	3	31%
5	23	200	5	2%
6	27	200	5	3%
7	23	250	5	26%
8	27	250	5	33%

28.5 Failure Mode Effect Analysis – FMEA

Just as in the previous section about DOE, a full explanation of the method will not be given here; just a brief introduction of this method will be described.

This method was developed in order to reduce the risks in the U.S. aerospace industry in the 1950s and has gained an increasing interest in other sectors as well. Today we expect that a lot of companies in industry are using FMEA in their daily work. It is therefore important that molders also have knowledge about this method as they may be drawn into collaborations with customers that use the method.

With FMEA you will assess the following criteria:

1. What errors might occur during the design or manufacture
2. The cause of these errors
3. What the effects of those errors can generate

When the method is used to its full extent, the daily work is facilitated because:

- The quality of the decisions will increase
- Communication will be improved
- Time and money will be saved by reducing the time to launch new products
- Late changes in the process can usually be avoided and thus cost is reduced
- You will get a tool to help identify and measure improvements (systematic testing)
- You will get a tool ensuring that specifications will be met and early errors can be avoided
- FMEA will facilitate the implementation of statistical process control (SPC) if required
- You will normally get increased productivity and thus better profitability

For molders, FMEA can be a great tool when it comes to troubleshooting and process optimization. The persons who should be involved in an FMEA project among molders are:

- Designers, who will have a given place within the FMEA group, since they usually are those who set the requirements
- Process engineers or machine setters, who have the knowledge of how the material is to be processed
- Mold makers, who have to adjust problems concerning the gate, cooling channels, venting and ejection, etc.
- Quality engineers, who are responsible for making sure that quality control is done often enough and performed in a correct manner

When developing new products, the FMEA process is divided into different phases:

- Functional FMEA, where the preliminary and general requirements are specified

Chapter 28 – Problem Solving and Quality Management

- Design FMEA, where the specifications of the various components are defined
- Process FMEA, where the manufacturer's process is specified and optimized

Table 28.4 The different phases of FMEA.

Functional FMEA	Design FMEA	Process FMEA
■ Will be performed when the first specification has been done ■ Will be the first rough analysis of the overall requirements and the resulting consequences ■ Will be good for the decision making to the selection of the concept or system solution	■ When the concept or system is selected the analysis will be done on several occasions during the design phase ■ The failure modes and failure effects of the component and materials will be analyzed and addressed	■ Analysis will be performed in the production process to identify any fault possibilities ■ Will be used in an existing process or when modifying a new process ■ The process is divided into sub-steps and each step is reviewed

28.5.1 General Concepts of FMEA

Below is an example of how to work using FMEA.

Figure 28.10 This figure describes some common concepts of FMEA:
1) Failure mode: The car has a flat tire
2) Failure cause: There is a nail in the tire
3) Failure effect: The car cannot be used

1. **Flat tire = Failure mode**   2. **Nail = Failure cause**
3. **The car cannot be used = Failure effect**

Tables are often used within FMEA analysis in order to create a risk assessment of the failures. Table 28.5 is such a table. The first eight columns represent the part of the table that is defined as the Quantitative Analysis while the next four are the risk assessment and defined as the Qualitative Analysis.

In the first column of the quantitative part of the table, the level of *Probability (P)* is entered.

This is supposed to give information on the error's probability and is appraised according to the following criteria:

Probability that the error will occur	Value *(P)*
Very unlikely	1
Small risk	2–3
Medium risk	4–6
High risk	7–8
Very high risk	9–10

The next column is the *Severity (S)*, which indicates how serious the failure is, appraised according to the following criteria:

Severity of the problem	Value *(S)*
No noticeable effect	1
Negligible impact. The user will probably be annoyed	2–3
Significant impact, i.e. noise or functional impairment	4–6
Considerable inconvenience requiring repair	7–8
Very serious. Risk of injury or violation of law	9–10

Finally the column *Detectability (D)* shows how easily the errors can be discovered:

Detectability of the problem	Probability of detection	Value *(D)*
Errors that are always noticed	>99.9%	1
Errors that are always detected during quality controls	>99%	2–3
Low probability of detection	90–99%	4–6
Very low probability of detection	50–90%	7–8
Unlikely that the error will be discovered	<50%	9–10

The *Risk level* is obtained by multiplying $P \times S \times D$ and is used to set the priority of the *Actions* that need to be taken. In the example with the car, the tire needs to be repaired. In Table 28.5 we have listed the *Failure Mode Effect Analysis* for the car:

Table 28.5 Table showing error effect analysis (FMEA).

No.	Item	Phase	Failure mode	Failure effect	Failure cause	Probability (P) (estimate)	Severity (S)	Detection (D)	Risk level	Actions
1	Car	Consumer	Flat tire	The car cannot be used	Nail	5	8	1	40	Repair
2										

CHAPTER 29
Troubleshooting – Causes and Effects

29.1 Molding Problems

In the previous chapters we dealt with defects caused by bad material. In this chapter we will discuss process-related errors. These can generally be divided into the following main groups:

1. Fill ratio, which means unfilled or overfilled parts
2. Surface defects
3. Strength problems
4. Dimensional problems
5. Production problems

In general, process problems belong to several of the main groups.

In order to identify and classify a problem and then find possible causes for it, you should ask the following questions:

1. What kind of problem is it?
2. What has changed?
3. When did this happen?
4. Where do the error/errors occur:
   - On the part/parts (the same place or randomly)?
   - In the production cycle?
5. How often does it occur?
6. How serious is it?

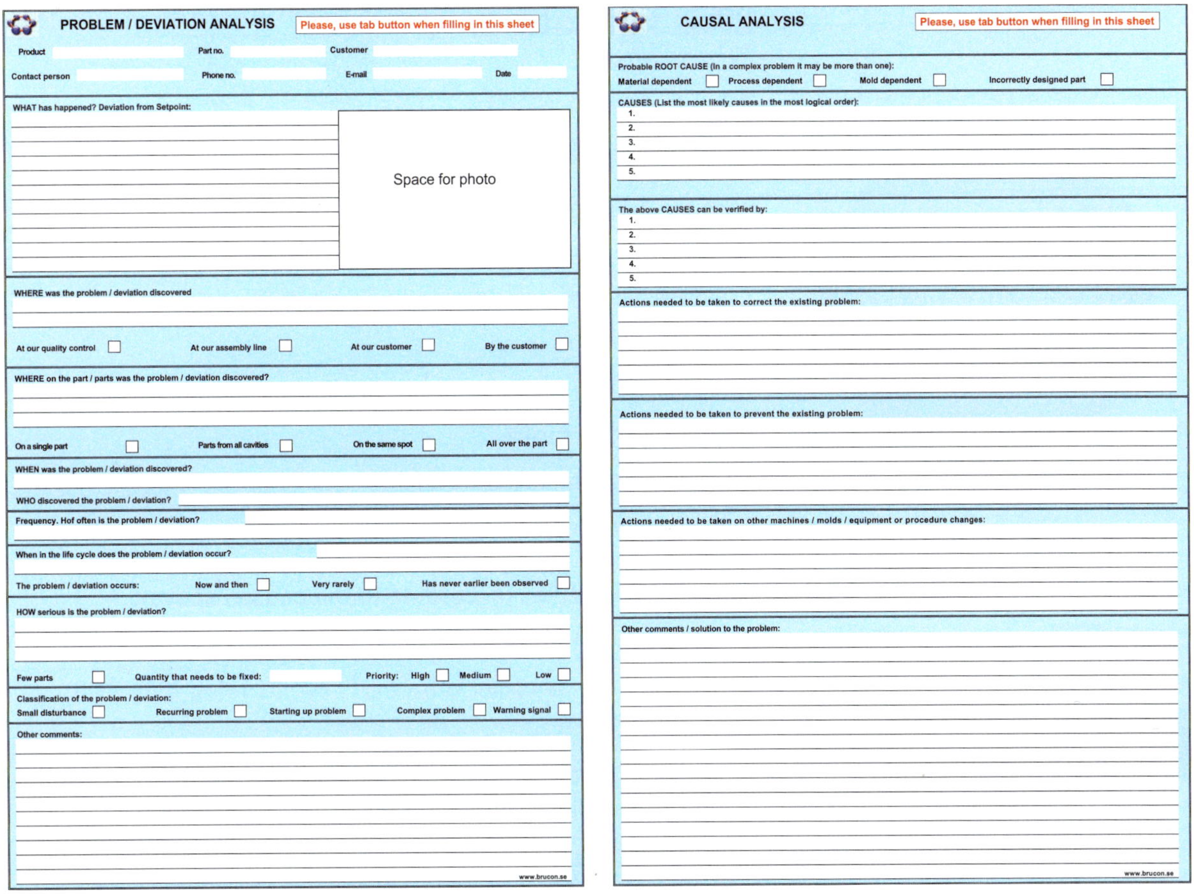

PROBLEM / DEVIATION ANALYSIS

Please, use tab button when filling in this sheet

Product | Part no. | Customer

Contact person | Phone no. | E-mail | Date

WHAT has happened? Deviation from Setpoint:

Space for photo

WHERE was the problem / deviation discovered

At our quality control ☐ At our assembly line ☐ At our customer ☐ By the customer ☐

WHERE on the part / parts was the problem / deviation discovered?

On a single part ☐ Parts from all cavities ☐ On the same spot ☐ All over the part ☐

WHEN was the problem / deviation discovered?

WHO discovered the problem / deviation?

Frequency. Hof often is the problem / deviation?

When in the life cycle does the problem / deviation occur?

The problem / deviation occurs: Now and then ☐ Very rarely ☐ Has never earlier been observed ☐

HOW serious is the problem / deviation?

Few parts ☐ Quantity that needs to be fixed: Priority: High ☐ Medium ☐ Low ☐

Classification of the problem / deviation:

Small disturbance ☐ Recurring problem ☐ Starting up problem ☐ Complex problem ☐ Warning signal ☐

Other comments:

www.brucon.se

CAUSAL ANALYSIS

Please, use tab button when filling in this sheet

Probable ROOT CAUSE (In a complex problem it may be more than one):

Material dependent ☐ Process dependent ☐ Mold dependent ☐ Incorrectly designed part ☐

CAUSES (List the most likely causes in the most logical order):

1.
2.
3.
4.
5.

The above CAUSES can be verified by:

1.
2.
3.
4.
5.

Actions needed to be taken to correct the existing problem:

Actions needed to be taken to prevent the existing problem:

Actions needed to be taken on other machines / molds / equipment or procedure changes:

Other comments / solution to the problem:

www.brucon.se

Figure 29.1 Forms for the analysis of problems in Excel format are available at www.brucon.se.

We will now describe a wide range of common and uncommon errors that can occur during the injection-molding process. We have also tried to list the causes of the most probable ones in a logical order, based on a large number of troubleshooting guides issued by leading plastic suppliers.

NOTE: When troubleshooting, it is important that the material supplier's process recommendations for the relevant material are available to adjust any incorrect settings.

Chapter 29 – Troubleshooting – Causes and Effects

29.2 Fill Ratio

29.2.1 Short Shots – The Part Is Not Completely Filled

Possible causes (listed in the most likely order):

1. Inadequate dosing; there is no cushion
2. Too-low hold pressure or wrong switch point
3. Too-low injection speed
4. Too-long injection time or wrong switch point
5. Faulty back flow valve
6. Insufficient venting (air traps)
7. Insufficient melt flow (too-high melt viscosity)

Suggested remedies (according to the causes above):

1. Increase the dosing. Check the material transport
2. Increase the switch point or the hold pressure (set the injection pressure to max)
3. Increase the injection speed to get faster mold filling
4. Increase the fill time and adjust the switch point
5. Replace the faulty back flow valve
6. Improve the venting:
   - Reduce the clamping force
   - *Workshop action:*
     Increase venting channels
7. Increase the melt flow (decrease the melt viscosity):
   - Increase if possible the melt temperature (start by checking the melt temperature with a pyrometer)
   - Increase the mold temperature
   - Change if possible to an easy-flow resin

Figure 29.2 The picture shows an unfilled part.

29.2.2 Flashes

Possible causes (listed in the most likely order):

1. Insufficient clamping force:
   - Too-low clamping pressure
   - Not enough clamping force on the injection machine
2. Too-high injection pressure or hold pressure
3. Too-high injection speed
4. Too-high melt flow (too-low melt viscosity)
5. Mold problems or a faulty design:
   - Too weak plates or too large diameter on the centering ring hole in the fixed mold half
   - Damaged parting line
   - The venting channels are damaged or worn

Figure 29.3 The figure shows flashes in the partijng line.

Suggested remedies (according to the causes above):

1. Increase the clamping pressure or machine size
2. Reduce the injection pressure or the hold pressure
3. Reduce the injection speed
4. Reduce the melt or mold temperature (check the temperatures with a pyrometer)
5. *Workshop action required (see also Chapter 16)*

29.2.3 Sink Marks

Possible causes (listed in the most likely order):

1. Insufficient material volume or leaking back flow valve (the screw hits the bottom)
2. Insufficient injection pressure or hold pressure
3. Too-short hold pressure time
4. Too-high or too-low injection speed
5. Too-high melt temperature
6. Mold problems or a faulty design:
   - Incorrect gate location
   - Too small gates or runners
   - Too thick walls
   - Incorrect rib design

Figure 29.4 Part with sink marks.

Suggested remedies (according to the causes above):

1. Make sure that there is enough material or check the function of the back flow valve
2. Increase the hold pressure switch point or the hold pressure (by pressure profile?)
3. Increase the hold pressure time
4. Adjust the injection speed
5. Decrease the melt temperature (check first the melt temperature with a pyrometer)
6. Replace the back flow valve
7. *Workshop action required (see also Chapter 16)*

29.2.4 Voids or Pores

Possible causes (in the most likely order):

1. Too-low hold pressure or difference between the injection pressure and the hold pressure
2. Too-short hold pressure time
3. Incorrect hold pressure switch
4. Too-high injection speed
5. Too-high melt temperature
6. Leaking back flow valve
7. Too-low back pressure
8. Mold problems or a faulty design:
    - Incorrect gate location
    - Too small gates or runners
    - Too thick walls

A further example can be seen in Figure 29.25.

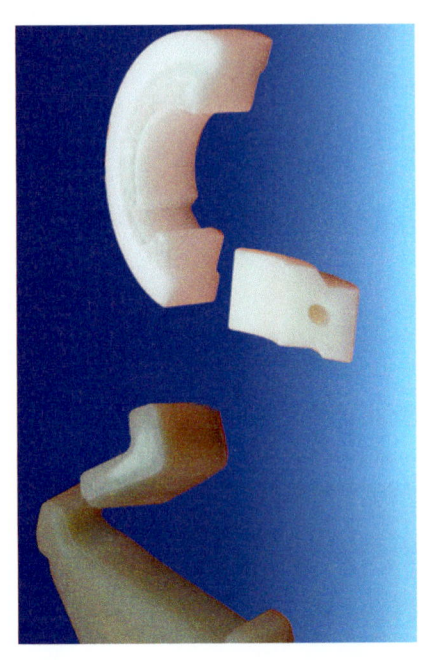

Figure 29.5 When sawing parts you can sometimes find voids in acetal parts (to the right) or the micro-pores in glass fiber reinforced polyamide (to the left).

Suggested remedies (according to the causes above):

1. Increase the hold pressure
2. Increase the hold pressure time
3. Adjust the hold pressure switch
4. Reduce the injection speed
5. Decrease the melt temperature (check first the melt temperature with a pyrometer)
6. Replace the back flow valve
7. Increase the back pressure
8. *Workshop action required (see also Chapter 16)*

29.3 Surface Defects

29.3.1 Burn Marks

29.3.1.1 Discoloration, Dark Streaks, or Degradation

Possible causes (listed in the most likely order):

1. Thermal degradation in the cylinder or nozzle
2. Thermal degradation in a hot runner mold

Suggested remedies (according to the causes above):

1. Check the melt with a pyrometer.
   If the melt is discolored:
   - Decrease if possible the melt temperature
   - Check if overheating occurs due to incorrectly installed or faulty temperature sensor in one of the heating bands (use the pyrometer between them)
   - Reduce the friction heat: Reduce the screw rotation speed and/or the back pressure
   - Check the hold-up time for the resin in the cylinder. If it is too long, change to a smaller cylinder or machine. Also check any dosing time delay
   - Change to an open nozzle if a shut-off nozzle is used
   - Check and clean the screw if there is any screw deposit

2. *Workshop action required (see also Chapter 16):*
   If a hot runner mold is used and the melt is not discolored:
   - Check the temperature in the manifold and nozzles
   - Reduce if possible the temperature
   - Check if there are any hold-up spots

Figure 29.6 Dark streaks on an acetal part.

29.3.1.2 Black Specks

Possible causes (listed in the most likely order):

1. Already found in the virgin resin granules or in regrind
2. Thermal degradation in the cylinder, nozzle, or back flow valve
3. Thermal degradation in the hot runner
4. Incorrectly cleaned plastification unit (screw and cylinder after material change)

Figure 29.7 Black specks on the surface.

Suggested remedies (according to the causes above):

1. Check the granules or regrind material visually
2. Check the melt temperature with a pyrometer, and if the melt contains black specks:
   - Decrease if possible the melt temperature
   - Check if there are any hold-up spots in the cylinder wall, between the cylinder and nozzle, on the screw, back flow valve, and screw tip
   - Check the hold-up time for the resin in the cylinder. If it is too long, change to a smaller cylinder or machine. Also check any dosing time delay
   - Change to an open nozzle if a shut-off nozzle is used
3. *Workshop action required (see also Chapter 16):*
   If the melt is not discolored: Check if there are any hold-up spots in the hot runner system
4. Clean the screw with the help of a wire brush

29.3.1.3 Splays or Silver Streaks (Partly over the Surface)

Possible causes (listed in the most likely order):

1. Gas bubbles that have been smeared on the surface during the filling:
   - Due to thermal degradation
   - Due to moisture in the material
   - Due to high shear in the nozzle, runners, or gate
2. Contamination due to inappropriate masterbatch in the resin

Suggested remedies (according to the causes above):

1. Check the melt temperature with a pyrometer.
   If the melt is "cooking":
   - Decrease if possible the melt temperature and reduce the screw rotation speed and the back pressure
   - Check if there are any hold-up spots in the cylinder wall, between the cylinder and the nozzle, on the screw, back flow valve, or screw tip
   - Check the hold-up time for the resin in the cylinder. If it is too long, change to a smaller cylinder or machine. Also check any dosing time delay
   - Change to an open nozzle if a shut-off nozzle is used
   - If the melt does not cook: Check if there are any hold-up spots (see also Chapter 16)

Figure 29.8 Silver streaks starting from the gate are signs of too high shear.

2. Reduce the shear:
   - Reduce the injection speed
   - *Workshop action required (see also Chapter 16):*
     Round sharp corners in runners or gates

3. Check the granules or regrind material visually or change the masterbatch

29.3.1.4 Diesel Effect – Entrapped Air

Possible causes (listed in the most likely order):

Degradation due to compressed air in the cavity (insufficient venting)

Suggested remedies (according to the causes above):

Eliminate the degradation:
- Reduce the injection speed
- Reduce the clamping pressure
- *Workshop action required (see also Chapter 16):*
  Improve the venting

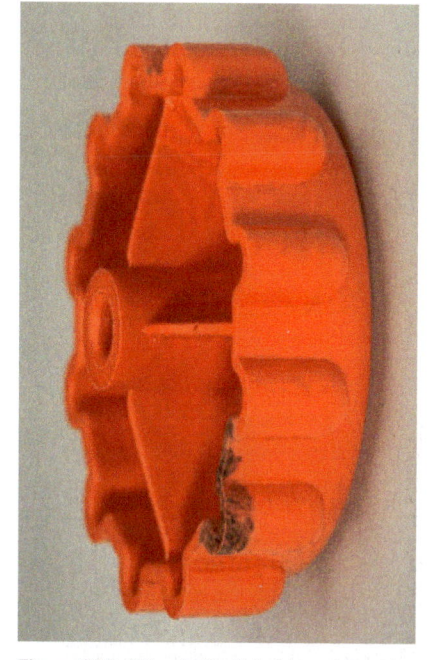

Figure 29.9 "Diesel effect" that has occurred when entrapped air has been compressed and heated.
[Photo: DuPont]

29.3.2 Splays or Silver Streaks (All over the Surface)

Possible causes (listed in the most likely order):

1. Gas or steam bubbles that have been smeared over the surface during the filling due to moisture in the material (visible all over the surface and on parts from all cavities)
2. Gas or steam bubbles that have been smeared over the surface during the filling due to thermal degradation or too high shear

Figure 29.10 Splays all over the surface on a polyamide part.

Suggested remedies (according to the causes above):

1. Check the melt temperature with a pyrometer.
   If the melt is "cooking":
   - Dry the material in a dehumidifying dryer (also non-hygroscopic materials need to be dried if there is condensation on the surface)
   - Check that the right drying temperature and time have been used
   - Check that the recommended maximum moisture content has been achieved
   - Check that the dried material has not come into contact with surrounding air (use a closed transport system between the dryer and the machine)
2. If the melt does not "cook": See Section 29.3.1.3 Splays or Silver Streaks

29.3.3 Color Streaks – Bad Color Dispersion

Possible causes (listed in the most likely order):

1. Uneven mixing of color pigments in the polymer
2. Uneven orientation of pigments during the filling
3. Color change due to thermal degradation

Figure 29.11 Blue color streaks on a light part due to unsufficient mixing of the pigments.

Suggested remedies (according to the causes above):

1. Check the melt temperature with a pyrometer.
   If the melt temperature is OK:
   - Increase the back pressure
   - Reduce the screw rotation speed
   - Change to a screw with mixer head

2. If masterbatch has been used:

- See (1) above
- Change to a masterbatch with smaller pigment size or different carrier

3. See Section 29.3.1.1 Discoloration

29.3.4 Color Streaks – Unfavorable Pigment Orientation

Possible causes (listed in the most likely order):

1. Metallic pigments visible in the weld-lines
2. Uneven orientation of pigments during the filling of the cavities

Suggested remedies (according to the causes above):

1. Change the filling sequence:

- Increase/decrease the injection speed
- Change the location of the weld-lines (see Section 29.3.9 Weld-Lines)
- Change to non-metallic pigments
- If there are high surface requirements painting will be a better alternative

2. If masterbatch is used:

- Change the filling sequence (see above)
- Change to a masterbatch with smaller pigment size or different carrier

Figure 29.12 Dark weld-lines on a silver colored part is almost impossible to get rid of.

29.3.5 Surface Gloss – Matte/Shiny Surface Variations

Possible causes (listed in the most likely order):

1. The surface is differently embossed by the cavity wall due to pressure variations
2. Wall temperature variations in the cavity

Suggested remedies (according to the causes above):

1. Improve the embossing of the wall:

- Increase the melt temperature
- Increase the mold temperature
- Increase the hold pressure or adjust the hold pressure switch
- Increase the hold pressure time

Figure 29.13 In this picture you can see the surface gloss difference with a blank triangle on the otherwise matte surface. This depends on the filling of the part being mainly done by the hold pressure, i.e. because of incorrect hold pressure switch point (too early).

Chapter 29 – Troubleshooting – Causes and Effects

2. Reduce the temperature variations:
   - Increase, decrease, or use a profile of the injection speed
   - Improve the mold temperature control (more cooling channels or mold heaters)

29.3.6 Surface Gloss – Corona Effect

Possible causes (listed in the most likely order):

1. Uneven filling sequence
2. Too small gate

Suggested remedies (according to the causes above):

1. Decrease the injection speed
2. *Workshop action required (see Chapter 16):*
   Increase the size of the gate

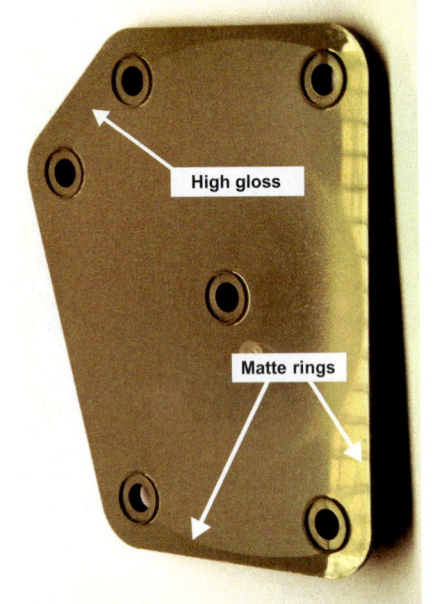

Figure 29.14 The corona effect is visible as matte rings around the gate, located in the hole in the middle, on the high-gloss polished surface.

29.3.7 Splays, Stripes, and Blisters

Possible causes (listed in the most likely order):

1. Small air bubbles entrapped in the melt are pressed against the surface during the filling process and form white or silver stripes
2. Turbulence at the filling due to differences in wall thickness, especially at ribs, bumps, or depressions

Suggested remedies (according to the causes above):

1. Reduce the risk of air entrapment in the surface:
   - Increase or decrease the injection speed
   - Decrease the decompression (suck-back)
   - Decrease the screw rotation speed
   - Check the back flow valve
   - Improve the venting

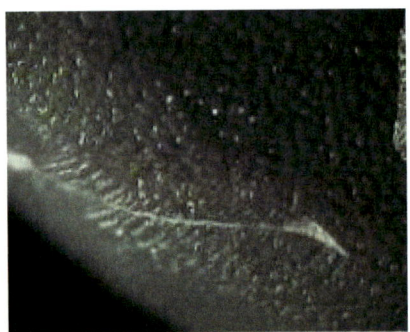

Figure 29.15 A white air stripe on a textured surface.

2. Reduce the risk of turbulence:
   - Round sharp corners (connection of ribs)
   - Reduce countersinking or engraving depth

29.3.8 Glass Fiber Streaks

Possible causes (listed in the most likely order):
1. The glass fiber in the melt orients toward the surface during the filling process and forms white or silver colored streaks
2. The melt solidifies too fast and causes the glass fibers not to be completely enclosed by the plastic
3. The glass fibers cause uneven shrinkage in the flow and cross directions
4. Runners or gates have sharp corners (too small radius), resulting in high shear
5. The gate is too small relative to the volume of the cavity, resulting in high shear

Suggested remedies (according to the causes above):
1. Increase the melt flow:
   - Increase the melt temperature if possible
   - Increase the mold temperature
2. Extend the solidification sequence:
   - Increase the injection speed
3. Reduce too big differences in mold shrinkage:
   - Increase the hold pressure
   - Increase the hold pressure time
4. Increase the radius in runners or gates
5. Increase the size of runners or gates

Figure 29.16 The glass fiber streaks are in a ring around the gate.

29.3.9 Weld-Lines (Knit-Lines)

Possible causes (listed in the most likely order):
1. The flow fronts have time to cool before they meet
2. The flow fronts do not merge well enough
3. The air does not have time to leave the cavity
4. You have chosen a particularly sensitive material
5. Mold problems or a faulty design (gate location)

Figure 29.17 The picture shows a bad weld-line on a hinge of a plastic box.

Suggested remedies (according to the causes above):

1. Improve the merging by temperature:
   - Increase the melt temperature
   - Increase the mold temperature
2. Improve the merging by pressure and time:
   - Increase the hold pressure
   - Increase the hold pressure time
3. Improve the venting
4. Replace if possible impact-modified resins with unmodified resins
5. *Workshop action required (see also Chapter 16)*

29.3.10 Jetting

Possible causes (listed in the most likely order):

1. Incorrect gate location, i.e. injecting directly into an empty space without breaking the plastic jet against an opposite wall or core
2. The plastic jets do not melt and merge together with the rest of the melt

Suggested remedies (according to the causes above):

1. Break the plastic jet:
   - Insert a core in the path of the jet
   - Increase the size of the gate
   - Change the angle of the gate
   - Move the gate so the plastic jet will break against the opposite wall
2. Improve the merging:
   - Increase the melt temperature
   - Increase the mold temperature
   - Reduce the injection speed
   - Profile the injection speed: Slow $\Rightarrow$ fast

Figure 29.18 Here you can see three jets starting from the gates in the larger hole.

29.3.11 Delamination

Possible causes (listed in the most likely order):

1. Too high shear of the melt flow
2. Too sharp corners in the gate or the part
3. Inappropriate masterbatch
4. Too high content of regrind
5. Remains of material from previous run are left in the cylinder of the machine

Figure 29.19 The delaminated material builds a thin skin on the surface of the part.

Suggested remedies (according to the causes above):

1. Reduce the shear by increasing the fluidity of the melt:
   - Increase the melt temperature
   - Increase the mold temperature
2. Reduce the shear at sharp corners:
   - Reduce the injection speed
   - Increase the radius of the corners
3. Select a masterbatch with the same carrier as the polymer
4. Reduce the content of regrind
5. Clean the screw and cylinder more efficiently

29.3.12 Record Grooves (Orange Peel)

Possible causes (listed in the most likely order):

1. The melt solidifies too quickly and the following liquid material flows over and forms a "wave", after which the process is repeated
2. The melt is not packed enough against the mold wall
3. Different surface finish on the mold halves, i.e. textured on one half and polished on the other

Suggested remedies (according to the causes above):

1. Reduce the cooling of the melt:
   - Increase the mold temperature
   - Increase if possible the melt temperature
   - Increase the injection speed
2. Increase the hold pressure
3. Select the same surface finish on both mold halves

Figure 29.20 This picture shows the so-called record groove effect on a black acetal surface.

Chapter 29 – Troubleshooting – Causes and Effects

29.3.13 Cold Slug

Possible causes (listed in the most likely order):

1. The material freezes in the nozzle
2. None or incorrectly located cold slug pocket in the runner
3. The melt flows into the fixed half during the opening or closing phase of the injection-molding cycle

Suggested remedies (according to the causes above):

1. Increase the nozzle temperature
2. Locate the cold slug pocket opposite the sprue in the mold
3. Reduce the risk of melt leakage into the mold:
   - Increase the decompression (suck-back)
   - Reverse the injection unit during the opening and closing phase
   - Increase the injection speed

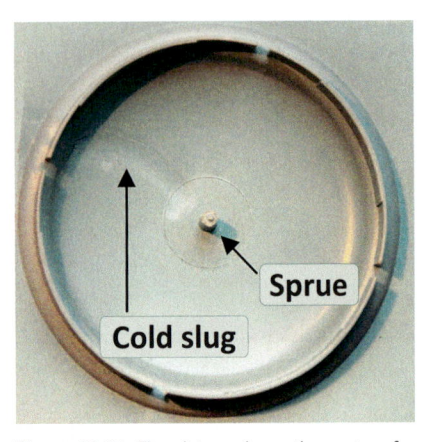

Figure 29.21 The picture shows the center of a hubcap. The gate is located on the opposite side. During the opening and closing phase, melt material has flowed into the cavity due to the injection unit abutting the mold.

29.3.14 Ejector Pin Marks

Possible causes (listed in the most likely order):

1. The part sticks too tightly in the cavity
2. The part is not cold enough (stiff) at the ejection
3. Mold problems or a faulty design

Suggested remedies (according to the causes above):

1. Reduce the mold shrinkage:
   - Reduce the hold pressure
   - Reduce the hold pressure time
   - Increase the release agent (surface lubrication) in the resin
   - Use a release spray (initially)
2. Eject or cool the part more efficiently:
   - Increase or decrease the ejection speed
   - Reduce the mold temperature
   - Increase the hold pressure time or the cooling time
3. *Workshop action required (see also Chapter 16):*
   - Increase the draft angles in the cavity
   - Change the size or design of the ejector pins

Figure 29.22 Here there are visible ejector pin marks looking like white crescent moons. You can also see a clear sink mark.

29.3.15 Oil Stain – Brown or Black Specks

Possible causes (listed in the most likely order):

1. Leaking cooling fluid when an oil temperature control unit is used
2. Leaking hydraulic oil hoses (cores)
3. Lubrication drops from the mold
4. Contamination from the gripper of the robot
5. Micro-cracks in the walls or plates of the mold

Suggested remedies (according to the causes above):

1. Check hoses
2. Check hose connections
3. Clean the mold
4. Clean the gripper of the robot
5. *Workshop action required (see also Chapter 16):*
   Repair the mold

Figure 29.23 Here there are brown greasy oil stains on a white plastic cap.

29.3.16 Water Stain

Possible causes (listed in the most likely order):

1. Leaking temperature-control hoses in the mold
2. Leaking gaskets in the mold
3. Cracks in the plates of the mold

Suggested remedies (according to the causes above):

1. Check hose connections and hoses
2. Check O-rings and gaskets in the mold
3. *Workshop action required (see also Chapter 16):*
   Repair the mold

Figure 29.24 Here there is a diagonal mark on the surface that was formed when plastic melt came in contact with water in the cavity.

Chapter 29 – Troubleshooting – Causes and Effects

29.4 Poor Mechanical Strength

29.4.1 Bubbles or Voids inside the Part

Possible causes (listed in the most likely order):

1. Gas or steam bubbles inside the part due to moisture in the resin (visible in parts from all cavities)
2. Air bubbles in the melt
3. Pores or voids due to bad packing of the part

Suggested remedies (according to the causes above):

1. Check the melt temperature with a pyrometer, and if the melt is "cooking":
   - Dry the resin in a dry air dryer (also non-hygroscopic resin needs to be dried if there is condensation on the surface
   - Check that correct drying temperature and time have been used
   - Make sure that the dried resin does not come in contact with the surrounding air in the room by using a closed transport system
2. Decrease the decompression (suck-back) by speed and/or length
3. See Section 29.2.4 Voids or Pores

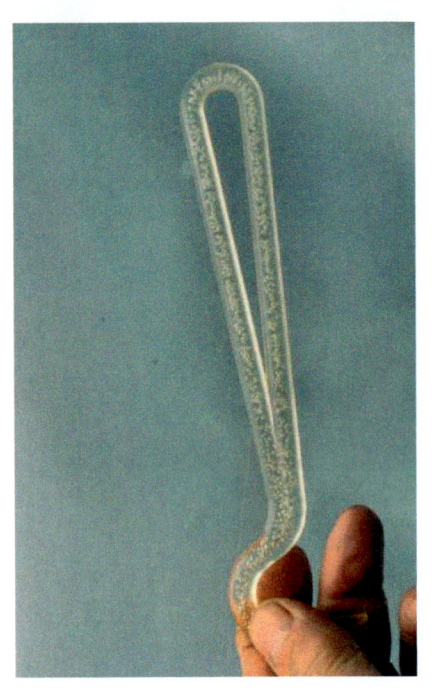

Figure 29.25 The picture shows a hoof pick in polycarbonate with steam bubbles inside the wall due to incorrect drying.

29.4.2 Cracks

Possible causes (listed in the most likely order):

1. The parts stick too tightly in the cavities
2. Stress cracking in amorphous plastics
3. Excessive load in the gate area
4. Fractural impression due to too small radius (sharp corners)

Suggested remedies (according to the causes above):

1. Reduce hold pressure or hold pressure time
2. Eliminate the risk of stress cracking:
   - Avoid constant load
   - Avoid exposure to solvents
   - Coat the surface with a protective layer, e.g. siloxane
3. *Workshop action required (see also Chapter 16):*
   - Move the gate
   - Add or increase the radius of the corners

Figure 29.26 The picture shows a mug in SAN that has cracked when ejected from the mold.

29.4.3 Unmelts (Also Called Pitting)

Possible causes (listed in the most likely order):

1. Too-low melt temperature
2. Incorrect temperature profile on the cylinder
3. The cylinder capacity in relation to the shot volume is on the low side
4. Too-low back pressure or too-high screw rotation speed
5. Too large granules in regrind
6. The material is contaminated with another polymer

Suggested remedies (according to the causes above):

1. Increase the melt temperature
2. Choose to move back the melting zone:
   - Straight or falling temperature profile
   - Preheat the material in a dryer
   - Increase the hopper temperature
3. Increase the time for the material to melt:
   - Increase the cycle time
   - Change to a larger cylinder or machine
4. Increase the back pressure
5. Reduce the screw rotation speed
6. Reduce the content of regrind or use a more efficient grinder
7. Examine the resin visually

Figure 29.27 Here you can see an acetal part colored by a black masterbatch. An incorrect cylinder temperature profile caused unmelted granules in the part. [Photo: DuPont]

Figure 29.28 This is a conveyor link that has failed under normal stress conditions due to exposure to a strong acid.

29.4.4 Brittleness

Possible causes (listed in the most likely order):

1. The resin has too-high moisture content when molded
2. Thermal degradation of the resin in the cylinder
3. The part had not shrunk enough
4. The part has an incorrect design (too sharp radius)
5. The part is made of a too brittle material
6. The material has been exposed to unfavorable chemicals

Suggested remedies (according to the causes above):

1. If the moisture content of the resin exceeds the recommended one: Dry the resin in a dry air dryer
2. Check that:
   - The recommended hold-up time of the resin in the cylinder has not been exceeded
   - There are no hold-up spots in the cylinder and nozzle
3. Increase hold pressure and hold pressure time
4. Increase the corner radius at gates and runners
5. Increase if possible the toughness of the material:
   - Condition polyamide parts in water after the molding
   - Choose an impact-modified or highly viscous grade of the material
6. Avoid chemicals that degrade the used material

29.4.5 Crazing

Possible causes (listed in the most likely order):

1. The material has been overloaded above its elongation of yield
2. Some materials (mainly amorphous such as polystyrene, PMMA, and polycarbonate) are particularly sensitive to crazing

Suggested remedies (according to the causes above):

1. Reduce the load:
   - Do not exceed the elongation at yield
   - Redesign the part
2. Replace the material with a less-sensitive semi-crystalline one such as polypropylene, acetal, or polyamide

Figure 29.29 The picture shows a pipe fitting that got crazing when a metal pipe was pressed into the plastic.

29.4.6 Problems with Regrind

Possible causes (listed in the most likely order):
1. The regrind has too large particles
2. The regrind has changed due to thermal degradation
3. The regrind is not dry enough
4. The glass fiber length in the regrind is too short (the reinforcing effect i.e. tensile strength and stiffness have decreased)

Suggested remedies (according to the causes above):
1. Reduce the variation of particle size:
   - Use a more efficient grinder
   - Screen away too large and small particles
2. Improve the quality of the melt:
   - Do not regrind parts with burns or discoloration
   - Do not regrind parts that have been produced in wet resin
   - Reduce the content of regrind to maximum 30%
3. Dry the regrind. NOTE: Regrind may often need to be dried for much longer time than recommended for virgin resin. However, you should not use the recommended drying temperature for a long drying time. You should use 10–20 °C lower temperature but still ensure that you reach the recommended moisture content of the material.
4. Increase the glass fiber length in the parts by:
   - Reduce if possible the screw rotation speed when dosing
   - Reduce the regrind content in virgin material

Figure 29.30 Here, a natural colored resin has been mixed with green masterbatch and green regrind. Some of the ground particles are both too big and show clear signs of degradation (black), which can cause strength problems.

29.5 Dimensional Problems

29.5.1 Incorrect Shrinkage

Possible causes (listed in the most likely order):
1. Incorrect processing parameters:
   - Hold pressure
   - Hold pressure time
   - Mold temperature
2. Insufficient packing of the part:
   - Leaking back flow valve (the screw hits the bottom)
   - Too small gates or runners
3. Incorrect shrinkage calculation during tool making
4. Material replacement with one with different shrinkage

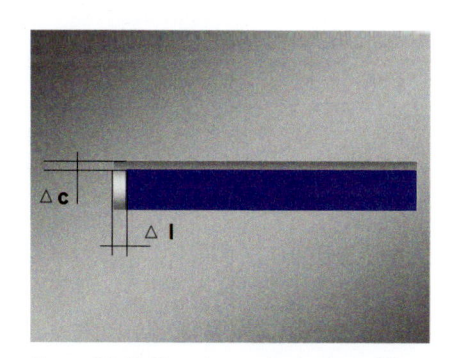

Figure 29.31 Most thermoplastics have different shrinkage in the flow and cross directions.

Suggested remedies (according to the causes above):

1. Compensate by:
   - Increase/decrease the hold pressure if the shrinkage is too big/low
   - Increase/decrease the hold pressure time if the shrinkage is too big/low
   - Decrease/increase the mold temperature time if the shrinkage is too big/low (NOTE: Take into account the post-shrinkage, which will increase when the mold temperature is decreased for semi-crystalline plastics)

2. Adjust by:
   - Replace the back flow valve
   - *Workshop action required (see also Chapter 16):*
     - Increase the dimensions of gates or runners
     - Compensate for the real shrinkage by adjusting the dimensions in the cavity

3. Replace, if possible, the material with another one that has the correct shrinkage

Figure 29.32 Different thermoplastics can manage different tolerance requirements.

29.5.2 Unrealistic Tolerances

Possible causes (listed in the most likely order):

1. The tolerances are set routinely and are not really needed
2. The tolerances are needed for the function of the part but cannot be fulfilled due to:
   - Unrealistic tolerances for tool making
   - Unrealistic tolerances in the injection-molding process, e.g. shrinkage
   - Too big warpage
   - Too big tolerances of the plastic raw material, e.g. glass fiber content
   - Too big swelling due to moisture absorption of the material in the part, e.g. polyamide
   - Too big coefficient of thermal expansion of the material in the part

Suggested remedies (according to the causes above):

1. Change in the drawing

2. Adjust:
   - The tolerances in the mold or redesign the part
   - See Section 29.5.1 Incorrect Shrinkage
   - See Section 29.5.3 Warpage
   - Replace if possible with a more dimensional material that withstands tighter tolerances
   - Replace with a material with less coefficient of thermal expansion or redesign the part

29.5.3 Warpage

Possible causes (listed in the most likely order):

Internal stresses due to:

1. Variations in the mold temperature (e.g. between mold halves or between mold half and cores)
2. Insufficient shrinkage compensation
3. Anisotropy of the plastic raw material, e.g. glass fiber
4. Varying wall thickness, e.g. ribs

Suggested remedies (according to the causes above):

1. Balance the mold temperature:
   - Test different temperatures on the mold halves and/or cores
   - Use a cooling fixture if nothing else works

2. Improve the shrinkage compensation:
   - Fix a leaking back flow valve (no cushion)
   - Increase the hold pressure
   - Increase the hold pressure time

3. Replace glass fiber reinforced materials with glass beads or mineral-filled materials

4. *Workshop action required (see also Chapter 16):*
   Adjust the mold

Figure 29.33 This picture shows a panel that should be flat but warps heavily.

Chapter 29 – Troubleshooting – Causes and Effects

29.6 Production Problems

29.6.1 Part Sticks in the Cavity

Possible causes (listed in the most likely order):

1. Too much shrinkage compensation, e.g. too hard packed
2. Incorrect temperature control of the mold
3. Mold problems or a faulty design

Suggested remedies (according to the causes above):

1. Reduce the shrinkage compensation:
   - Reduce the hold pressure
   - Reduce the hold pressure time
   - Reduce the injection speed
   - Use mold release spray temporarily
2. Adjust the temperature settings:
   - Reduce the mold temperature and eventually also the melt temperature
   - Increase the cooling time temporarily
3. *Workshop action required (see also Chapter 16):*
   - Increase the draft angle
   - Reduce the depth of the surface pattern
   - Reduce too big undercut
   - Repair damages in the surface

Figure 29.34 The part sticks in the fixed mold half. It does not follow the moving half with the ejector pins.

29.6.2 Part Sticks on the Core

Possible causes (listed in the most likely order):

1. Too much shrinkage compensation, e.g. too hard packed
2. The core is too hot or too cold
3. Vacuum can occur particularly on thin-walled parts
4. Mold problems or a faulty design

Suggested remedies (according to the causes above):

1. Reduce the shrinkage compensation:
   - Reduce the hold pressure
   - Reduce the hold pressure time
   - Reduce the injection speed
   - Use mold release spray temporarily

Figure 29.35 The part has followed the movable mold half but the ejectors pins do not have enough power for ejection of the part.

2. Adjust the temperature settings:
   - Reduce or increase the temperature of the core
   - Increase (temporarily) or reduce the cooling time
3. Same remedies as points (1) and (2) above
4. *Workshop action required (see also Chapter 16):*
   - Increase the draft angle
   - Reduce the depth of the surface pattern
   - Reduce too big undercuts
   - Increase the ejection length
   - Repair damage in the surface

29.6.3 Part Sticks on the Ejector Pins

Possible causes (listed in the most likely order):
1. Unfavorable ejection sequence
2. Mold problems or a faulty design
3. The part is not stiff enough at ejection

Suggested remedies (according to the causes above):
1. Adjust the ejection sequence
   - Increase if possible the ejection length
   - Increase if possible the ejection speed
   - Eject twice
   - Use the help of compressed air
   - Use a picking robot
2. *Workshop action required (see also Chapter 16):*
   - Too short length of the ejector pins due to mechanical restriction
   - Increase the number of ejector pins
3. Increase the cooling time

Figure 29.36 The part has been ejected from the cavity but is hanging on the ejector pins.

Chapter 29 – Troubleshooting – Causes and Effects

Figure 29.37 When the sprue sticks in the mold you need to anneal a brass screw and let it melt into the sprue.
When the screw has cooled down and the plastic has solidified, you can pull out the sprue carefully by using a pair of pliers.

29.6.4 Sprue Sticks in the Mold

Possible causes (listed in the most likely order):

1. The sprue has been shrinkage compensated too much
2. The nozzle is frozen
3. The nozzle is leaking (drooling)
4. The nozzle diameter is too big
5. The nozzle radius is damaged or too big
6. Mold problems or a faulty design

Suggested remedies (according to the causes above):

1. Reduce the shrinkage compensation:
   - Reduce the hold pressure or hold pressure time
   - Reduce the injection speed
   - Use mold release spray temporarily
2. Increase the nozzle temperature
3. Reduce the risk of leakage:
   - Reduce the nozzle temperature
   - Increase decompression (suck-back)
   - Reverse the injection unit during the opening and closing phase
4. Replace the nozzle with a smaller one, i.e. radius about 0.5 mm smaller than the sprue bush
5. Repair the nozzle
6. *Workshop action required (see also Chapter 16)*:
   - Too small taper on the sprue
   - Increase the undercut of the puller (cold slug pocket)
   - Increase the corner radius (R) between the sprue and the runner to $R = 0.5$ times the wall thickness
   - Improve the fit between the sprue bush in the mold and the nozzle

29.6.5 Stringing

Possible causes (listed in the most likely order):

1. The sprue pulls a thread from the nozzle
2. A thread from the previous shot hangs and enters the cavity

Suggested remedies (according to the causes above):

1. Eliminate the risk for stringing:
   - Increase the decompression (suck-back)
   - Adjust the nozzle temperature either up or down
2. Same remedies as point (1) above

Figure 29.38 You can get stringing problems both in semi-crystalline and amorphous thermoplastics.

CHAPTER 30
Statistical Process Control (SPC)

Statistical process control is a method that has long been used in the engineering industry to improve the quality of produced products. Regarding the production of plastic products it has not yet been put into use on a large scale. SPC among molders is, however, largely increasing. In this chapter we will give the reader an orientation on the principles and different concepts used. This chapter has been developed together with Nielsen Consulting (www.nielsenconsulting.se), a Swedish consultant who specializes in SPC training and has contributed with text and images.

30.1 Why SPC?

SPC is a very useful and profitable method because it:

- **Creates customer value**, i.e. improves the function or extends the lifetime of the customer's product
- **Reduces rejects** by focusing on the tolerance center (see terms below) instead of the tolerance limits
- **Prevents failure** of the products because actions are taken at the right time
- **Reduces the need for final inspection**, i.e. deliveries with high capability (see terms below) do not need a final inspection
- **Promotes customer relations** because capability allows the customer to take the goods without performing an incoming inspection
- **Detects machine failure** at an early stage and thus becomes an aid in state-based maintenance
- **Reduces inventory costs** by error-free deliveries and allows a reduced inventory
- **Can reduce stress** in production as the need to measure and control the process is reduced
- **Can facilitate price discussions** as accurate deliveries usually mean more satisfied customers
- **May increase staff engagement** through increased understanding of the process as it is easier to see patterns and trends in the process
- **Provides a uniform approach** when there is no room for "sole and absolute discretion"
- **Is a tool in the Lean process** that provides continuous improvement with focus on satisfying the customer

30.2 Definitions in SPC

30.2.1 Normal Distribution (Gaussian Dispersion)

This is the way in which the measured values, in most cases, will be distributed as a result of the random spread around its mean value (highest point of the hump); see Figure 30.1.

Notice that most of the measured values are around the hump and that there are fewer values closer to the periphery. It is in other words not very likely that you at all, under random sampling, will find some details at the periphery. It is not enough that the details that you happen to measure fall within the tolerance range. To see the Gaussian dispersion many details need to be measured, and it may be time consuming. But there is a shortcut by using the standard deviation!

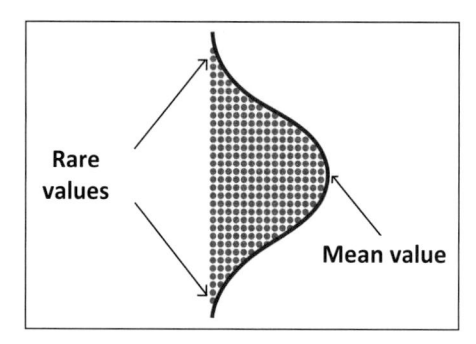

Figure 30.1 Normal distribution and mean value.

30.3 Standard Deviations

30.3.1 One Standard Deviation

This is a statistical function that will be used to calculate the normal distribution. It is calculated by measuring the distance from the mean value (the point at the far right of the curve in Figure 30.2) to the point where the curve changes direction and begins to bulge outward. This distance represents one standard deviation. This means that you do not need to measure several hundred details to find out how much the machine or process varies. You can instead calculate the spread by using the standard deviation. In probability theory the standard deviation is represented by the Greek letter sigma: σ.

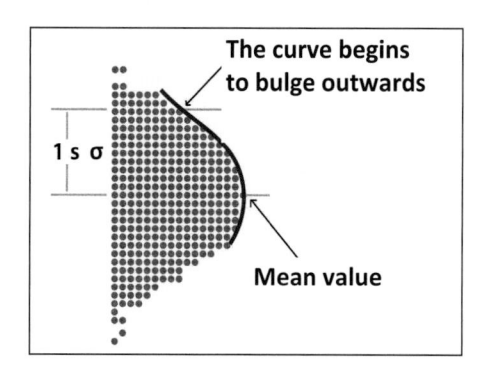

Figure 30.2 One standard deviation.

30.3.2 Six Standard Deviations (Six Sigma)

To calculate the normal distribution, you multiply the standard deviation by six and thereby get the normal distribution curve. In other words, if you had continued to measure details, they would have filled the curves width, but now we instead calculate the width of the curve; see Figure 30.3.

The normal distribution is thus based on a standard deviation and consists of six such standard deviations. The six standard deviations capture approximately 99.73% of the outcome. It also means that 0.27% of the outcome is not within the normal distribution curve.

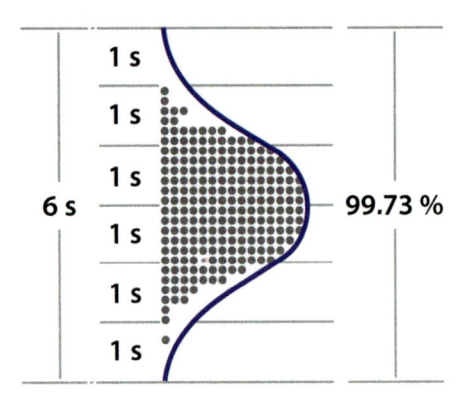

Figure 30.3 Six standard deviations.

30.3.3 Control Limits

The control limits are important parts of the statistical control process. They have nothing to do with the tolerance limits, as the control limits are there to alert you when the process changes its behavior.

An important principle is that the control limits are used together with the mean values within a control chart in order to control the process. This is unlike the tolerance limits, which are used together with measured single points in order to determine if the part in question is approved or not.

The control limits are used to center the process around the target value, which generally is used in the same way as the tolerance middle. The control limits will also show where the limits of the stable process are specified. This means that you basically do not have any reason to react until the control chart signals that something has happened. A typical control chart is an XR chart, where the process mode and dispersion are monitored by means of sample groups and control limits. When a point falls outside of the control limits in the X part, this means that the process mode has changed; see Figure 30.4.

When a point falls outside the control limit of the R part, the process distribution has changed; see Figure 30.5.

30.3 Standard Deviations

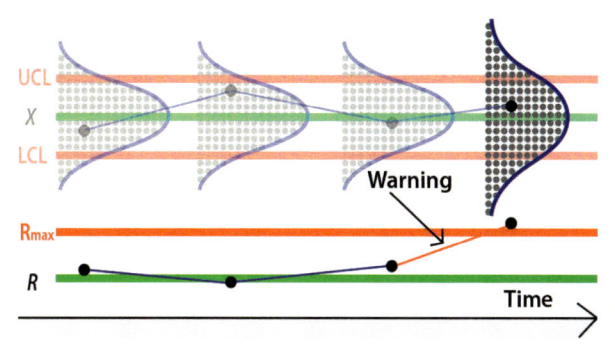

Figure 30.4 An XR diagram with a value outside the upper control limit.

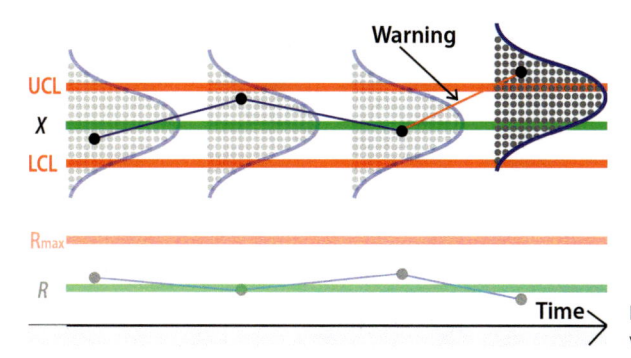

Figure 30.5 An XR diagram with a value outside R-max in the RX diagram.

How Are the Control Limits Determined?

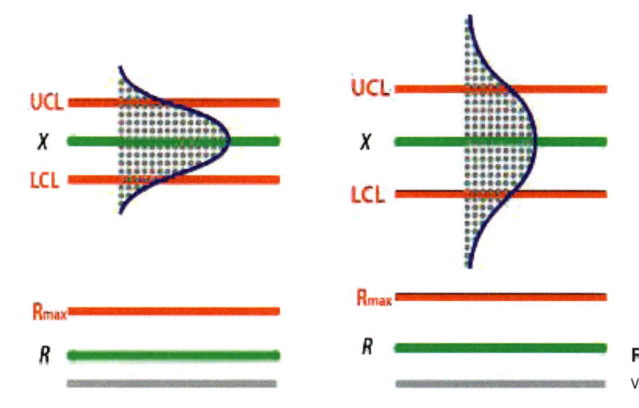

Figure 30.6 An XR diagram with a value outside R-max in the RX diagram.

The best will be to let the control limits adapt to the process. A smaller process distribution gives a narrower control range, while a wider distribution gives a larger control range.

A common myth is that the machine setter will then adjust the process too often, but in practice it is usually the other way around.

The process will be adjusted less often compared to not using SPC. By allowing the control limits to monitor the process you will react neither too early nor too late when the process changes its behavior.

Other Ways to Set the Control Limits

In some cases, it may cause difficulties to let the control limits adapt to the process. An example of this is when the process uses tools that cannot be easy to adjust. Examples of such tools are molds for injection molding. As these molds often provide a very small process distribution and thus a narrow control range, it may therefore be appropriate to release the control limits from the process and instead lock them at a certain distance from the limits of tolerance; see Figure 30.7.

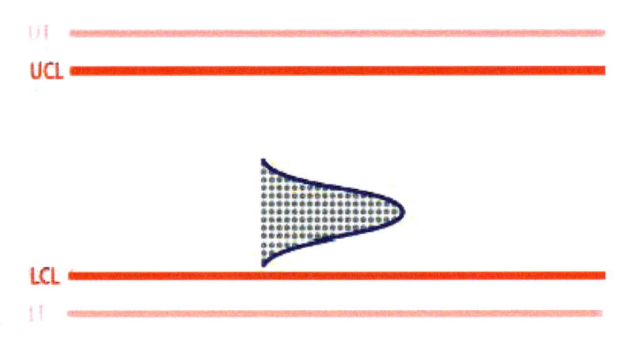

Figure 30.7 An XR diagram with a value outside R-max in the RX diagram.

30.3.4 Target Value

If you have a shaft that is to be installed in a molded slide bearing, initial lubrication will prolong the service life of the bearing. A large shaft diameter (in the upper tolerance range) will allow a smaller amount of grease to fit between the shaft diameter and the hole diameter. This gives a poorer lubrication effect and thus results in faster wear and shorter life span compared to an optimum fit.

A small shaft diameter in turn causes play between the shaft and the hole. This play tends to increase rapidly with time, causing a shorter life span.

The function will be the best at the target value, which in this case is in the middle of the tolerance range; see Figure 30.8. For single-sided properties such as warpage, surface finish, and strength, the target value is instead 0. By using statistical process control you will be allowed to center the process toward the target value.

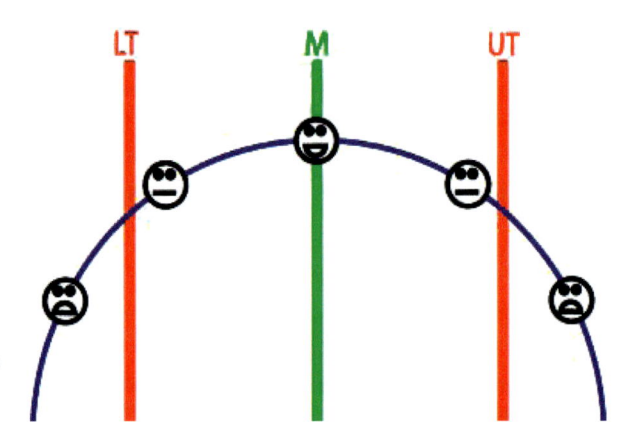

Figure 30.8 The smiley symbols represent the shaft and shows that the position on the target value will give the best result.

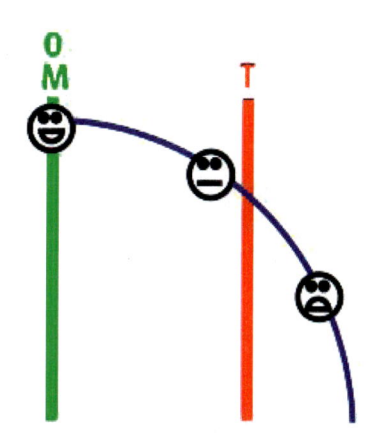

Figure 30.9 The smiley symbols represent the surface finish and show that the best value is equal to 0.

30.3.5 Target Value Centering (TC)

TC is the distance from the target value T to the mean value of the machine or process distribution expressed as a percentage of the tolerance range; see Figure 30.10. Nowadays this is not so commonly used, but previously the maximum deviation was often set to e.g. TC ± 15%.

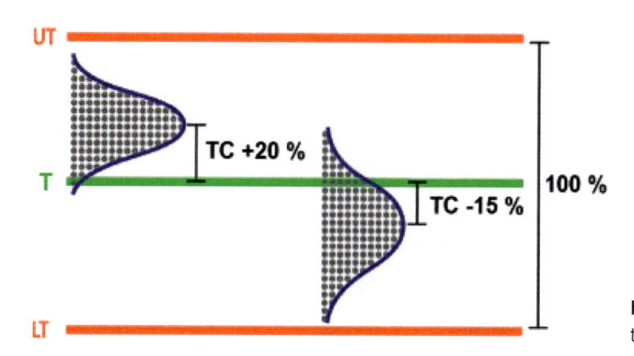

Figure 30.10 Centering toward the target value.

30.3.6 Capability Machine (Cm)

The Cm value describes the machine capability and refers to the number of times the machine Cm distribution fits within the tolerance range. The higher the value of the Cm ratio, the better is the machine.

If the Cm value is for example 2.5, the value of the machine's distribution fits 2.5 times within the tolerance width, while a Cm of 1.0 means that it will only fit once.

Note that if the distribution lies off-center it has the same size, which means the same Cm value. The Cm value does not take into account the distribution's position relative to the upper or lower tolerance limit. It only shows the relationship between the distribution of the machine's value and the tolerance range.

Chapter 30 – Statistical Process Control (SPC)

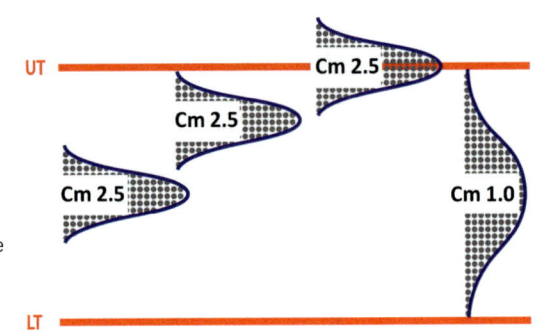

Figure 30.11 Different positions of the distribution of the machine capability Cm. The first curve is that of the best machine.

30.3.7 Capability Machine Index (Cmk)

In order to also add into consideration the machine's capability in relation to the tolerance limits, the Cmk value is used. This value describes the machine's capability in accordance with the corrected position. A high Cm value is not worth much if the machine settings are heavily off-center in relation to the center of the tolerance. A large Cmk ratio thus means that you have a good machine with a low distribution in relation to the tolerance range. It also means that it is correctly positioned in relation to the tolerance center. If the Cmk value is equal to the Cm value, the machine is set to produce exactly in the middle of the tolerance range; see Figure 30.12.

A common minimum value for machine capability is a Cmk value of 1.67.

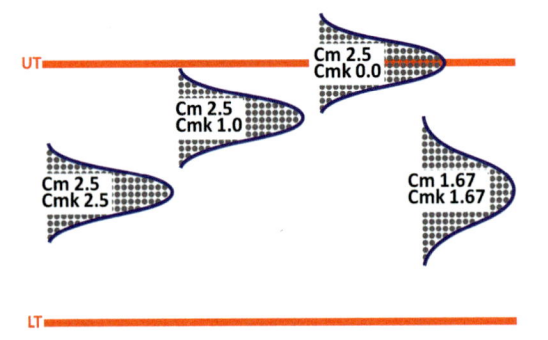

Figure 30.12 The Cmk value refers to the position of the machine capability.

30.3.8 Capability Process (Cp)

The Cp value describes the process capability and refers to the number of times that the process distribution fits within the tolerance width. The higher the Cp value, the better is the process.

If for instance the Cp value is 2.0 this means that the distribution of the process fits twice within the tolerance width, while a Cp value of 1.0 means that it only fits once. Note that if the distribution lies off-center it has the same size, which means the same Cp value. The Cp value does not take into account the distribution's position relative to the upper or lower tolerance limit. It only shows the relationship between the distribution of the process value and the tolerance range.

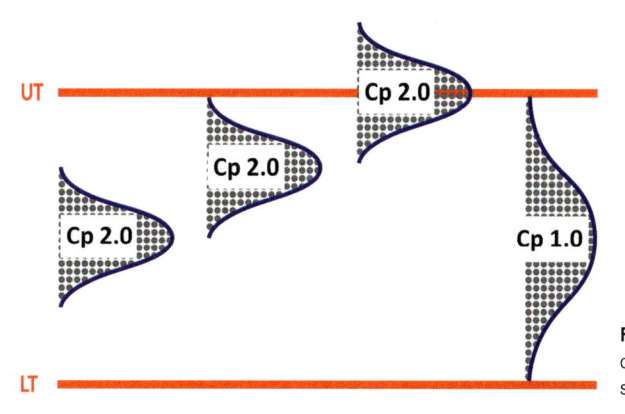

Figure 30.13 Different values for the capability process. The first curve shows the best settings.

30.3.9 Capability Process Index (Cpk)

In order to also add into consideration the process capability in relation to the tolerance limits, the Cpk value is used. This value describes the process capability in accordance with the corrected position. A high Cp value is not worth much if the process settings are heavily off-center in relation to the center of the tolerance. A large Cpk ratio thus means that you have a robust process with a low distribution in relation to the tolerance range. It also means that it is correctly positioned in relation to the tolerance center. If the Cpk value is equal to the Cp value, the process is set to produce exactly in the middle of the tolerance range; see Figure 30.14.

A common minimum value for process capability is a Cpk value of 1.33.

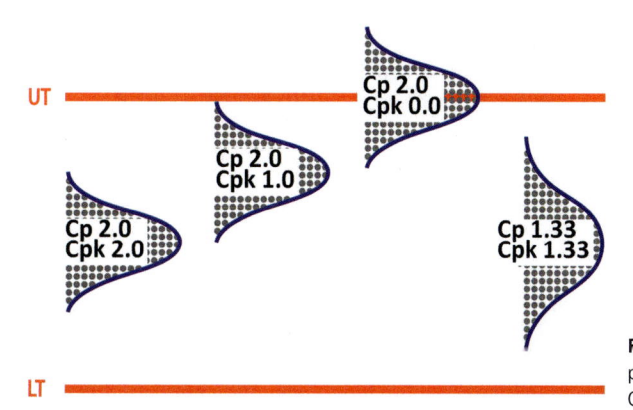

Figure 30.14 Various camps on process capability generally require Cpk = 1.33.

30.3.10 Six Important Factors

The following six factors are considered to cause the distribution in capability measurements:

1. The machine
   (i.e. how worn is it and what is the condition of the mold)

2. The method of measurement
   (i.e. the resolution of the testing device and its distribution)

3. The human behind the measured values
   (i.e. what is his/her experience, accuracy, and concentration)
4. The material
   (i.e. variations in melt viscosity and/or glass fiber content)
5. The environment
   (i.e. temperature, humidity, and how much the central cooling system varies)
6. The process
   (i.e. injection molding, extrusion, blow molding)

30.3.11 Machine Capability

The machine capability is measured using Cm and Cmk values. These values give a snapshot of how well a machine at this time can produce parts in relation to the tolerance limits.

Figure 30.15 shows a few different snapshots. When measuring the machine capability it is important not to change any settings, molds, materials, machine setters, or measurement methods as well as not have any interruptions. Of the factors described in Section 30.3.10 it is only the machine and the measurement that are allowed to influence the result.

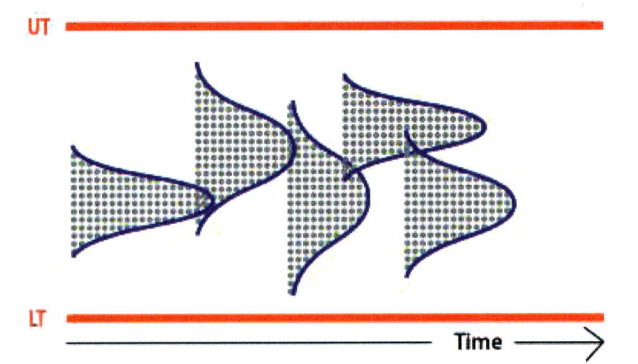

Figure 30.15 Here there are different measurement series that will display the machine capability.

30.3.12 Process Capability

Process capability is a long-term study measured in Cp and Cpk values that shows how successfully a process produces in relation to the tolerance limits. This is both during the period covered by the study, but also within the near future.

It is possible to express process capability as the sum of a number of machine capabilities over a long period of time. When measuring the process capability, everything that influences the process is to be included in the measurement. In other words: of the six factors above, all six are allowed to influence the result.

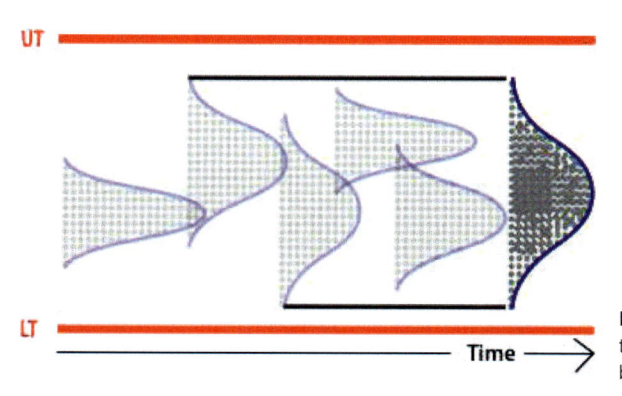

Figure 30.16 The process capability is the sum of a number of machine capabilities.

30.4 How SPC Works in Practice

30.4.1 Software

When SPC is introduced in your production process, you need computer software to calculate the various statistical values of the process. Figure 30.17 shows the interface of software named Microlog 32 by Fourtec (see www.fourtec.com).

In the RX chart in Figure 30.17 there is a target weight of 50.000 grams, an upper tolerance limit of 0.900 grams and a lower limit of 0.000 grams. The series is performed between the hours of 16:19 and 02:37. Each blue dot represents either a change of a process parameter or a disturbance in the process, and these should always be documented. By studying the curve we can see that the outcome is very well centered around the target value. The target value centering only has a deviation of 0.29%.

The process is very robust with a Cp value of 8.88 and a Cpk value of 8.82. In the chart in Figure 30.17 the setter has manually filled in each value, but this could easily have been done by a robot placing each part on a scale, which in turn has been connected to a computer.

The deviations must however be recorded manually.

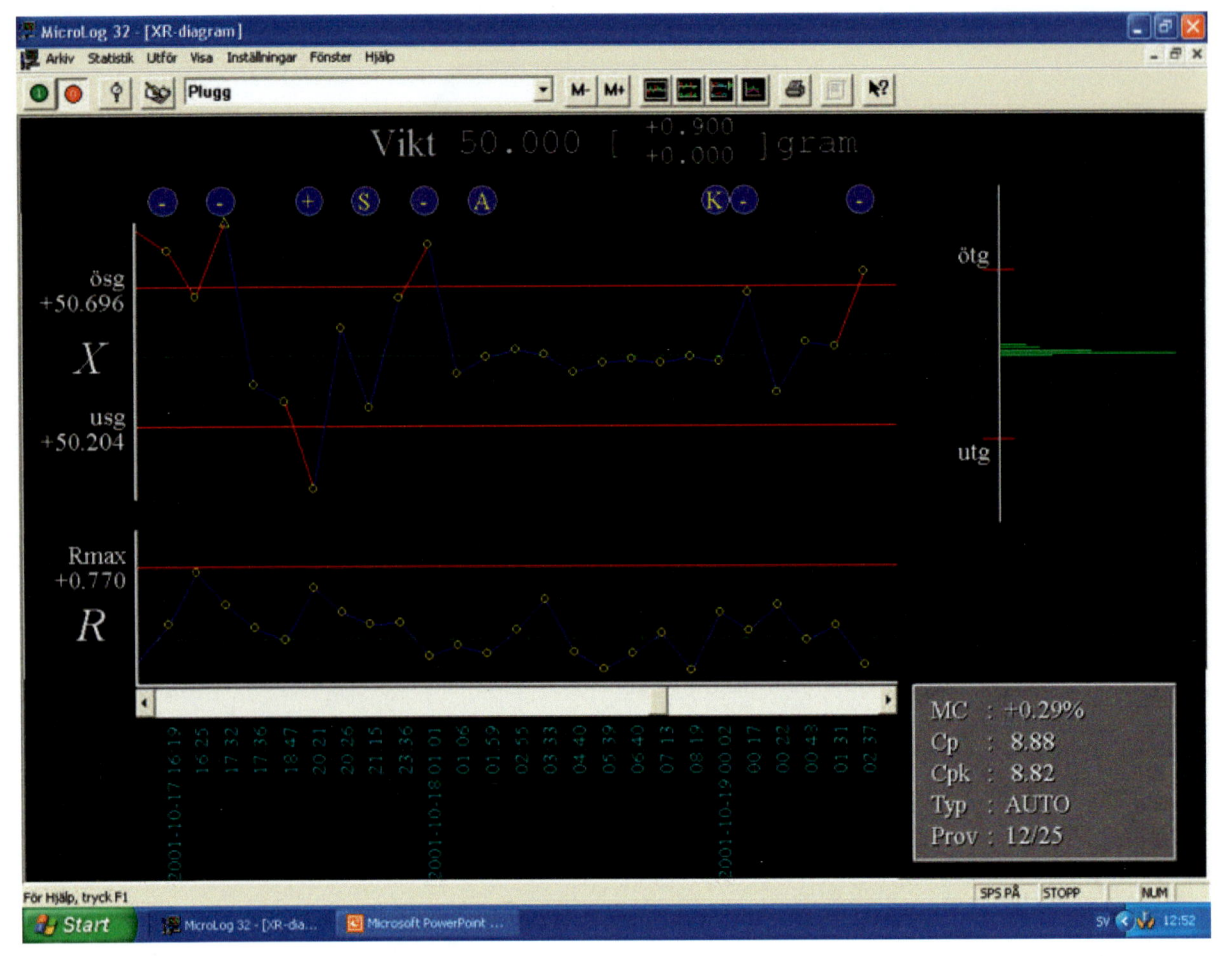

Figure 30.17 In the RX diagram here we can see the weight distribution of plastic parts produced by injection molding over a time frame of a few hours. The blue circles with a negative sign inside show when the hold pressure has been decreased. The circle with the positive signs inside shows when the hold pressure has been increased. The circles with an S and a K inside display when the process was interupted due to problems with the robot. The circle with an A inside marks the switch from a mixture of 30% regrind in the material to 100% virgin resin.

30.4.2 Process Data Monitoring

The interest in statistical process control is constantly increasing among molders. Engel, one of the leading injection-molding machine manufacturers, offers an SPC system used for online quality monitoring using up to 20 different process parameters. The system is integrated with the machine control system and can be supplemented on already delivered Engel machines. See www.engelglobal.com.

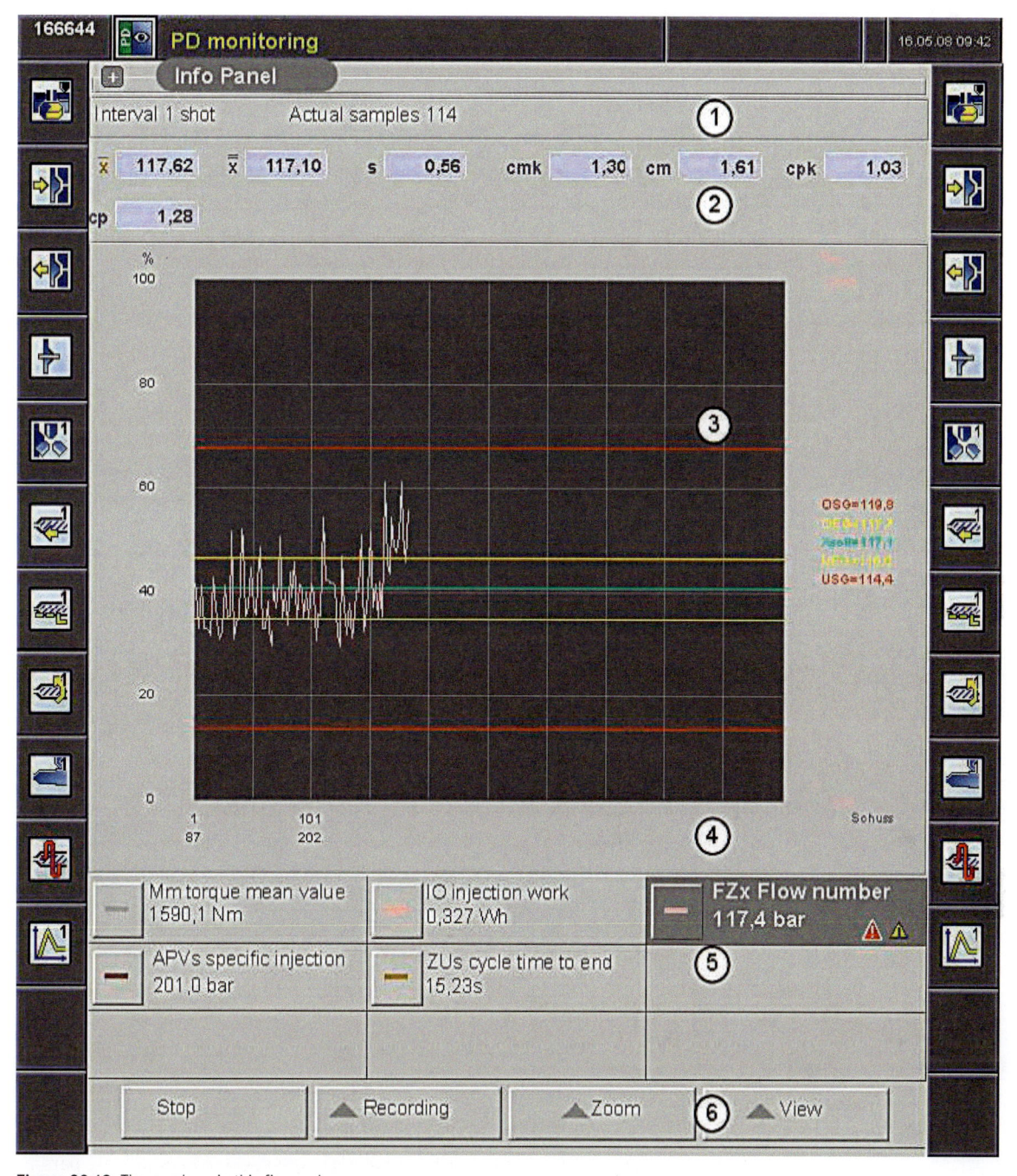

Figure 30.18 The numbers in this figure show:
- ❶ Measurement setting and current number of measurements
- ❷ Line for statistical characteristic values
- ❸ Horizontal graph for one process parameter each displaying:
    - Set value (blue)
    - Actual value (turquoise)
    - Mean value of sample (ocher)
    - Overall mean value (green)
    - Control limits (yellow)
    - Specification limits (red)
- ❹ Lines for measurement number and shot counter and/or time
- ❺ Curve legend with keys for curve-specific settings
- ❻ Function and menu keys

CHAPTER 31
Internet Links

The following companies have contributed with information and/or photos for this book and are highly recommended if you need more information about their products or services:

Company	Internet link	Product or service
Acron Formservice AB	www.acron-form.se	Rapid prototyping
Ad Manus Materialteknik AB	www.ad-manus.se	Training/testing/analysis
AD-Plast AB	www.ad-plast.se	Injection moldings
Arla Plast AB	www.arlaplast.se	Extrusion
Arta Plast AB	www.artaplast.se	Injection moldings
Clariant Sverige AB	www.clariant.com	Masterbatch
Digital Mechanics AB	www.digitalmechanics.se	Rapid prototyping
DuPont Engineering Polymers	plastics.dupont.com	Plastic raw material
DSM	www.dsm.com	Plastic raw material
DST Control AB	www.dst.se	Electro-optical systems
Engel Sverige AB	www.engelglobal.com	Injection molding machines
Erteco Rubber & Plastics AB	www.erp.se	Plastic raw material
European Bioplastics	www.european-bioplastics.org	European trade organization
Ferbe Tools AB	www.ferbe.se	Tool maker
Hordagruppen AB	www.hordagruppen.com	Blow moldings
IMCD Sweden AB	www.imcd.se	Plastic raw material
IfBB Institute for bioplastics and biocomposites	www.ifbb-hannover.de	Development of biopolymers
IKEM	www.ikem.se	Swedish trade organization (mobile apps)
K.D. Feddersen Norden AB	www.kdfeddersen.com	Plastic and machine distributor
Makeni AB	www.makeni.se	Injection moldings
Mape Plastic AB	www.mapeplastics.se	Plastic raw material
Mettler Toledo AB	www.se.mt.com	Equipment for analysis
Miljösäck AB	www.miljosack.se	Climate-smart plastic bags
Orthex Sweden AB	www.orthexgroup.se	Consumer products
Plastinject AB	www.plastinject.se	Injection moldings
Polykemi AB	www.polykemi.se	Plastic raw material
Polymerfront AB	www.polymerfront.se	Plastic raw material
Polyplank AB	www.polyplank.se	Recycled products
Protech AB	www.protech.se	Rapid prototyping equipment
Resinex Nordic AB	www.resinex.se	Plastic raw material
Rotationsplast AB	www.rotationsplast.se	Rotational moldings
Sematron AB	www.sematron.se	Vacuum forming
Stebro Plast AB	www.stebro.se	Injection moldings
Talent Plastics AB	www.talentplastics.se	Injection moldings and extrusion
Tojos Plast AB	www.tojos.se	Injection moldings
Vadstena Lasermärkning	www.lasermarkning.se	Laser marking equipment
Weland Medical AB	www.weloc.com	Plastic clips

Index

A

ABS 18
acetal 26
acrylonitrile-butadiene-styrene 18
actual value 302
additive manufacturing 154
additives 78
amorphous 5
analytical troubleshooting 301
angle tool 189
anisotropic behavior 241
assembly methods 260
atactic 77

B

back pressure 289
barrier screws 181
biocomposites 59, 68
biodegradable 59
bioplastics 59, 62
biopolyamide 65
biopolyester 63
blisters 326
blow molding 204
brainstorming 308
brittleness 334
bubbles 332
burn marks 321

C

cable production 193
CAMPUS 98
capability machine 347
capability process 348
cellulose 63
chemical properties 83
chroming 122
co-extrusion 191
cold slug 330
cold slug pockets 133
color streaks 324
compounding 195
control limits 344

cooling systems 134
cooling time 292
corner radius 252
corrugation 185
cost calculations 168
cracks 332
crazing 334
creep 242
cushion 295
cylinder 180

D

dark streaks 321
decompression 294
degradation 321
delamination 329
design of experiments 309
design rules 238
deviation 302
diesel effect 323
dimensional problems 335
discoloration 321
dosing length 294
dosing time 292
draft angles 138
dry air dryer 279
drying 278

E

ejector pin marks 330
ejector systems 137
electrical properties 84, 96
environmental factors 70
EPS 15
extrusion 177

F

factorial experiments 310
family molds 126
feeding 186
fill ratio 318
film blowing 191

flame retardancy 86
flammability 95, 107
flashes 319
flexural modulus 91
fluoropolymers 47

G

gas injection 116
gate 131
gate location 256
glass fiber streaks 327
glass transition temperature 5
granulation methods 78

H

HB rating 95
HDPE 8
heat deflection temperature 86, 93
heat stabilization 85
high-performance thermoplastics 46
hinge 198
hold pressure 287
hold pressure switch 294
hold pressure time 287
hot air dryer 279
hot plate welding 265
hot runner systems 132
"hot stamp" printing 118
hygroscopic 278

I

impact strength 92
infrared spectrophotometer 108
infrared welding 266
injection-molding cycle 114
injection-molding machine 110
injection-molding methods 110
injection-molding process 270
injection pressure 286
injection speed 286, 289
injection unit 112

Index

in-mold decoration 120
isotactic 77

J

jetting 328

L

laser marking 121
laser welding 266
LCP 50
LCPA 65
LDPE 8
liquid crystal polymer 50
LLDPE 8
locking unit 113

M

machine capability 350
masterbatch 80
material data 88
Material Data Center 98
material selection 209
MDPE 8
mechanical properties 79
melt index 97
melting point 5
melt temperature 282
metalizing 122
microtome analysis 109
mold 123
mold design 139
mold filling analysis 146
mold shrinkage 97, 284
molds with melt cores 128
monofilament 194
monomer 76
muffing 203
multi-component injection molding 115
multi-component molds 127

N

nonlinear 240
nozzle diameter 278
nylon 23

O

oil stain 331
orange peel 329

P

PA 23
painting 121
PBT 29
PC 33
PE 7
PEEK 53
PEI 54
peripheral speed 280, 291
PET 29
PEX 8
PHA 66
physical properties 80
PLA 64
plastic 1
PMMA 21
polyamide 23
polybutylene terephthalate 29
polycarbonate 33
polyester 29
polyether ether ketone 53
polyetherimide 54
polyethylene 7
polyethylene terephthalate 30
polylactide 64
polymer 1
polymerization 76
polymethyl methacrylate 21
polyoxymethylene 26
polyphenylene sulfide 52
polyphenylsulfone 57
polypropylene 11
polystyrene 15
polysulfone 56
polytetrafluoroethylene 47
polyvinylchloride 13
POM 26
pores 320
post-shrinkage 284
PP 11
PPA 48
PPS 52
PPSU 57

printing 117
problem analysis 306
process capability 350
processing data 298
process parameters 297
production problems 338
profile 196
prototype molds 145
PS 15
PSU 56
PTFE 47
PTT 65
PVC 13
pyrometer 282

Q

quality control 102

R

record grooves 329
recycling 73
regrind 335
reject 170
relaxation 242
replacement cost calculation 175
requirement specifications 210
ribs 254
riveting 268
rotating cores 125
rotational molding 206
rotational welding 264
runner systems 130

S

SAN 17
SBS 38
scanning electron microscope 108
screen printing 119
screw diameter 292
screw rotation 292
screw rotation speed 290
sealing lip 197
SEBS 38
semi-crystalline 5
service temperature 93
setpoint 302

shrinkage 284, 335
silver streaks 322, 324
single-screw extruder 181
sink marks 319
Six Sigma 344
sliding joint 198
snap-fit joint 199
specific heat 6
specific volume 6
spiral forming 202
splays 322, 324, 326
stack molds 125
standard deviations 343
starch 63
statistical design of experiments 309
statistical process control 342
stiffness 89
stress concentration 252
stress-strain curve 90
stringing 341
stripes 326
styrene-acrylonitrile 17
suck-back 294
surface treatment 117
syndiotactic 77

T

tampon printing 119
target value 346
target value centering 347
temperature profile 281
tensile modulus 91
tensile strength 88
testing methods 105
thermal properties 85
thermoplastic elastomers 36
thermosets 3
three-plate molds 124
tolerances 258, 336
total shrinkage 284
toughness 89
TPC-ET 42
TPE-A 44
TPE-E 42
TPE-O 36
TPE-S 38
TPE-U 41
TPE-V 39
TPO 36
TPU 41
TPV 39
troubleshooting 316
twin-screw extruder 182
two-component molding 115
two-plate molds 123

U

UHMWPE 8
UL service temperature 93
ultrasonic welding 262
unmelts 333

V

vacuum forming 207
venting systems 136
venting zone 180
vibration welding 263
visual inspection 104
voids 320, 332
V rating 95

W

wall thickness 251
warpage 337
water injection molding 116
water stain 331
weather resistance 81
weld lines 257
weld-lines 327
winding 188

Y

yield stress 90